INTERPRETING ENGINEERING DRAWINGS

METRIC EDITION

C. Jensen
TECHNICAL DIRECTOR
IN THE PROVINCE OF ONTARIO

R. Hines
ELECTRO MECH. SPECIALIST
C.G.E.

 VAN NOSTRAND REINHOLD COMPANY
NEW YORK CINCINNATI TORONTO LONDON MELBOURNE

Copyright ©1979 by Litton Educational Publishing, Inc.

Library of Congress Catalog Card Number 77-29111
ISBN 0-442-24136-4

Printed in the United States of America.

Published in 1979 by Van Nostrand Reinhold Company
A division of Litton Educational Publishing, Inc.
135 West 50th Street, New York, NY 10020, U. S. A.

Van Nostrand Reinhold Limited
1410 Birchmount Road
Scarborough, Ontario M1P 2E7, Canada

Van Nostrand Reinhold Australia Pty. Ltd.
17 Queen Street
Mitcham, Victoria 3132, Australia

Van Nostrand Reinhold Company Limited
Molly Millars Lane
Wokingham, Berkshire, England

16 15 14 13 12 11 10 9 8 7 6 5 4 3 2 1

Library of Congress Cataloging in Publication Data

Jensen, Cecil Howard, 1925-
   Interpreting engineering drawings.

   Includes index.
   1. Engineering drawings. I. Hines, Raymond D.
joint author. II. Title.
T379.J45 1978   604'.2'5   77-29111
ISBN 0-442-24136-4

PREFACE

New technological developments have created a need for well-prepared instructional material for persons both entering and engaged in technical positions in industry. INTERPRETING ENGINEERING DRAWINGS provides instructional material in the correct sequence for students, apprentices, journeymen, and others who need to know how to read and interpret engineering drawings.

One of the most significant challenges currently encountered by industry throughout the Western Hemisphere is the adoption and implementation of the metric system of measurement. This text reflects and supports this adaptation through the exclusive use of metric units in all applicable areas. New drafting conventions introduced by advances in new production techniques and simplified drafting standards are used throughout the text. The use of a second color provides improved readability, contrast, and emphasis to the written material and illustrations.

The written material and illustrations preceding the drawing assignments contain related technical information and principles of drafting necessary to interpret the new material on each drawing. In some instances, shop practices are defined and explained to clarify the meanings of operational notes which appear on shop drawings.

Selected tables and charts taken from handbooks and manufacturers' catalogs are in the back of the book. These comprise the reference material needed to solve the problems. Further information can be obtained by writing to the American Society of Mechanical Engineers, United Engineering Center, 345 East 47th Street, New York, NY 10017 and the Canadian Standards Association, 178 Rexdale Boulevard, Rexdale, Ontario, Canada, M9W 1R3.

To distinguish between the illustrations used in the explanatory material and the assignments, the illustrations are given a unit and figure number while the assignments are given a drawing number preceded by the letter "A." *Figure 1-1,* for example, is the first illustration in the book and *Drawing A-1* is the first assignment.

Mr. Cecil H. Jensen, Technical Director of the McLaughlin Collegiate and Vocational Institute, Oshawa, Ontario, Canada, has over twenty-three years of teaching experience in mechanical drafting. He is the successful author of many technical books including: *Drafting Fundamentals, Engineering Drawing and Design,* and *Home Planning and Design.* Before he began teaching, Mr. Jensen spent several years in industrial design. He has also been responsible for the supervision of the teaching of technical courses for the General Motors apprentices in Oshawa, Canada.

Mr. Raymond D. Hines contributes thirty-five years of experience in the field of electro-mechanical design to this text. He is currently employed as a Design Specialist by the General Electric Company, Guelph, Ontario, Canada, where he was recently responsible for the internal design layout of the largest high-voltage power transformer built in North America. He is intensely interested in the advancement of the drafting design profession through the latest methods of drawing presentations.

CONTENTS

APPENDIX

Table

UNIT 1

BASES FOR INTERPRETING DRAWINGS

Engineering or *technical drawings* furnish a description of the shape and size of an object. Other information necessary for the construction of the object is given in a way that renders it readily recognizable to anyone familiar with engineering drawings.

Pictorial drawings are similar to photographs because they show objects as they would appear to the eye of the observer, figure 1-1. Such drawings, however, are not often used for technical designs. The drawings used in industry must clearly show the exact shape of objects. This cannot usually be accomplished in just one pictorial view, because many details of the object may be hidden or not clearly shown when the object is viewed from only one side.

For this reason the drafter must show a number of views of the object as seen from different directions. These views, referred to as front view, top view, right-side view, etc., are systematically arranged on the drawing sheet and projected from one another, figure 1-2, page 2. This type of projection is called *orthographic projection*. The ability to understand and visualize an object from these views is essential in the interpretation of engineering drawings.

The principles of orthographic projection can be applied in four angles, or systems: first-, second-, third-, and fourth-angle projection. However, only two systems, first- and third-angle projection are used. *Third-angle projection*, commonly referred to as *American Projection*, is used in the United States, Canada, and many other countries. *First-angle projection*, commonly referred to as *European Projection*, which is described in detail in Unit 16, is used mainly in Europe. Because world trade has brought about the exchange of both engineering drawings and products, drafters are now called upon to communicate in both types of orthographic projection, as well as pictorial representations.

THIRD-ANGLE PROJECTION

The third-angle system of projection is used almost exclusively on mechanical engineering drawings because it permits each facet of the object to be drawn in true scale and without distortion of form along all

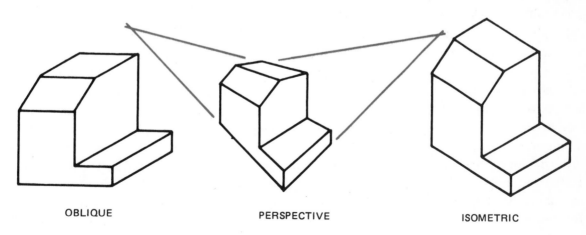

OBLIQUE   PERSPECTIVE   ISOMETRIC

Fig. 1-1 Pictorial drawings

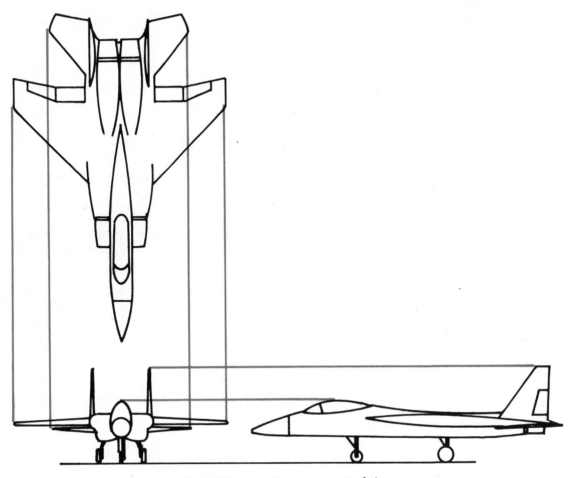

Fig. 1-2 Systematic arrangement of views

dimensions. This system of projection also involves much less time than other methods.

Three views are usually sufficient to describe the shape of an object. The most commonly used views are the front, top, and right-side. In third-angle projection the object may be assumed to be enclosed in a glass box, figure 1-3(B). A view of the object drawn on each side of the box represents that which is seen when looking perpendicularly at each face of the box. If the box were unfolded as if hinged around the front face, the desired orthographic projection would result, figures 1-3(C) and 1-3(D). These views are identified by names as shown. With reference to the front view:

- The top view is placed above

- The bottom view is placed underneath
- The left view is placed on the left
- The right view is placed on the right
- The rear view is placed at the extreme left or right as may be found convenient.

When referring to the overall size of an object, the terms width, height, and depth are used to describe the views. The terms length and thickness are not recommended because they do not apply in all cases.

It must be clearly understood that these terms are used to describe the shape of the views. The longest or smallest dimension can be the width, depth, or height dimension.

The height of the object is seen in the front, right-side, left-side, and rear views.

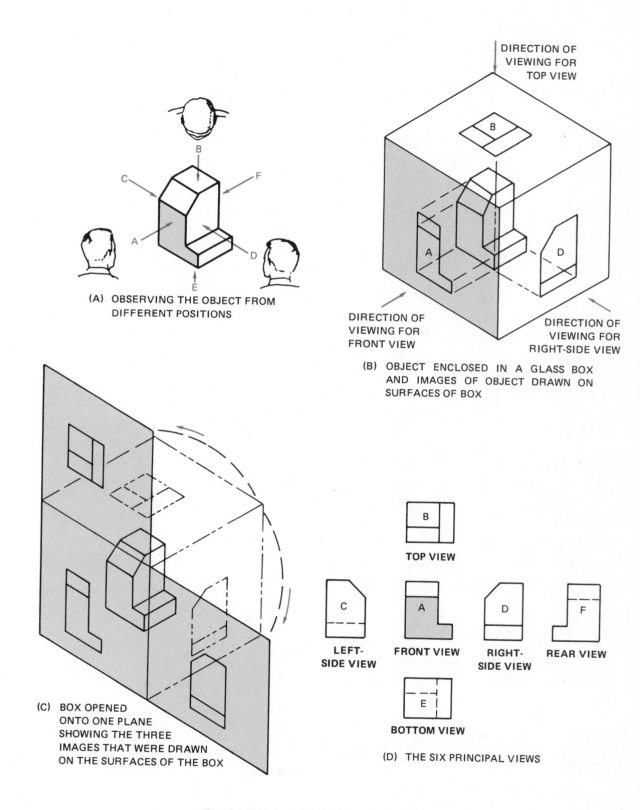

(A) OBSERVING THE OBJECT FROM DIFFERENT POSITIONS

DIRECTION OF VIEWING FOR TOP VIEW

DIRECTION OF VIEWING FOR FRONT VIEW

DIRECTION OF VIEWING FOR RIGHT-SIDE VIEW

(B) OBJECT ENCLOSED IN A GLASS BOX AND IMAGES OF OBJECT DRAWN ON SURFACES OF BOX

(C) BOX OPENED ONTO ONE PLANE SHOWING THE THREE IMAGES THAT WERE DRAWN ON THE SURFACES OF THE BOX

TOP VIEW

LEFT-SIDE VIEW

FRONT VIEW

RIGHT-SIDE VIEW

REAR VIEW

BOTTOM VIEW

(D) THE SIX PRINCIPAL VIEWS

Fig. 1-3 Third-angle orthographic projection

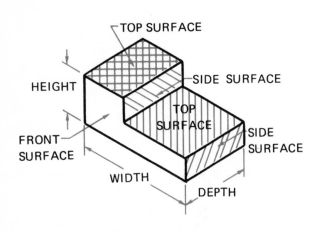

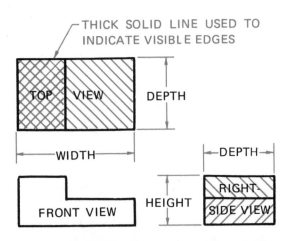

(A) PICTORIAL DRAWING
(ISOMETRIC PROJECTION)

(B) ORTHOGRAPHIC PROJECTION DRAWING
(THIRD-ANGLE PROJECTION)

Fig. 1-4 A simple object shown in orthographic and pictorial form

The width of the object is depicted in the front, top, bottom, and side views. Seldom are more than three views necessary to completely describe the shape of an object. Therefore, the simple object shown in figure 1-4 can be used to illustrate position of these principal dimensions.

In figure 1-4 the object is shown in both orthographic projection and pictorial form. The drawing uses each view to represent the exact shape and size of the object and the relationship of the three views to one another. This principle of projection is used in all mechanical drawings. The isometric drawing shows the relationship of the front, top, and right-side surfaces.

ISO PROJECTION SYMBOL

Since two types of projection, first- and third-angle, are used on engineering drawings,

and since each has the same units of measurements, it is necessary to be able to identify the type of projection. The *International Standards Organization* (ISO) has recommended that the symbol shown in figure 1-5 be shown on all drawings. Its preferred location is the lower right-hand corner of the drawing, adjacent to the title block, figure 1-6. To aid the reader in learning the language of industry, some drawings in this text have been drawn in first- as well as in third-angle projection. The ISO symbol will indicate the type of projection used.

TITLE BLOCK

All drawings should have some form of title block, placed preferably in the lower right-hand corner. A title block may contain such information as:

- Name of the part

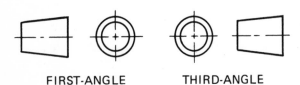

FIRST-ANGLE   THIRD-ANGLE

Fig. 1-5 ISO projection symbols

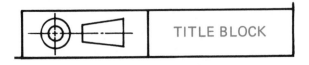

TITLE BLOCK

Fig. 1-6 Locating ISO symbol on drawing paper

- Order number
- Date
- Number of the drawing
- Scale size used
- Name of the drafter
- Name of the drawing inspector
- Material to be used

DRAWING STANDARDS

The drawings and information shown throughout this text are based on the *American Standards Drafting Manual Y14* and the *Canadian Engineering Drawing Standards B 78.* In some areas of drawing practice, such as simplified drafting, national standards have not yet been established. The authors have, in such cases, adopted the practices used by leading industries in the United States and Canada.

VISIBLE LINES

A thick solid line is used to indicate the visible edges and corners of an object. Visible lines should stand out clearly in contrast to other lines, making the general shape of the object apparent to the eye.

LETTERING ON DRAWINGS

The most important requirements for lettering are legibility, reproducibility, and ease of execution. These requirements are best met by the style of lettering known as standard uppercase Gothic, as shown in figure 1-7.

ABCDEFGHIJKLMNOP
QRSTUVWXYZ
1234567890
VERTICAL

ABCDEFGHIJKLMNOP
QRSTUVWXYZ
1234567890
SLOPED

Fig. 1-7 Lettering for drawings

For these reasons it is used on mechanical engineering drawings. Vertical lettering is preferred, but sloping style may be used, though never on the same drawing as vertical lettering. Suitable lettering size for notes and dimensions is 3 mm. Larger characters are used for drawing titles and numbers. They are also used where it may be necessary to bring some part of the drawing to the attention of the reader.

SKETCHING

Sketching is a necessary part of the course of interpreting technical drawings since the skilled technician in the shop is frequently called upon to sketch and explain points to other people. Sketching also helps to develop a good sense of proportion and accuracy of observation. The most common types of sketching paper are shown in figure 1-8. Each square on the paper may represent 1 mm, 10 mm, 0.10 inch or 1 inch of actual object length. Figure 1-8 illustrates the use of graph paper for both pictorial and orthographic projection.

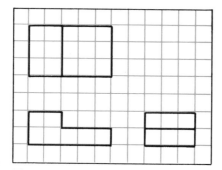

(A) COORDINATE SKETCHING PAPER USED FOR SKETCHING ORTHOGRAPHIC PROJECTION

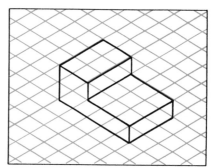

(B) ISOMETRIC SKETCHING PAPER USED FOR SKETCHING PICTORIAL DRAWINGS

Fig. 1-8 Sketching paper

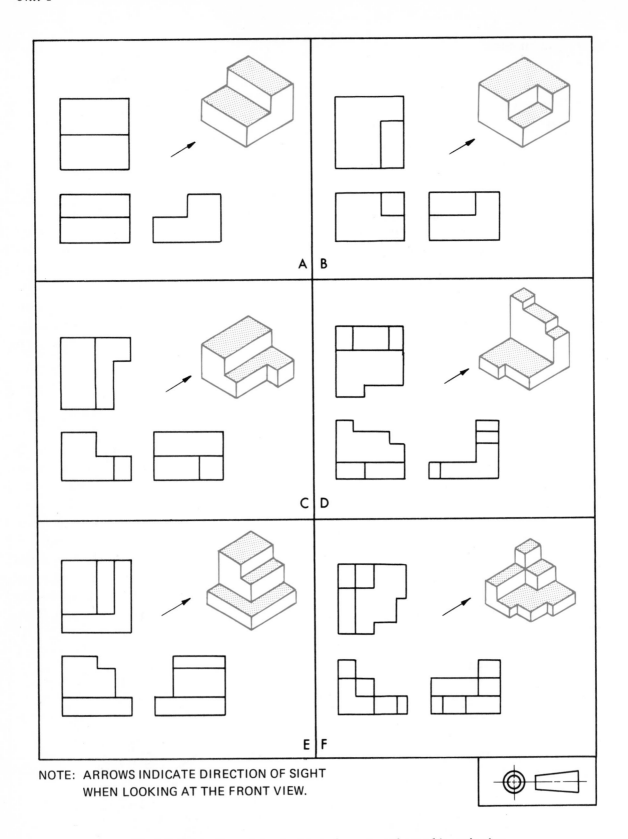

NOTE: ARROWS INDICATE DIRECTION OF SIGHT
WHEN LOOKING AT THE FRONT VIEW.

Fig. 1-9 Illustrations of simple objects drawn in orthographic projection

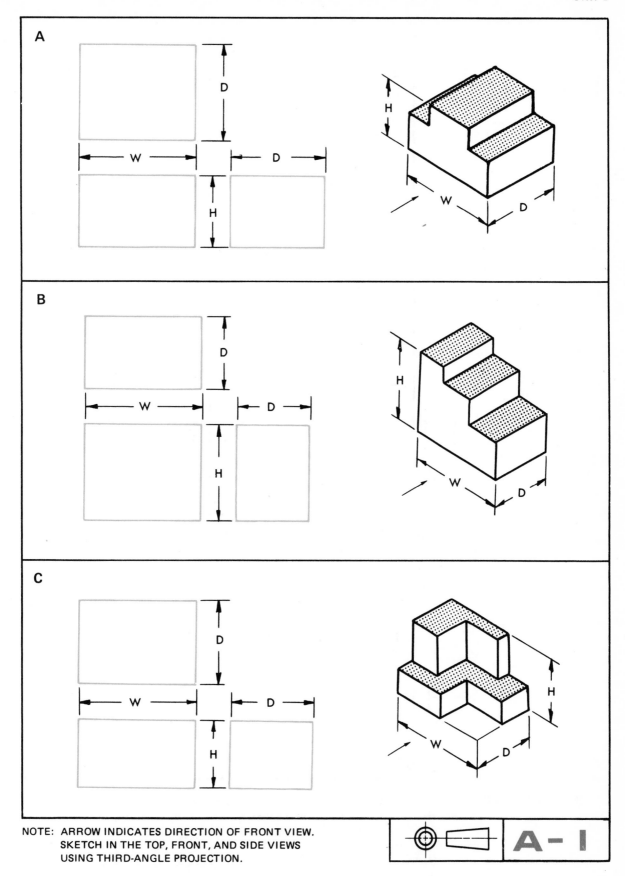

A

B

C

NOTE: ARROW INDICATES DIRECTION OF FRONT VIEW.
SKETCH IN THE TOP, FRONT, AND SIDE VIEWS
USING THIRD-ANGLE PROJECTION.

A-1

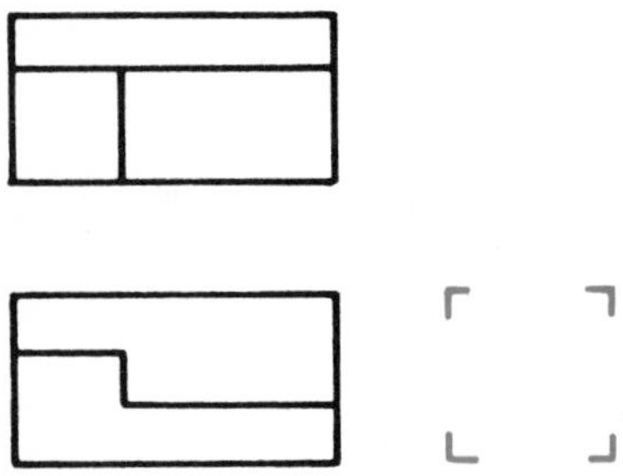

A

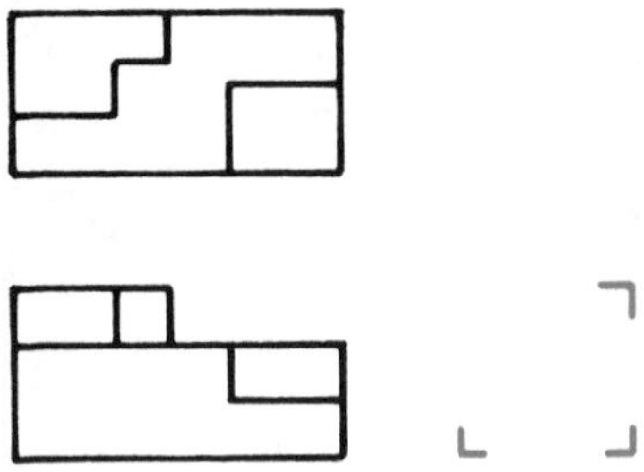

B

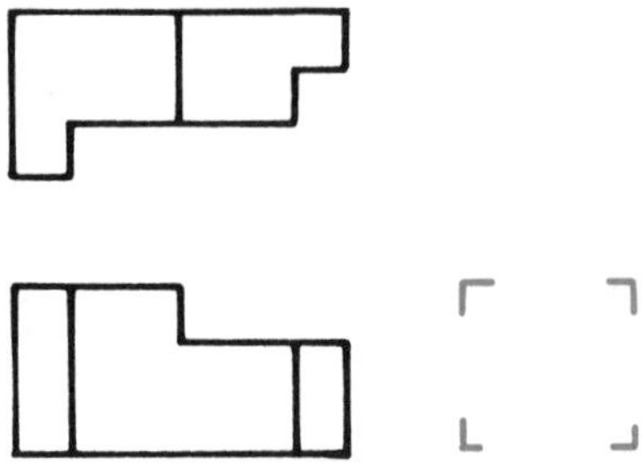

C

SKETCH THE SIDE VIEWS; THE TOP AND FRONT VIEWS ARE COMPLETE.

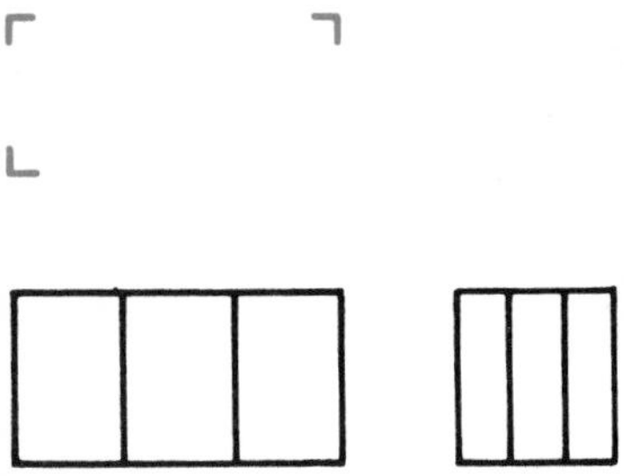

D

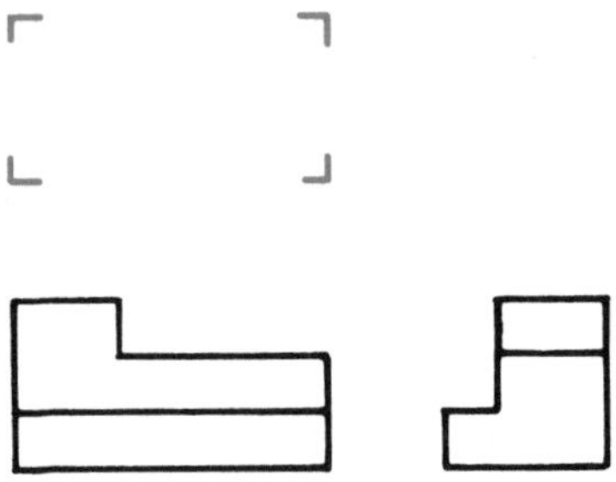

E

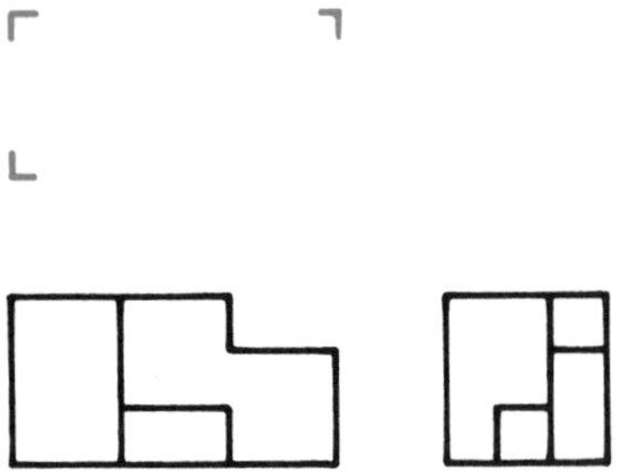

F

SKETCH THE TOP VIEWS; THE FRONT AND SIDE VIEWS ARE COMPLETE.

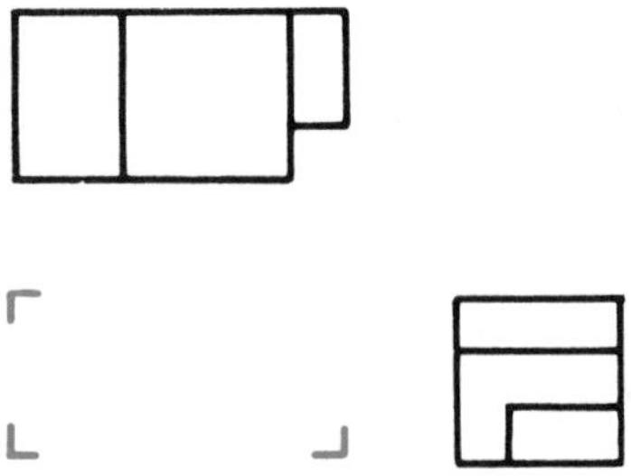

G

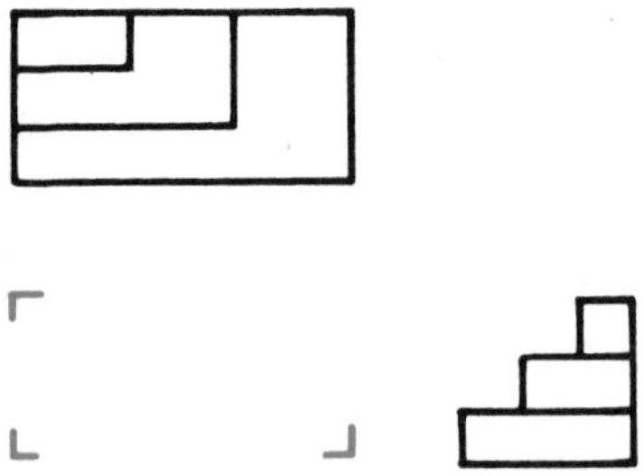

H

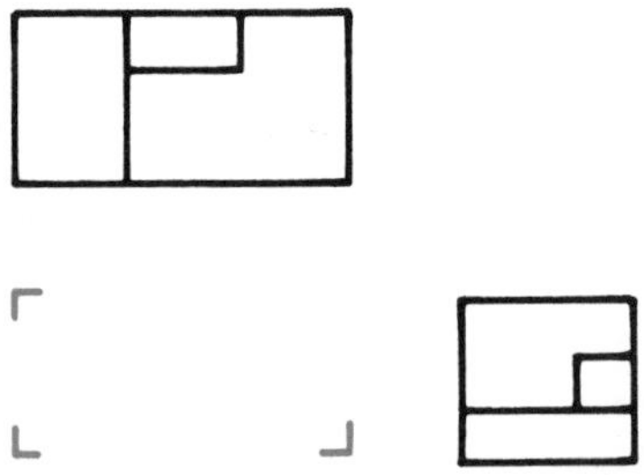

K

SKETCH THE FRONT VIEWS; THE TOP AND SIDE VIEWS ARE COMPLETE.

COMPLETION ASSIGNMENT: ALL SURFACES ARE
VERTICAL OR HORIZONTAL.

A-2

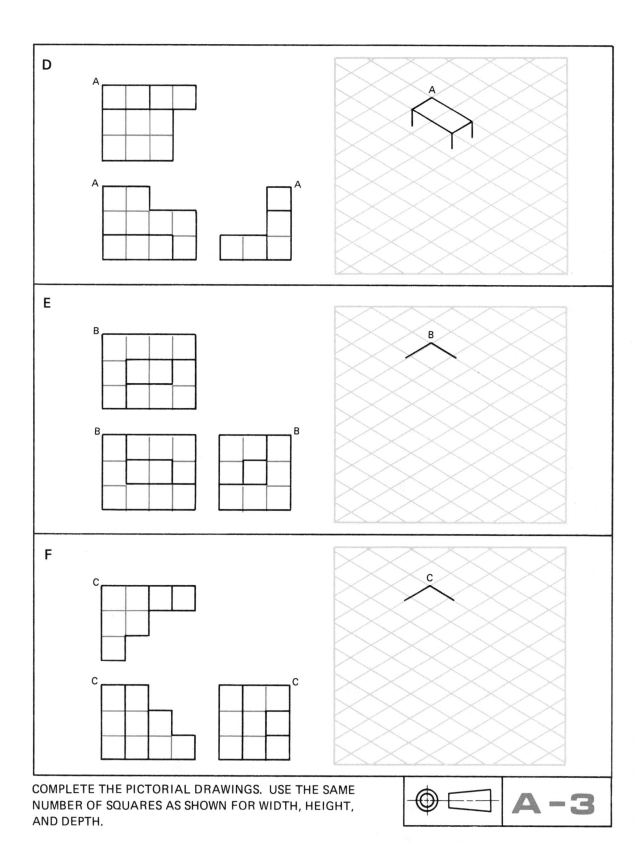

D

E

F

COMPLETE THE PICTORIAL DRAWINGS. USE THE SAME
NUMBER OF SQUARES AS SHOWN FOR WIDTH, HEIGHT,
AND DEPTH.

A-3

UNIT 2

WORKING DRAWINGS

A working drawing, sometimes referred to as a detail or assembly drawing, contains the complete information for the construction of an object or the assembly of a product. The information found on this drawing may be classified under three headings:

- *Shape.* The description of the shape of an object is indicated by the number and type of views and lines selected to completely show or describe the form of an object.

- *Dimensioning.* The description of the size and location of the shape of an object is indicated by lines and numerals used to specify the dimensions of an object.

- *Specifications.* Additional information such as general notes, material, heat treatment, machine finish, and applied finish are indicated by special entries written on the drawings.

This data is found on the drawing, in the title block.

DIMENSIONING

Dimensions are indicated on drawings by extension lines, dimension lines, leaders, arrowheads, figures, notes, and symbols. These lines and dimensions define such geometrical characteristics as lengths, diameters, angles, and locations, figure 2-1. The lines used in dimensioning are *thin* in contrast to the outline of the object. The dimension must be clear and concise, permitting only one interpretation. In general, each surface, line, or point is located by only one set of dimensions. Exceptions to these rules are arrowless and tabular dimensioning which are discussed in Unit 18.

Placement of Dimensions

The two most commonly used methods of dimensioning are the *unidirectional system* and the *aligned system.* The aligned system, which is read from the bottom and right side of the drawing, is outdated. The currently preferred method is the unidirectional system which is read from the bottom only, figure 2-2.

Dimension Lines

Dimension lines denote particular sections of the object. They should be drawn parallel to the section they define. Dimension lines terminate in arrowheads which touch an extension line and are broken in order to allow the insertion of the dimension. Unbroken lines are sometimes used as a simplified drafting practice. Where space does not permit the insertion of the dimension line and dimension between the extension lines, the dimension line may be placed outside the extension line. The dimension can also be placed outside the extension line if the space between the extension lines is limited. In restricted areas a dot may be used instead of two arrowheads. These methods are shown in figure 2-3, page 12.

Extension Lines

Extension lines denote the points or surfaces between which a dimension applies. They extend from object lines and are drawn perpendicularly to the dimension lines as shown in figures 2-1 and 2-2. A small gap is left between the extension line and the outline to which it refers.

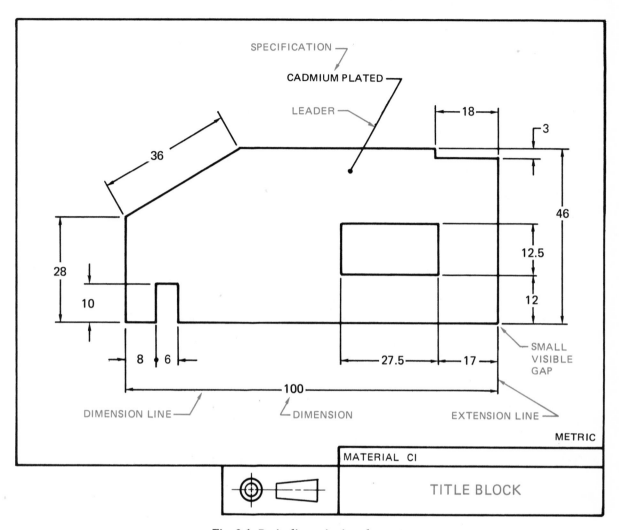

Fig. 2-1 Basic dimensioning elements

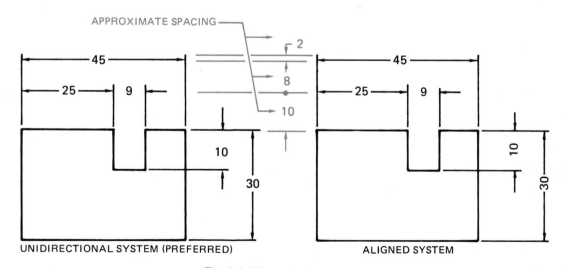

Fig. 2-2 Dimensioning systems

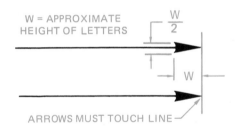

(A) SIZE OF ARROWHEADS

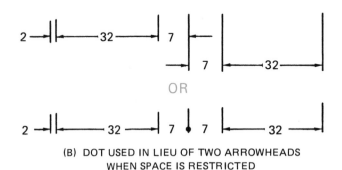

OR

(B) DOT USED IN LIEU OF TWO ARROWHEADS WHEN SPACE IS RESTRICTED

Fig. 2-3 **Arrowheads**

Where extension lines cross arrowheads or dimension lines close to arrowheads, a break in the extension line is permitted, figure 2-4.

Leaders

Leaders are used to direct dimensions or notes to the surface or points to which they apply, figure 2-5. A leader consists of a line with or without short horizontal bars adjacent to the note or dimension and an inclined portion which terminates with an arrowhead touching the line or point to which it applies. A leader may terminate with a dot when it refers to a surface within the outline of a part.

Hidden Lines

Most objects drawn in engineering offices are complicated and contain many surfaces and edges. Many features (lines, holes, etc.) cannot be seen when viewed from the outside of the piece. These hidden

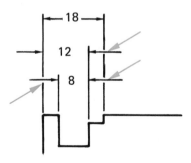

Fig. 2-4 **Breaks in extension lines**

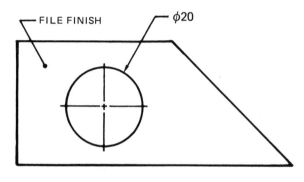

Fig. 2-5 **Using a leader for dimensions and notes**

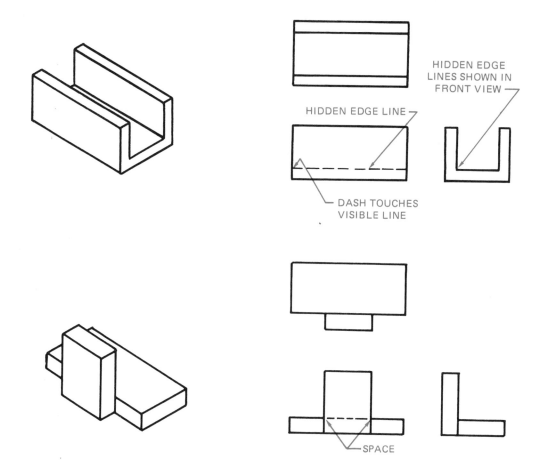

Fig. 2-6 Hidden lines

edges are shown on the drawings by a series of short dashes which are referred to as a *hidden line*. The length of the dashes varies slightly in relationship to the size of the drawing. Hidden lines are usually required on the drawing to help show the true shape of the object, but may be omitted when not required to preserve the clarity of the drawing.

Hidden lines should always start and end with a dash except when such a dash would form a continuation of a visible detail line. Dashes should always join at corners. Illustrations of these hidden line techniques are shown in figure 2-6.

UNITS OF MEASUREMENT

Although the metric system of dimensioning has become the official standard of measurement, most drawings in existence in North America are dimensioned in inches or feet and inches. For this reason, drafters and persons involved in the reading of engineering drawings should be familiar with all the dimensioning systems they may encounter.

The assignment following the instructional material in each unit will use metric dimensions when possible. When the metric units of measurement are used the drawing will prominently display the word METRIC and state the units of measurement shown on the drawing. Drawings not showing the METRIC designation are dimensioned in inches. However, all drawings accompanying the text material are dimensioned in millimetres unless otherwise stated.

The SI (Metric) System

The standard metric units on engineering drawings are the millimetre for linear measure and the micrometre for surface roughness. For architectural drawings metre and millimetre units are used, figure 2-7.

In metric dimensioning, as in decimal-inch dimensioning, numerals to the right of the decimal point indicate the degree of precision.

Whole dimensions do not require a zero to the right of the decimal point.

2 not 2.0
10 not 10.0

A millimetre value of less than one is shown with a zero to the left of the decimal point.

0.2 not .2 or .20
0.26 not .26

Commas should not be used to separate groups of three numbers in either inch or metric values. A space should be used in place of the comma.

32 541 not 32,541
2.562 826 6 not 2.5628266

Identification. A metric drawing should include a general note, such as UNLESS OTHER-WISE SPECIFIED DIMENSIONS ARE IN MILLIMETRES. A metric drawing should in addition, be identified by the word METRIC prominently displayed near the title block.

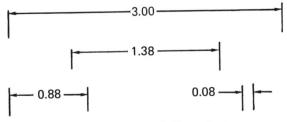

Fig. 2-8 Decimal-inch dimensioning

English System

The Decimal-Inch System. In the decimal-inch system parts are designed in basic decimal increments, preferably 0.02 inches and are expressed as two place decimal numbers. Using the 0.02 module, the second decimal place (hundredths) is an even number or zero, figure 2-8. Sizes other than these, such as 0.25, are used when essential to meet design requirements. When greater accuracy is required, sizes are expressed as three or four place decimal numbers such as 1.875.

The Fractional-Inch System. In this system, sizes are expressed in common fractions, the smallest division being 64ths, figure 2-9. Sizes other than common fractions are expressed as decimals.

The Feet and Inches System. This system is used in architectural drawing and for large structural and installation drawings. Dimensions greater than 12 inches, with parts of an inch given as common fractions. Inch

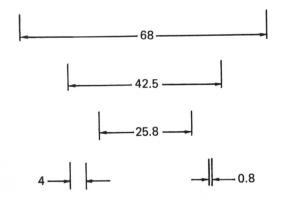

Fig. 2-7 Metric (millimetre) dimensioning

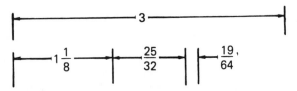

Fig. 2-9 Fractional-inch dimensioning

marks are not used, and a zero is added to indicate no full inches. A dash mark is placed between the foot and inch values, figure 2-10.

CHOICE OF DIMENSIONS

The choice of the most suitable dimensions and dimensioning methods will depend, to some extent, on whether the drawings are intended for unit production or mass production.

Unit production refers to cases in which each part is to be made separately, using general-purpose tools and machines. Details on custom-built machines, jigs, fixtures, and gauges required for the manufacture of production parts are made in this way. Frequently, only one of each part is required.

Mass production refers to parts produced in quantity, where special tooling is usually provided. Most part drawings for manufactured products are considered to be for mass-produced parts.

Functional dimensioning should be expressed directly on the drawing, especially for mass-produced parts. This will result in the selection of datum features on the basis of function and assembly for unit-produced parts, it is generally preferable to select datum features on the basis of manufacture and machining.

Nonfunctional dimensions should be selected on the basis of facilitating production and inspection rather than to facilitate tool and gauge design.

BASIC RULES FOR DIMENSIONING

- Place dimensions between the views when possible.
- Place the dimension line for the shortest width, height, and depth, nearest the outline of the object. Parallel dimension lines are placed in order of their size, making the longest dimension line the outermost.

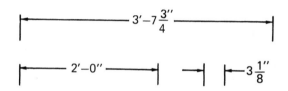

Fig. 2-10 Feet and inch dimensioning

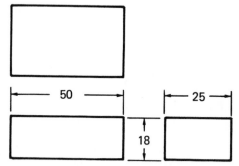

(A) PLACE DIMENSIONS BETWEEN VIEWS.

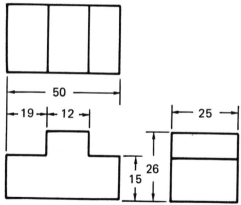

(B) PLACE SMALLEST DIMENSION NEAREST THE VIEW BEING DIMENSIONED.

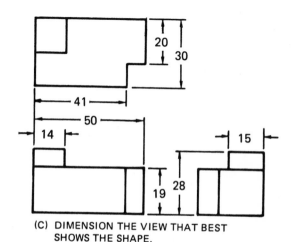

(C) DIMENSION THE VIEW THAT BEST SHOWS THE SHAPE.

Fig. 2-11 Basic dimensioning rules

- Place dimensions near the view that best shows the characteristic contour or shape of the object. In following this rule, dimensions will not always be between views.
- On large drawings, dimensions can be placed on the view to improve clarity.
- Use only one system of dimensions, either the unidirectional or the aligned, on any one drawing.

INFORMATION SHOWN ON ASSIGNMENT DRAWINGS

Circled numbers and letters shown in red are to be followed on the drawing assignments so that questions may be asked about lines and surfaces without involving a great number of descriptive items. They are learning aids and, as the course progresses, are omitted from the more advanced problems. To simplify the drawings, the actual working drawing is drawn in black. The information which is used in the developing of interpreting technical drawings is shown in red and would not appear on working drawings found in industry.

REFERENCES AND SOURCE MATERIALS

1. *American Drafting Standards Manual Y14*, (ASME).
2. Canadian Standard Association, *Bulletin 78.2*.

QUESTIONS

1. What is the name of the object?
2. What is the drawing number?
3. How many pieces are to be made?
4. Of what material is the part made?
5. What is the overall width?
6. What is the overall depth?
7. What is the overall height?
8. Which line or surface in the side view represents surface (F) in the top view?
9. Which line or surface in the side view represents surface (E) in the top view?
10. Which line or surface in the side view represents surface (G) in the top view?
11. Which line or surface in the side view represents surface (L) of the front view?
12. What is the vertical height in the side view from the surface represented by line (P) to that represented by line (Q)?
13. What is the height of the step in the side view from the bottom of the part to the surface represented by surface (E)?
14. Which two letters in the top view represent distance (V) in the side view?
15. The dimensions of which two letters in the top view represent distance (W) in the side view?
16. Which line or surface in the side view represents surface (M) in the front view?
17. What is the height of line (N)?
18. Which line or surface in the front view represents the surface (R) in the side view?
19. Which line or surface in the top view represents surface (L)?
20. Which line or surface in the front view represents surface (F)?
21. Which line or surface in the front view represents surface (E)?
22. Which line or surface in the top view represents surface (M)?
23. What type of line is (T)?
24. What type of line is (Y)?
25. What units of measurement are used on this drawing?
26. Calculate dimensions (B), (C), (D), and (W).

METRIC	
QUANTITY	2
MATERIAL	MS
SCALE	1:1
DRAWN	DATE

COUNTER CLAMP BAR

A-4

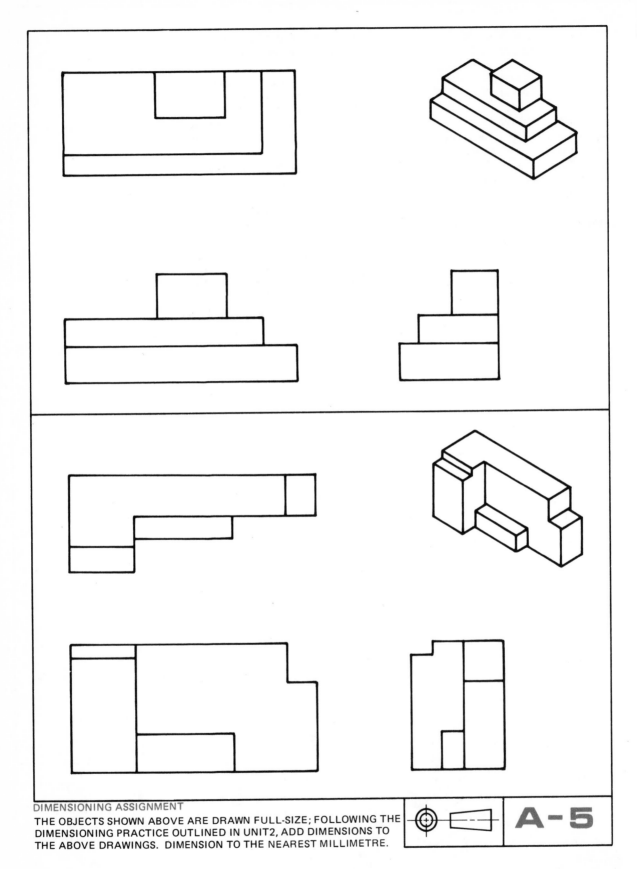

DIMENSIONING ASSIGNMENT
THE OBJECTS SHOWN ABOVE ARE DRAWN FULL-SIZE; FOLLOWING THE
DIMENSIONING PRACTICE OUTLINED IN UNIT2, ADD DIMENSIONS TO
THE ABOVE DRAWINGS. DIMENSION TO THE NEAREST MILLIMETRE.

A-5

DRAWING REPRODUCTION

The original drawings made by the drafter are kept in the engineering department at all times. Copies or prints of the originals are sent to the shop. For many years the most popular reproduction method was the blueprint. The demand for a faster and more versatile type of reproduction during World War II led to the introduction of the diazo process (white prints) of duplicating drawings, figure 3-1.

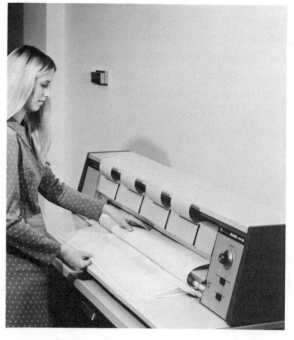

(B) A WHITEPRINT DIAZO MACHINE IN USE

Fig. 3-1B The diazo method of reproduction

Today, the demand for more economical and versatile methods of reproducing drawings and the expense of maintaining filing systems has brought about the introduction of new methods of reproducing drawings, two of which are *microfilming* and *photoreproduction*.

The use of microfilm has been steadily increasing in recent years. Microfilm has a 35 mm image set into an aperture card, figure 3-2. Using this technique, the drawing may also be stored on a roll of 35 mm film.

Microfilm enables the drawing to be stored in a fraction of the floor space needed for filing the full-size paper originals. It can be referenced on the screen of a reader or, if a print is required, microfilm can be enlarged and reproduced on a reader-printer, figure 3-3, page 20.

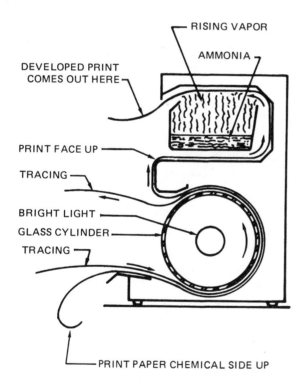

RISING VAPOR

AMMONIA

DEVELOPED PRINT COMES OUT HERE

PRINT FACE UP

TRACING

BRIGHT LIGHT

GLASS CYLINDER

TRACING

PRINT PAPER CHEMICAL SIDE UP

ROLLERS MOVE THE TRACING AND PRINT AROUND THE LIGHT, AND MOVE THE PRINT PAST THE RISING AMMONIA VAPOR.

(A) THE DIAZO PROCESS

Fig. 3-1A The diazo method of reproduction

Photoreproduction is one of the newest and most versatile methods for reproducing engineering drawings, figure 3-4. One such printer offers, in addition to full-scale printing, five reduction settings by which a drawing, or part of a drawing, can be reduced in size. Drawing reductions allow significant advantages: less paper consumption, lower handling and mailing costs, and smaller filing space requirements.

Because photoreproduction utilizes photography, it is suitable for scissors and paste-up drafting. This technique allows the drafter to incorporate previously printed or drawn material into his own work.

Fig. 3-4 Photoreproduction process. This machine will reproduce, reduce, fold, and sort prints.

Fig. 3-2 Aperture cards

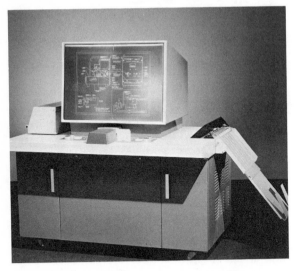

Fig. 3-3 Reader-printer for microfilm

ABBREVIATIONS USED ON DRAWINGS

Abbreviations and symbols are used on drawings to conserve time and space, but only when their meaning is perfectly understood. A few of the more common abbreviations and symbols are shown in table 8 of the appendix. A complete set of abbreviations for drawing use may be found in ANSI Y1.1 or CSA Z-85, *Abbreviations For Use On Drawings And In Text.*

Another advantage is that since the drawing is photographed and prints or tracings can be made from the original work, the drafter can draw on practically any type of paper. However, in industrial practice drawings are usually done on translucent sheets.

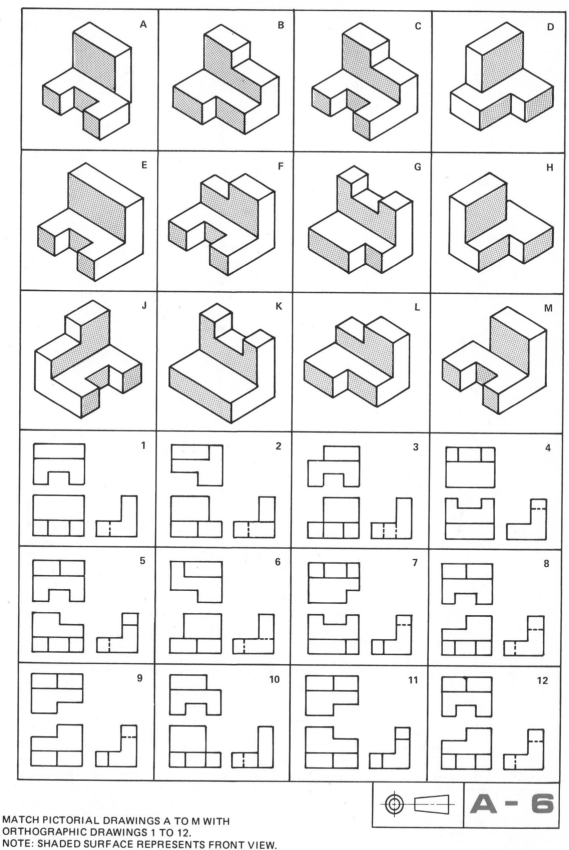

MATCH PICTORIAL DRAWINGS A TO M WITH
ORTHOGRAPHIC DRAWINGS 1 TO 12.
NOTE: SHADED SURFACE REPRESENTS FRONT VIEW.

A-6

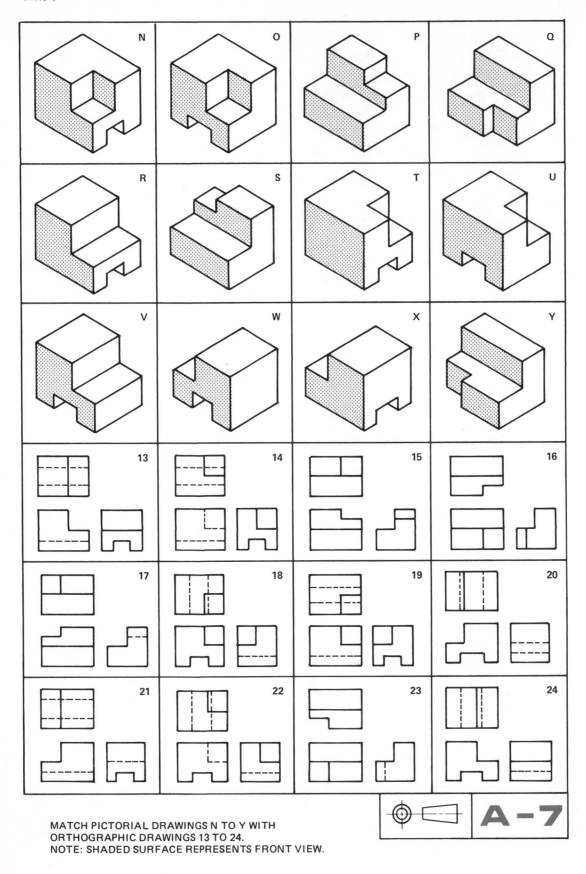

MATCH PICTORIAL DRAWINGS N TO Y WITH
ORTHOGRAPHIC DRAWINGS 13 TO 24.
NOTE: SHADED SURFACE REPRESENTS FRONT VIEW.

A-7

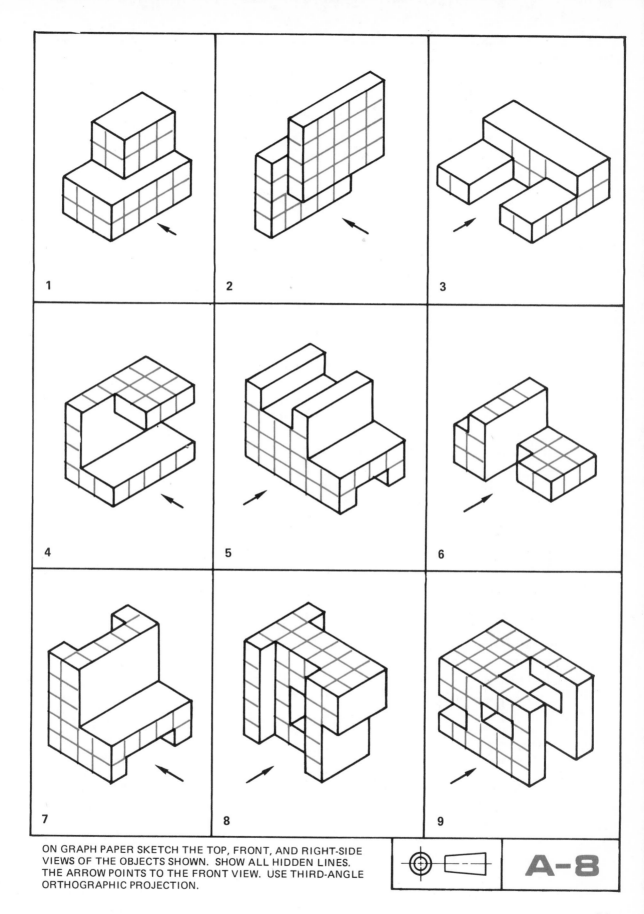

1

2

3

4

5

6

7

8

9

ON GRAPH PAPER SKETCH THE TOP, FRONT, AND RIGHT-SIDE
VIEWS OF THE OBJECTS SHOWN. SHOW ALL HIDDEN LINES.
THE ARROW POINTS TO THE FRONT VIEW. USE THIRD-ANGLE
ORTHOGRAPHIC PROJECTION.

A-8

23

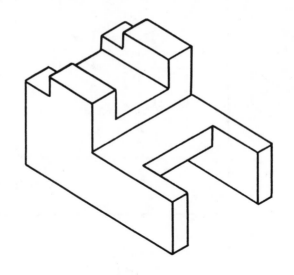

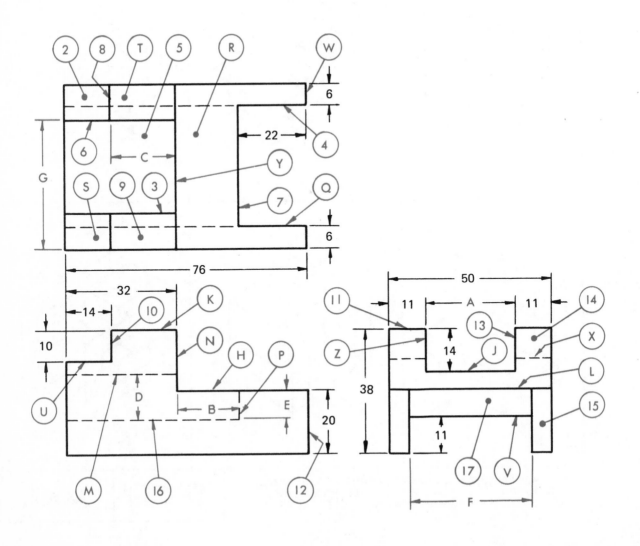

QUESTIONS

1. What is the object?
2. What is the drawing number?
3. How many castings are required?
4. What material is the part made of?
5. What is the overall width?
6. What is the overall height?
7. What is the overall depth?
8. Calculate distances A through G.
9. Which line in the top view represents surface (P)?
10. Which line in the side view represents surface (5)?
11. Which line in the side view represents surface (R)?
12. Which surfaces in the top view does line (K) of the front view represent?
13. Which surface in the top view does line (M) in the front view represent?
14. Which line in the side view represents the same surface represented by line (M) of the front view?
15. What kind, or type of line, is line (M)?
16. Which front view line does line (X) in the side view represent?
17. Which front view line does line (Y) in the top view represent?
18. Which line in the front view does surface (15) in the side view represent?
19. Which front view line represents surface (R) in the top view?
20. Which surface in the side view represents line (N) of the front view?
21. Which line in the side view represents surface (2)?
22. Which surface in the side view does line (P) represent?
23. Which surface in the top view does line (11) represent?
24. Which line in the side view does line (3) in the top view represent?
25. Which line in the side view does line (16) in the front view represent?
26. Which surface in the side view does (W) represent?

ANSWERS

1 _____
2 _____
3 _____
4 _____
5 _____
6 _____
7 _____
8 A _____
  B _____
  C _____
  D _____
  E _____
  F _____
  G _____
9 _____
10 _____
11 _____
12 _____
13 _____
14 _____
15 _____
16 _____
17 _____
18 _____
19 _____
20 _____
21 _____
22 _____
23 _____
24 _____
25 _____
26 _____

METRIC

MATERIAL	MALLEABLE IRON
SCALE	NOT TO SCALE
DRAWN BY	DATE

FEED HOPPER   **A-9**

INCLINED SURFACES

If the surfaces of an object lie in either a horizontal or a vertical position, the surfaces appear in their true shape in one of the three views and these surfaces appear as a line in the other two views.

When a surface is sloped or inclined in only one direction, that surface is not seen in its true shape in the top, front, or side views. It is, however, seen in two views as a distorted surface. On the third view it appears as a line.

The true length of surfaces **A** and **B** in figure 4-1 is seen in the front view only. In the top and side views, only the width of surfaces **A** and **B** appears in its true size. The length of these surfaces is foreshortened.

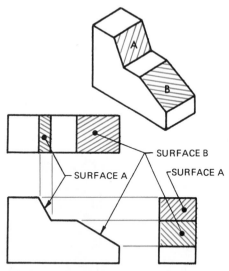

NOTE: THE TRUE SHAPES OF SURFACES A AND B DO NOT APPEAR ON THE TOP OR SIDE VIEWS.

Fig. 4-1 Inclined surfaces

Where an inclined surface has important features that must be shown clearly and without distortion, an auxiliary or helper view must be used. These views will be discussed in detail later in the book.

MEASUREMENT OF ANGLES

Some objects do not have all their features positioned in such a way that all surfaces can be in the horizontal and vertical plane at the same time. The design of the part may require that some of the lines in the drawing be shown in a direction other than horizontal or vertical. This will necessitate some lines to be drawn at an angle.

The amount of this divergence, or obliqueness, of lines may be indicated by either an offset dimension or an angle dimension as shown in figure 4-2.

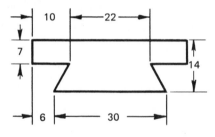

LINEAR MEASUREMENTS

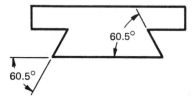

ANGLE MEASUREMENTS USING
DECIMAL DEGREES

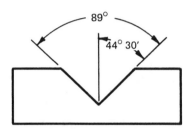

ANGLE MEASUREMENTS USING
DEGREES AND MINUTES

Fig. 4-2 Dimensioning angles

Angle dimensions may be expressed in degrees and decimal parts of a degree. They may be expressed in degrees, minutes, and seconds. The latter method is now preferred. The symbols for degrees (°), minutes ('), and seconds (") are included with the appropriate values.

For example, 2°; 30°; 28°10'; 0°15'; 27°13'15"; 0°0'30"; 0.25°; 30°0'0"; ± 0°2'30"; and 2.0° ± 0.5° are all correct forms.

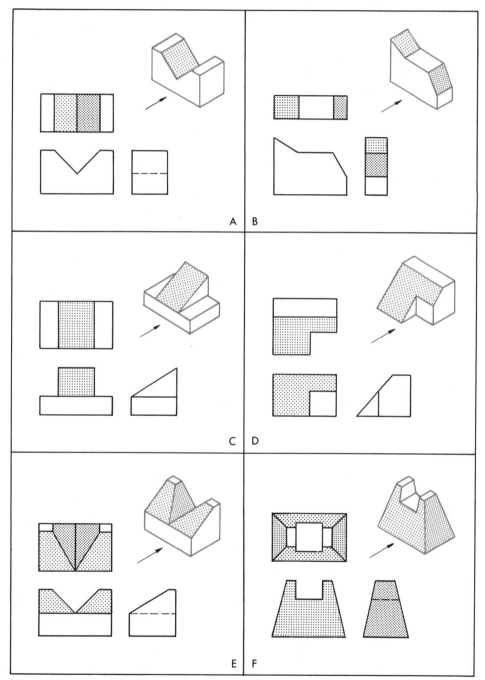

NOTE: ARROWS INDICATE DIRECTION OF SIGHT WHEN LOOKING AT THE FRONT VIEW.

Fig. 4-3 Illustrations of simple objects having inclined surfaces

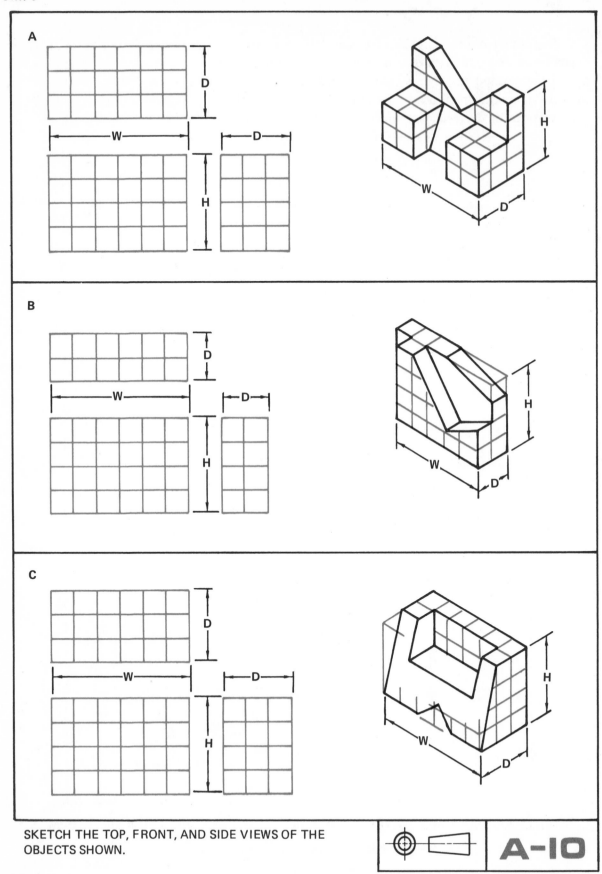

A

B

C

SKETCH THE TOP, FRONT, AND SIDE VIEWS OF THE OBJECTS SHOWN.

A-10

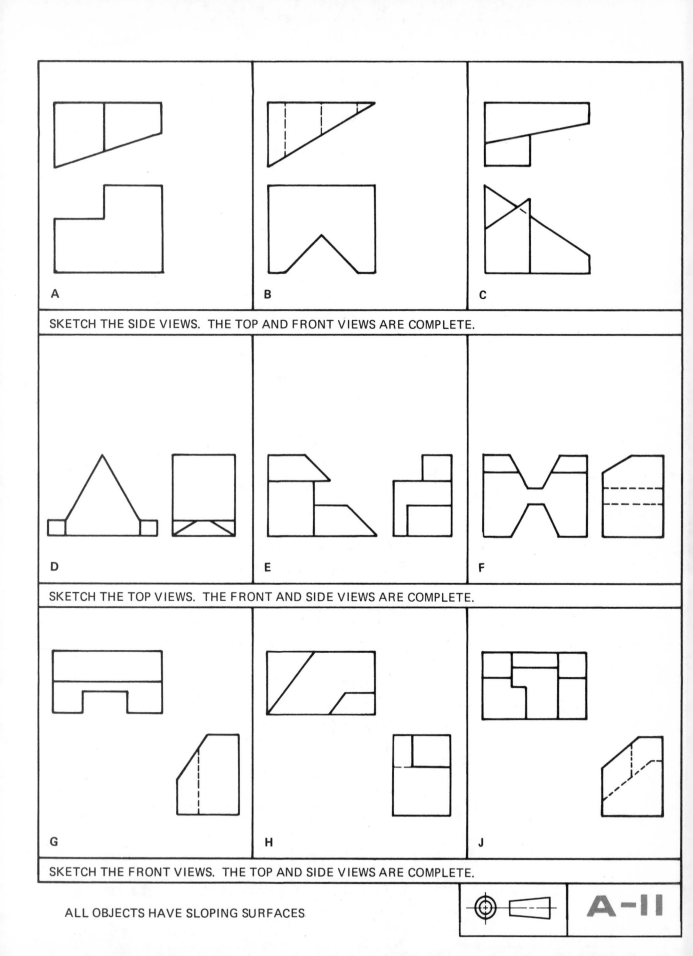

A

B

C

SKETCH THE SIDE VIEWS. THE TOP AND FRONT VIEWS ARE COMPLETE.

D

E

F

SKETCH THE TOP VIEWS. THE FRONT AND SIDE VIEWS ARE COMPLETE.

G

H

J

SKETCH THE FRONT VIEWS. THE TOP AND SIDE VIEWS ARE COMPLETE.

ALL OBJECTS HAVE SLOPING SURFACES

A-11

1. Calculate distances A to G.
2. At what angle is line ⑥ to the vertical?
3. At what angle is line ⑦ to the horizontal?
4. Locate surface ⑥ in the side view.
5. Locate surface ① in the side view.
6. Locate surface ⑥ in the top view.
7. Which lines in the side view are represented by line ② in the front view?
8. Locate ⑱ in the top view.
9. Locate surface ⑨ in the side view.
10. Locate surface ⑫ in the front view.
11. Locate surface ③ in the top view.
12. Which lines in the side view are represented by point ④ in the front view?
13. Which line in the side view is line ⑯ in the top view?
14. Locate surface ⑩ in the side view.
15. Locate surface ⑩ in the front view.
16. Locate surface ⑫ in the side view.
17. Locate surface ⑮ in the side view.
18. Which lines does point ④ represent in the top view?
19. Locate surface ⑥ in the top view.
20. Locate line ⑯ in the side view.
21. Locate line ㉕ in the top view.
22. Which line in the front view is surface ⑮ in the top view?

1 A _____
B _____
C _____
D _____
E _____
F _____
G _____
2 _____
3 _____
4 _____
5 _____
6 _____
7 _____
8 _____
9 _____
10 _____
11 _____
12 _____
13 _____
14 _____
15 _____
16 _____
17 _____
18 _____
19 _____
20 _____
21 _____
22 _____

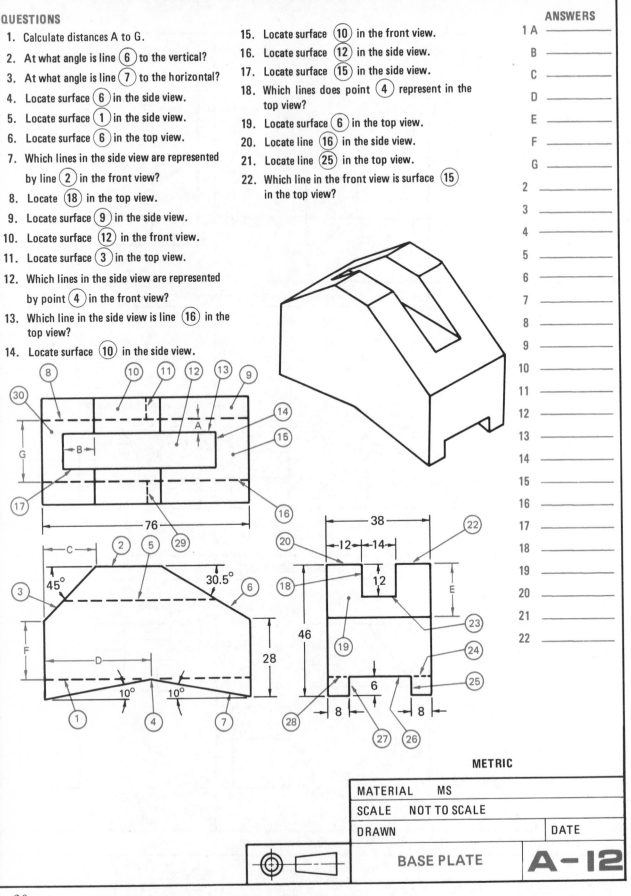

METRIC

MATERIAL	MS	
SCALE	NOT TO SCALE	
DRAWN		DATE
BASE PLATE		A-12

UNIT

5

CIRCULAR FEATURES

Typical parts with circular features are illustrated in figure 5-1. Note that the circular feature appears circular in one view only and that no line is used to indicate where a curved surface joins a flat surface. Hidden circles, like hidden flat surfaces, are represented on drawings by a hidden line.

CENTER LINES

A *center line* is drawn as a thin broken line of long and short dashes, spaced alternately. Center lines may be used to indicate center points, axes of cylindrical parts, and axes of symmetry. Solid center lines are often used as a simplified drafting practice; however, the interrupted line is preferred. Center lines should project for a short distance beyond the outline of the part or feature to which they refer. They may be lengthened for use as extension lines for dimensioning purposes; in this case the extended portion is not broken.

In end views of circular features, the point of intersection of two center lines is shown by two intersecting short dashes. However, for very small circles a solid unbroken line is recommended. These techniques are shown in figure 5-2, page 32.

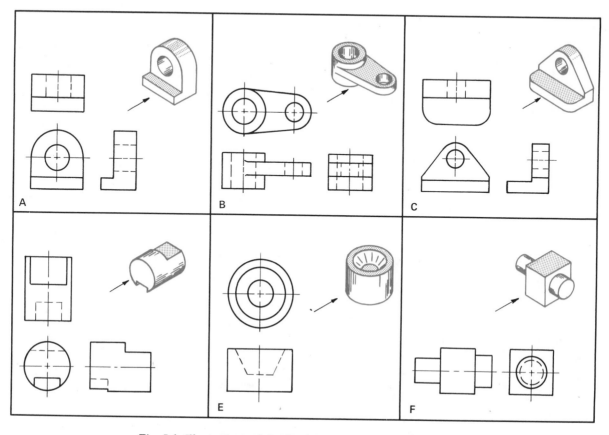

Fig. 5-1 Illustrations of simple objects having circular features

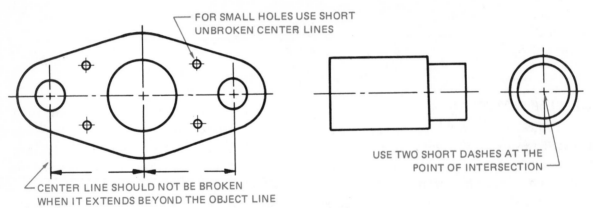

Fig. 5-2 Center line application

DIMENSIONING OF CYLINDRICAL FEATURES

Features shown as circles are normally dimensioned by one of the methods shown in figure 5-3. Where the diameters of a number of concentric cylinders are to be given, it may be more convenient to show them on the side view. The symbol ⌀ shown before

dimension or the abbreviation DIA is shown after the diametral dimension.

The radius of the arc is used in dimensioning a circular arc. The letter R accompanies the dimension to indicate that it is a radius. Approved methods for dimensioning arcs are shown in figure 5-4.

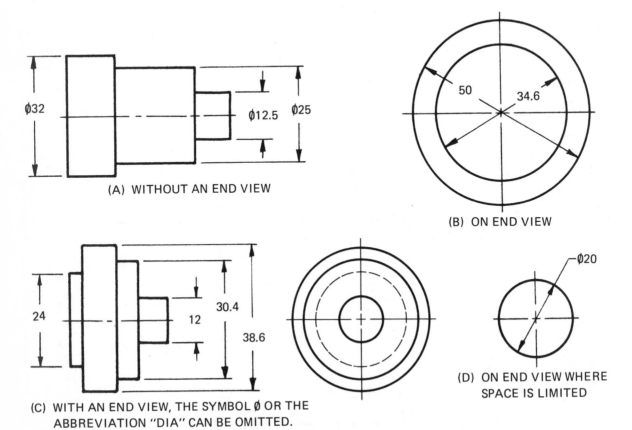

Fig. 5-3 Dimensioning diameters

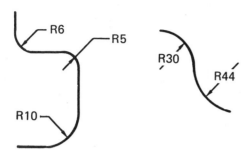

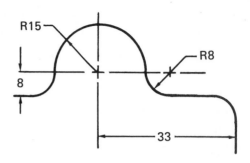

(A) DIMENSIONING RADII WHICH NEED NOT HAVE THEIR CENTER POINTS LOCATED

(B) RADII WITH LOCATED CENTERS

Fig. 5-4 Dimensioning radii

DIMENSIONING CYLINDRICAL HOLES

Specification of the diameter with a leader, as shown in figure 5-5, is the preferred method for designating the size of small holes. For larger holes, use one of the methods illustrated in figure 5-3. When the leader is used the symbol Ø precedes the abbreviation or DIA follows the size of the hole. The note end of the leader terminates in a short horizontal bar. When two or more holes of the same size are required, the number of holes is specified after the size. If a blind hole is required, the depth of the hole is included in the dimensioning note; otherwise, it is assumed that all holes shown are through holes. Terms such as drill, ream, bore, etc. should be avoided on drawings. The method of producing the hole should be left to the discretion of either the shop personnel or the planning department.

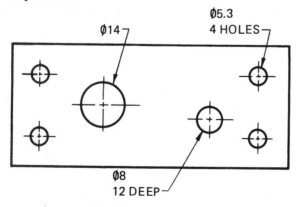

Fig. 5-5 Dimensioning cylindrical holes

DRILLING, REAMING, AND BORING

Drilling is the process of piercing a hole through a solid with a drill or the enlarging of a smaller hole. For some types of work holes must be drilled smooth, straight, and an exact size. On others, accuracy of location and size of the hole are not as important.

When accurate holes of uniform diameter are required, they are first drilled slightly undersize and then reamed. *Reaming* is the process of sizing a hole to a given diameter with a reamer in order to produce a hole which is round, smooth, and straight.

Boring is one of the more dependable methods of producing holes which are round and concentric. The term boring refers to the enlarging of a hole by means of a boring tool. The use of reamers is limited to the sizes of available reamers. Holes, however, may be bored to any size desired.

The degree of accuracy to which a hole is to be machined is specified on the drawing. The method of producing the hole is left to the production department, job planner, or machinist.

ROUNDS AND FILLETS

A round, or radius, or chamfer is put on the outside of a piece to improve its appearance and avoid forming a sharp edge that might chip off under a sharp blow or cause interference. A fillet is additional metal allowed in the inner intersection of two surfaces. This increases the strength of the object. A general note, such as ROUNDS AND FILLETS R3 or ROUNDS AND FILLETS R3 UNLESS OTHERWISE SHOWN is used on the drawing instead of individual dimensions.

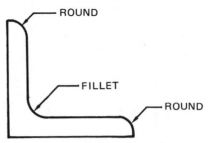

Fig. 5-6 Fillets and rounds

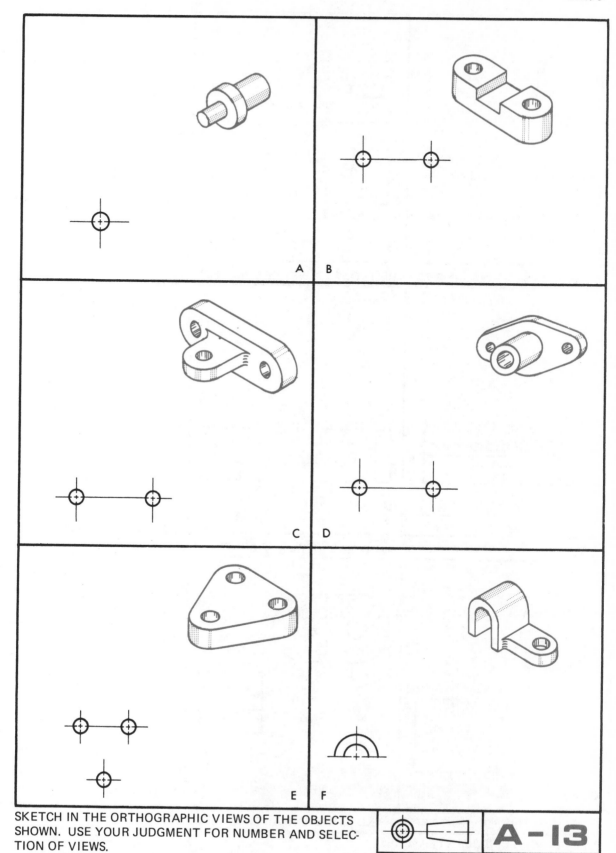

A B

C D

E F

SKETCH IN THE ORTHOGRAPHIC VIEWS OF THE OBJECTS
SHOWN. USE YOUR JUDGMENT FOR NUMBER AND SELEC-
TION OF VIEWS.

A-13

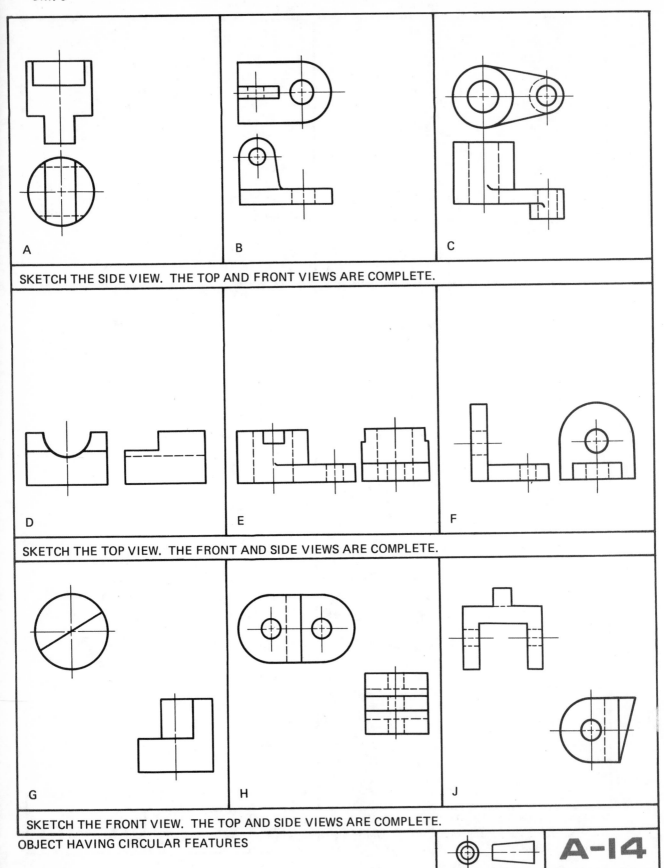

SKETCH THE SIDE VIEW. THE TOP AND FRONT VIEWS ARE COMPLETE.

A

B

C

D

E

F

SKETCH THE TOP VIEW. THE FRONT AND SIDE VIEWS ARE COMPLETE.

G

H

J

SKETCH THE FRONT VIEW. THE TOP AND SIDE VIEWS ARE COMPLETE.

OBJECT HAVING CIRCULAR FEATURES

A-14

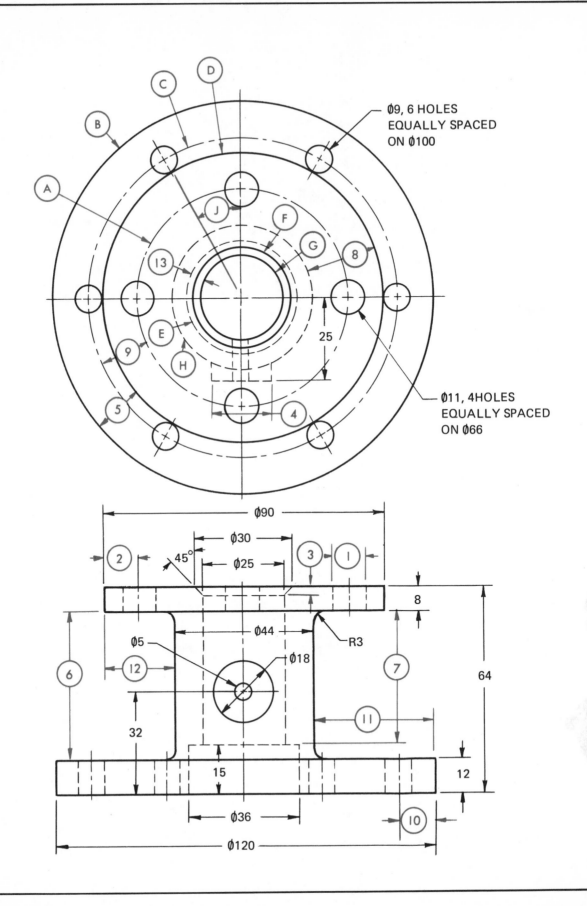

Ø9, 6 HOLES
EQUALLY SPACED
ON Ø100

Ø11, 4HOLES
EQUALLY SPACED
ON Ø66

25

Ø90

Ø30

Ø25

45°

Ø44

R3

Ø5

Ø18

Ø36

Ø120

8

64

12

32

15

A B C D F G H J E

13 9 5 4 8

2 3 1 6 12 7 11 10

38

1. What are the diameters of circles (A) to (H) ?

2. How many holes are in the bottom surface?
3. How many holes are in the top surface?
4. How deep is the Ø25 hole?

5. What is angle (J) ?

6. How thick is the largest flange?
7. What size bolts would be used for the top flange? Allow for 1 mm clearance.
   (Refer to the Capscrew Chart in the appendix.)
8. What size bolts would be used for the bottom flange? Allow for 1 mm clearance.
   (Refer to the Capscrew Chart in the appendix.)

9. Calculate distances (1) to (13) .

ANSWERS

1 A _____

  B _____

  C _____

  D _____

  E _____

  F _____

  G _____

  H _____

2 _____

3 _____

4 _____

5 _____

6 _____

7 _____

8 _____

9 (1) _____

(2) _____

(3) _____

(4) _____

(5) _____

(6) _____

(7) _____

(8) _____

(9) _____

(10) _____

(11) _____

(12) _____

(13) _____

	METRIC	
MATERIAL	CAST IRON	
SCALE	NOT TO SCALE	
DRAWN		DATE

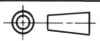

COUPLING   **A-I5**

UNIT 6

DRAWING TO SCALE

When objects are drawn at their actual size, the drawing is called full-scale or scale 1:1. Many objects, however, including buildings, ships, and airplanes, are too large to be drawn full-scale. Therefore, they must be drawn to a *reduced scale*. An example would be the drawing of a house to a scale of 1:50 (metric) or 1:48 (1/4″ = 1 foot) in the inch-foot scale.

Frequently, small watch parts and similar objects are drawn larger than their actual size in order to clearly define their shapes. This is called drawing to an *enlarged scale*. The minute hand of a wrist watch, for example, could be drawn to scale 5:1.

Many mechanical parts are drawn to half scale, 1:2, and fifth scale, 1:5. Notice that the scale of the drawing is expressed in the form of a ratio. The left side of the ratio represents a unit of measurement of the size drawn, the right side, the measurement of the actual object. Thus 1 unit of measurement on the drawing equals 5 units of measurement on the actual object. Many of the common scale ratios are shown in figure 6-1.

Drafting scales are constructed with a variety of combined scale ratios marked on their surfaces. This combination of scale ratios spares the drafter the inconvenience of calculating the size to be drawn when working to a scale other than full-size. The more common drafting scales are shown in figure 6-2.

SI (Metric Scales)

The millimetre is the linear unit of measurement for mechanical drawings. Scale multipliers and divisors of 2 and 5 are recommended, resulting in the scales shown in figure 6-1. The

Metric Scale (Millimetres)		100:1 50:1 20:1 10:1 5:1 2:1	1:1	1:2 1:5 1:10 1:20 1:50 1:100
Inch Scales	Decimal-inch	10:1 4:1 2:1	1:1	1:2 1:3 1:4
	Fractional-inch		1:1	1:2 1:4 3:4
	Civil Engineer's Scale		1:1	1:20 1:30 1:40 1:50 1:60
Foot Scales			1:1	$\frac{1}{8}''$ = 1 FOOT $\frac{1}{4}''$ = 1 FOOT 1″ = 1 FOOT 3″ = 1 FOOT

Fig. 6-1 Common drafting scales used on drawings

numbers shown indicate the size difference between the drawings and the actual part. For example, the ratio 10:1 on the drawing means the drawing is ten times the actual size of the part, whereas a ratio of 1:5 on the drawing means the object is five times larger than it is shown on the drawing.

The units of measurement for architectural drawings are the metre and millimetre. The same scale multipliers and divisors used for mechanical drawings are used for architectural drawings.

The Inch Scales. There are three types of scales showing values that are equal to 1

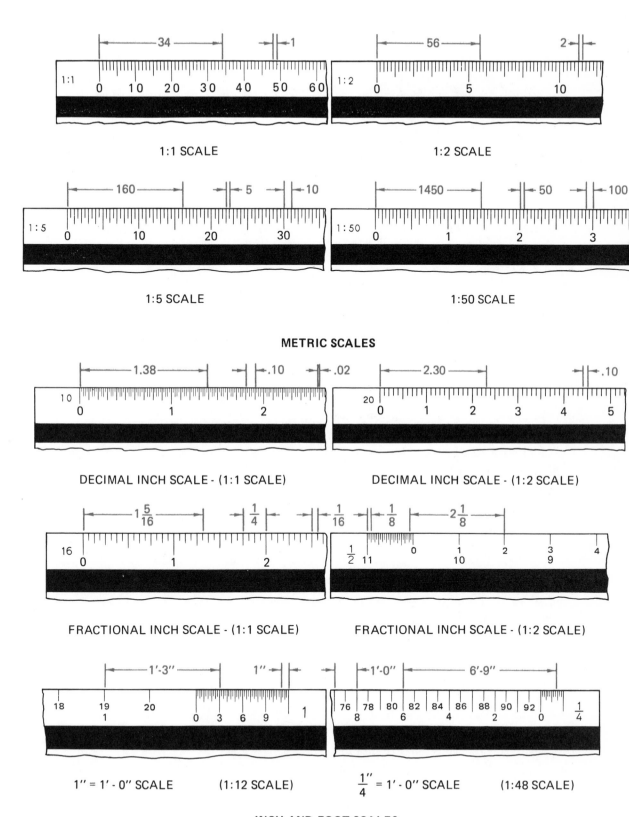

METRIC SCALES

DECIMAL INCH SCALE - (1:1 SCALE)   DECIMAL INCH SCALE - (1:2 SCALE)

FRACTIONAL INCH SCALE - (1:1 SCALE)   FRACTIONAL INCH SCALE - (1:2 SCALE)

1″ = 1′ - 0″ SCALE   (1:12 SCALE) $\frac{1″}{4}$ = 1′ - 0″ SCALE   (1:48 SCALE)

INCH AND FOOT SCALES

Fig. 6-2 Drafting scales

inch. They are the *decimal-inch scale,* the *fractional-inch scale,* and the *civil engineer's scale.* The civil engineer's scale has divisions of 10, 20, 30, 40, 50, 60, and 80 parts to the inch.

On fractional-inch scales, multipliers or divisors of 2, 4, 8, and 16 are used, offering such scales as full-size, half-size, quarter-size, etc.

The Foot Scales. These scales are used mostly in architectural work. The main difference between foot and inch scales is that in the foot scale each major division represents a foot, not an inch, and the end units are subdivided into inches or parts of an inch. The more common scales are the 1/8 inch = 1 foot; 1/4 inch = 1 foot; 1 inch = 1 foot; and 3 inches = 1 foot scales.

MACHINE SLOTS

Slots are used chiefly in machines to hold parts together. The two principal types are *T slots* and *dovetails,* figure 6-3.

A dovetail is a groove or slide whose sides are cut on an angle. This forms an interlocking joint between two pieces, enabling the slot to resist pulling apart in any direction other than along the lines of the dovetail slide itself.

The dovetail is commonly used in the design of slides, including lathe cross slides, milling machine table slides, and other sliding parts.

REFERENCE DIMENSIONS

When a reference dimension is shown on a drawing for information only and is not necessary for the manufacture of the part, it must be clearly labelled. The approved method for indicating reference dimensions on a drawing is the enclosure of the dimensions inside parentheses, figure 6-4. Previously, reference dimensions were indicated by placing the abbreviation REF after or below the dimension.

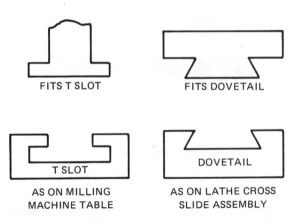

Fig. 6-3 Typical machine slots

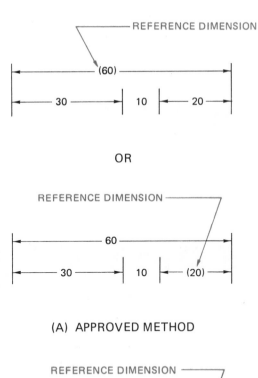

(A) APPROVED METHOD

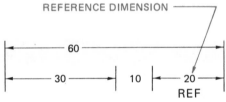

(B) FORMER METHOD

Fig. 6-4 Reference dimensions

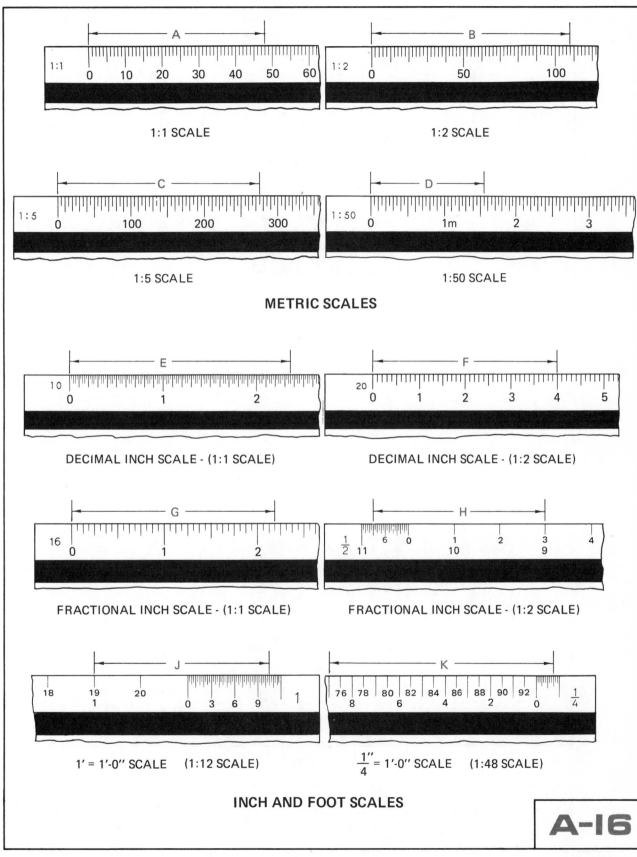

1:1 SCALE

1:2 SCALE

1:5 SCALE

1:50 SCALE

METRIC SCALES

DECIMAL INCH SCALE - (1:1 SCALE)

DECIMAL INCH SCALE - (1:2 SCALE)

FRACTIONAL INCH SCALE - (1:1 SCALE)

FRACTIONAL INCH SCALE - (1:2 SCALE)

1' = 1'-0'' SCALE   (1:12 SCALE)

$\frac{1''}{4}$ = 1'-0'' SCALE   (1:48 SCALE)

INCH AND FOOT SCALES

A-16

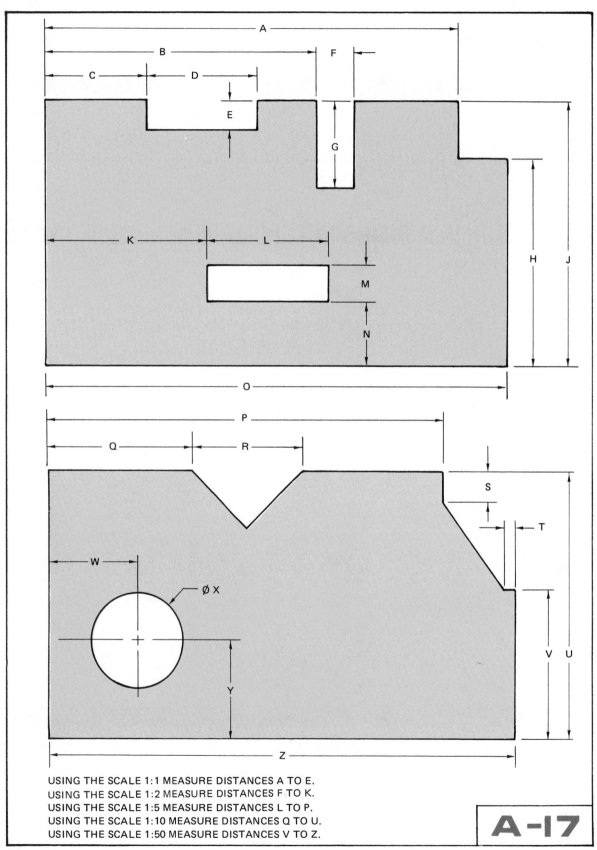

USING THE SCALE 1:1 MEASURE DISTANCES A TO E.
USING THE SCALE 1:2 MEASURE DISTANCES F TO K.
USING THE SCALE 1:5 MEASURE DISTANCES L TO P.
USING THE SCALE 1:10 MEASURE DISTANCES Q TO U.
USING THE SCALE 1:50 MEASURE DISTANCES V TO Z.

A -17

QUESTIONS

1. In which view is the shape of the dovetail shown?
2. In which view is the shape of the T slot shown?
3. How many rounds are shown in the top view?
4. In which view is a fillet shown?
5. Which line in the top view represents surface (R) of the side view?
6. Which line in the front view represents surface (R)?
7. Which line in the top view represents surface (L) of the side view?
8. Which line in the front view represents surface (L)?
9. Which line in the side view represents surface (A) on the top view?
10. Which dimension in the front view represents the width of surface (A)?
11. What type of lines are (B), (J), and (K)?
12. How far apart are the two invisible edge lines of the side view?
13. What dimension indicates how far line (J) is from the base of the slide?
14. How wide is the opening in the dovetail?
15. Which two lines in the top view indicate the opening of the dovetail?
16. At what angle to the horizontal is the dovetail cut?
17. In the side view, how far is the lower left edge of the dovetail from the left side of the piece?
18. What are the lengths of dimensions (Y), (V), and (X)?
19. What is the height of the dovetail?

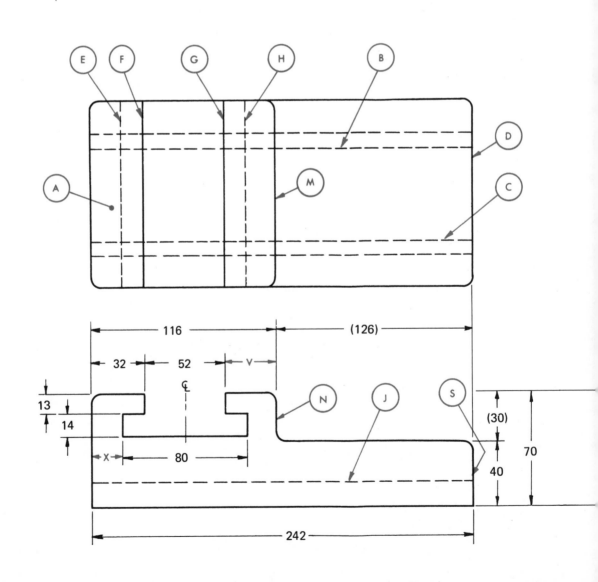

46

20. How much material remains between the surface represented by line (Q) and the top of the dovetail after the cut has been taken?
21. What is the vertical distance from the surface represented by line (Q) to that represented by line (T)?
22. Which dimension represents the distance between lines (F) and (G)?
23. What is the overall height of the T slot?
24. How wide is the metal at the sides of the 52 mm opening of the T slot?
25. What is the width of the bottom of the T slot?
26. What is the height of the opening of the bottom of the T slot?
27. What is the horizontal distance from line (N) to line (S)?
28. What is the unit of measurement for the angles shown?
29. How many reference dimensions are shown on the drawing?
30. What is the size of the largest reference dimension?

ANSWERS

1 _____	16 _____
2 _____	17 _____
3 _____	18 Y _____
4 _____	V _____
5 _____	X _____
6 _____	19 _____
7 _____	20 _____
8 _____	21 _____
9 _____	22 _____
10 _____	23 _____
11 B _____	24 _____
J _____	25 _____
K _____	26 _____
12 _____	27 _____
13 _____	28 _____
14 _____	29 _____
15 _____	30 _____

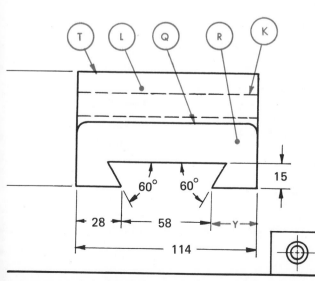

METRIC	
MATERIAL	CAST IRON
SCALE	NOT TO SCALE
DRAWN	DATE
COMPOUND REST SLIDE	A-18

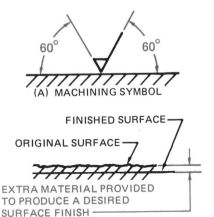

(A) MACHINING SYMBOL

FINISHED SURFACE

ORIGINAL SURFACE

EXTRA MATERIAL PROVIDED
TO PRODUCE A DESIRED
SURFACE FINISH

(B) MEANING

Fig. 7-1 Machining symbol

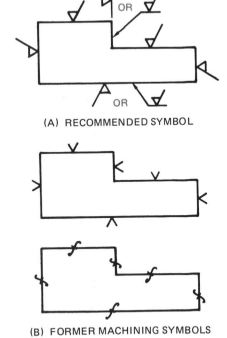

(A) RECOMMENDED SYMBOL

(B) FORMER MACHINING SYMBOLS

Fig. 7-2 Application of machining symbols

MACHINING SYMBOLS

When preparing working drawings of parts to be cast or forged, the drafter must indicate part surfaces which require machining or finishing. This requirement is indicated by adding a machining allowance symbol to the surface or surfaces which must be finished by the removal of material, figure 7-1. Figure 7-2 shows the current machining symbol and those which were formerly used on drawings. This information is essential in order to alert the patternmaker and diemaker to provide extra metal on the casting or forging to allow for the finishing process. Depending on the material to be cast or forged, between 2 and 4 mm are usually allowed on small castings and forgings for each surface that requires finishing, figure 7-3.

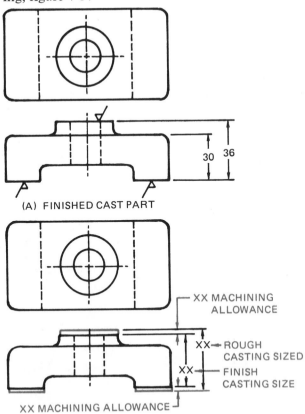

30   36

(A) FINISHED CAST PART

XX MACHINING
ALLOWANCE

XX ROUGH
CASTING SIZED

XX FINISH
CASTING SIZE

XX MACHINING ALLOWANCE

(B) CASTING WITH EXTRA METAL
ALLOWED FOR MACHINING

Fig. 7-3 Allowances for machining

Like dimensions, machining symbols are not duplicated on the drawing. They should be used on the same view as the dimensions that give the size or location of the surfaces. The symbol is placed on the line representing the surface or on a leader or an extension line locating the surface. The symbol and the inscription should be oriented so they may be read from the bottom or right-hand side of the drawing.

Where all the surfaces are to be machined, a general note, such as FINISH ALL OVER may be used and the symbols on the drawings omitted.

The old machining symbols shown in figure 7-2(B) are found on many drawings in use today. When called upon to make changes or revisions on a drawing already in existence, a drafter must adhere to the drawing conventions shown on that drawing.

The machining symbol does not indicate surface finish quality. The surface texture symbol which is defined and discussed in Unit 9 is used to signify the desired surface finish quality.

Indication of Machining Allowance

When the value of the machining allowance must be specified, it is indicated to the left of the symbol, figures 7-4 and 7-5. This value is expressed in millimetres or inches depending on which units of measurement are used on the drawing.

Removal of Metal Prohibited

A surface from which the removal of metal is prohibited is indicated by the symbol shown in figure 7-6. This symbol may also be used in a drawing relating to a production process. It indicates that a surface must be left the way it is affected by a preceding manufacturing process, regardless of the removal of material or other changes.

NOT-TO-SCALE DIMENSIONS

When a dimension on a drawing is altered, making it not-to-scale, a straight freehand line is drawn below the dimension to indicate that the dimension is not drawn to scale, figure 7-7.

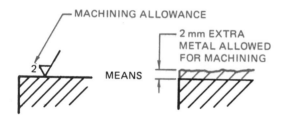

Fig. 7-4 Indication of machining allowance

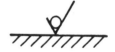

Fig. 7-6 Symbol for removal of material not permitted

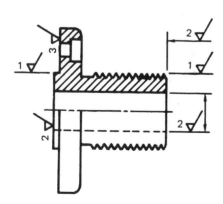

Fig. 7-5 Indicating machining allowance on drawing

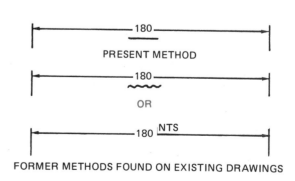

Fig. 7-7 Indicating dimensions that are not to scale

This is a change from earlier methods of indicating not-to-scale dimensions. Formerly, a wavy line below the dimension or the letters NTS beside the dimension were used to indicate not-to-scale dimensions.

DRAWING REVISION

Drawing revisions are made to accommodate improved manufacturing methods, reduce costs, correct errors, and improve design. A clear record of these revisions must be registered on the drawing.

All drawings must carry a change or revision table down the top right-hand side of the drawing. Additionally, provision may be made for recording a revision number or symbol, the date, the drafter's name or initials, and approval of the change. Should the drawing revision cause a dimension or dimensions to be different from the scale indicated, the dimensions which are not-to-scale should be indicated. Typical revision tables are shown in figure 7-8.

When many revisions are needed, a new drawing should be made. The words RE-DRAWN AND REVISED should appear in the revision column of the new drawing when this is done.

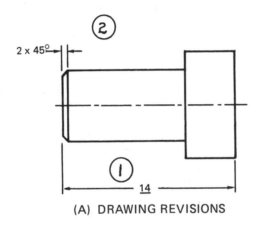

(A) DRAWING REVISIONS

REVISIONS			
Rev.	Description	Date	Approved
1	Length was 20	Jan. 6/77	J. Campbell
2	Chamfer Added	Mar. 3/77	D. Arnold

(B) TYPICAL VERTICAL REVISION BLOCK
(SIZE MAY VARY)

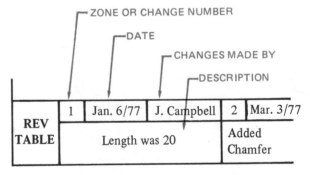

(C) HORIZONTAL REVISION BLOCK
(SIZE MAY VARY)

Fig. 7-8 Drawing revisions

BREAK LINES

Break lines, as shown in figure 7-9 are used to shorten the view of long uniform sections. They are also used when only a partial view is required. Such lines are used on both detail and assembly drawings. The thin line with freehand zigzags is recommended for long breaks, and the jagged line for wood parts. The special breaks shown for cylindrical and tubular parts are useful when an end view is not shown; otherwise, the thick break line is adequate.

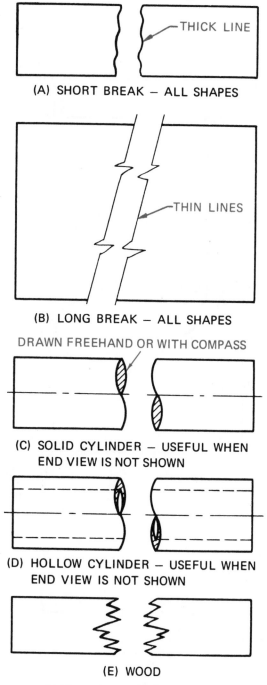

(A) SHORT BREAK – ALL SHAPES

(B) LONG BREAK – ALL SHAPES

(C) SOLID CYLINDER – USEFUL WHEN END VIEW IS NOT SHOWN

(D) HOLLOW CYLINDER – USEFUL WHEN END VIEW IS NOT SHOWN

(E) WOOD

Fig. 7-9 Conventional break lines

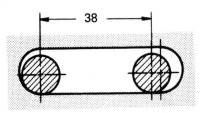

DETAIL OF 12 mm BOLTS IN SLOT

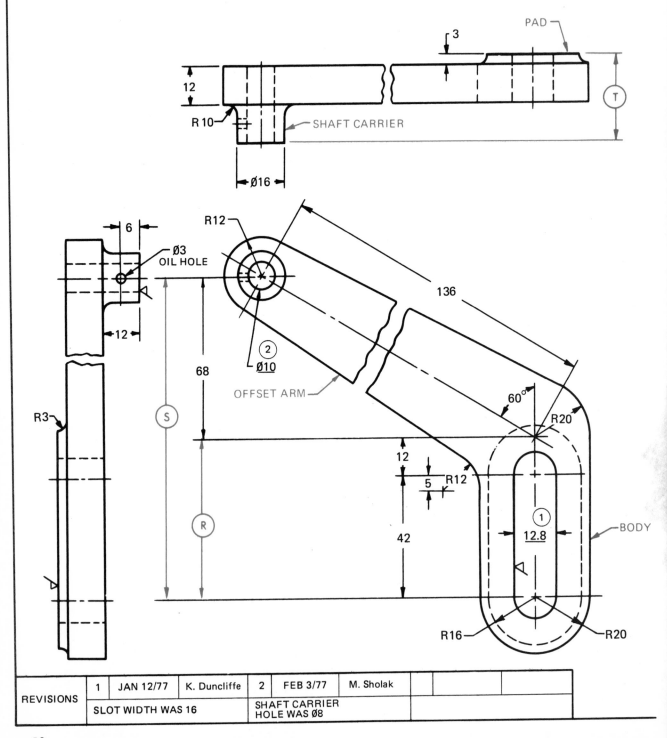

3

PAD

12

R 10

SHAFT CARRIER

Ø16

T

6

Ø3
OIL HOLE

12

R12

136

2
Ø10

OFFSET ARM

68

R3

S

60°

R20

12

5

R12

1
12.8

BODY

R

42

R16

R20

QUESTIONS

1. At what angle is the offset arm to the body of the piece?
2. What is the center-to-center measurement of the length of the offset arm?
3. Which radius forms the upper end of the offset arm?
4. Which radii form the lower end of the offset arm where it joins the body?
5. What is the width of the bolt slot in the body of the bracket?
6. What is the center-to-center length of this slot?
7. What was the slot width before revision?
8. Which radius forms the ends of the pad?
9. What is the overall length of this pad?
10. What is the overall width of the pad?
11. What is the radius of the fillet between the pad and the edge of the piece?
12. What is the diameter of the shaft carrier body?
13. What is the diameter of the shaft carrier hole?
14. What is the distance from the face of the shaft carrier to the face of the pad?
15. What is the radius of the inside fillet between the arm and the body of the piece?
16. If M12 bolts are used in holding the bracket to the machine base, what is the clearance on each side of the slot?
17. If the center-to-center distance of two M12 bolts which fit in the slot is 38 mm, how much play is there lengthwise in the slot?
18. What size oil hole is in the shaft carrier?
19. How far is the center of this hole from the face of the shaft carrier?
20. How thick is the combined body and pad?
21. Calculate distances (R), (S), and (T).
22. The hole in the shaft carrier was revised. What is the difference in size between the new and old hole?
23. How many dimensions are not to scale?
24. If 2 mm is allowed for each surface to be machined, what would be the overall thickness of the original casting?

ANSWERS

1 _____   21 (R) _____

2 _____   (S) _____

3 _____   (T) _____

4 _____   22 _____

5 _____   23 _____

6 _____   24 _____

7 _____

8 _____

9 _____

10 _____

11 _____

12 _____

13 _____

14 _____

15 _____

16 _____

17 _____

18 _____

19 _____

20 _____

METRIC

MATERIAL	CI	
SCALE	1:2	
DRAWN		DATE

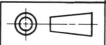

OFFSET BRACKET

A-19

UNIT

8

SECTIONAL VIEWS

Sectional views, commonly called *sections,* are used to show interior detail too complicated to be shown clearly and dimensioned by outside views and hidden lines. A sectional view is obtained by supposing the nearest part of the object has been cut or broken away on an imaginary cutting plane. The exposed or cut surfaces are identified by section lining or crosshatching. Hidden lines and details behind the cutting-plane line are usually omitted unless they are required for clarity. It should be understood that only in the sectional view is any part of the object shown as having been removed.

A sectional view frequently replaces one of the regular views. For example, a regular front view is replaced by a front view in section, as shown in figure 8-1.

The Cutting-Plane Line

A cutting-plane line indicates where the imaginary cutting takes place. The position of the cutting plane is indicated, when necessary, on a view of the object or assembly by a cutting-plane line as shown in figure 8-2. The ends of the cutting-plane line are bent at 90 degrees and terminated by arrowheads to indicate the direction of sight for viewing the section. Cutting planes are not shown on sectional views. The cutting-plane line may be

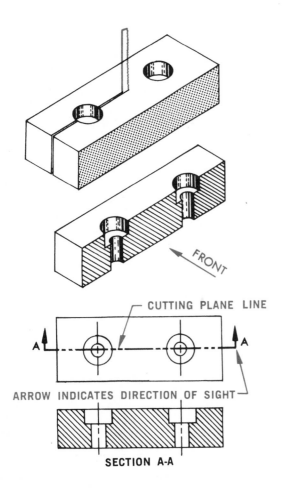

Fig. 8-1 A section drawing

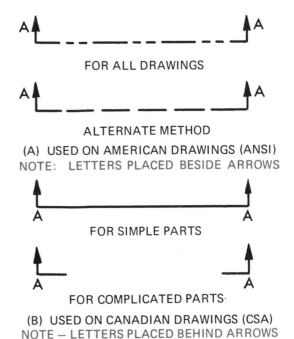

FOR ALL DRAWINGS

ALTERNATE METHOD

(A) USED ON AMERICAN DRAWINGS (ANSI)
NOTE: LETTERS PLACED BESIDE ARROWS

FOR SIMPLE PARTS

FOR COMPLICATED PARTS

(B) USED ON CANADIAN DRAWINGS (CSA)
NOTE — LETTERS PLACED BEHIND ARROWS

Fig. 8-2 Cutting-plane lines

omitted when it corresponds to the center line of the part or when only one sectional view appears on a drawing.

If two or more sections appear on the same drawing, the cutting-plane lines are identified by two identical large, single-stroke, Gothic letters. One letter is placed at each end of the line. Sectional view subtitles are given when identification letters are used and appear directly below the view, incorporating the letters at each end of the cutting-plane line thus: SECTION A-A, or abbreviated, SECT. A-A.

Section Lining

Section lining indicates the surface that has been cut and makes it stand out clearly. Section lines usually consist of thin parallel lines, figure 8-3, drawn at an angle of approximately 45 degrees to the principal edges or axis of the part.

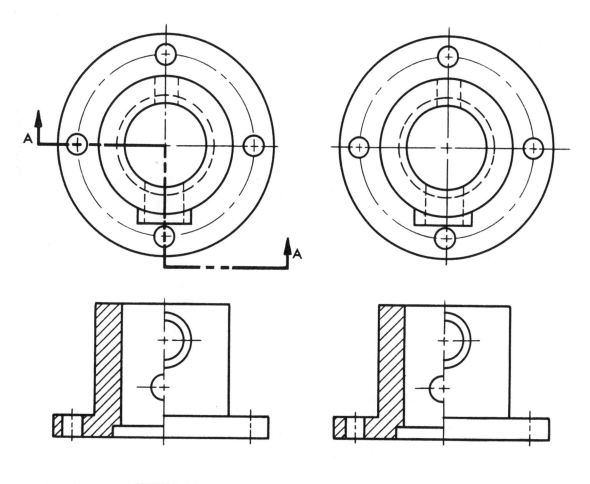

SECTION A-A

Letters, subtitle, and cutting-plane line used when more than one section view appears on a drawing or when they make the drawing clearer.

Letters, subtitle, and cutting-plane line may be omitted when they correspond with the center line of the part and when there is only one section view on the drawing.

Fig. 8-3 Identification of cutting plane and section view

When it is desirable to indicate differences in materials, other symbolic section lines are used, such as those shown in figure 8-4. If the part shape would cause section lines to be parallel or nearly parallel to one of the sides or features of the part, an angle other than 45 degrees is chosen.

The spacing of the hatching lines is uniform to give a good appearance to the drawing. The pitch, or distance, between lines varies between 1 and 2.5 mm, depending on the size of the area to be sectioned. Section lining is similar in direction and spacing in all sections of a single component.

Wood and concrete are the only two materials usually shown symbolically. When wood symbols are used, the direction of the grain is shown.

TYPES OF SECTIONS

Full Sections

When the cutting plane extends entirely through the object in a straight line and the front half of the object is theoretically removed, a *full section* is obtained, figure 8-5. This type of section is used for both detail and assembly drawings. When the cutting plane divides the object into two identical parts, it is not necessary to indicate its location. However, the cutting plane may be identified and indicated in the usual manner to increase clarity.

Half Sections

A symmetrical object or assembly may be drawn as a *half section,* showing one-half up to the center line in section and the other

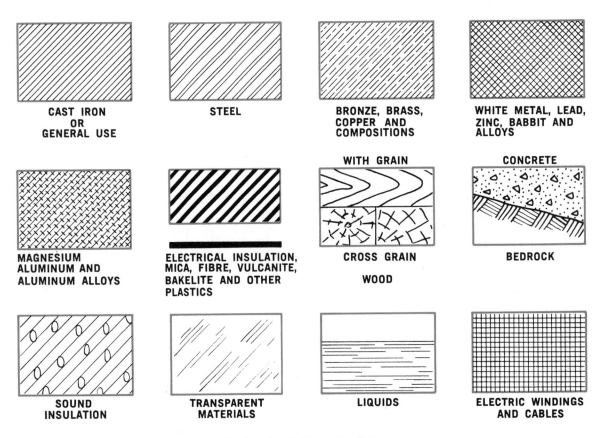

CAST IRON
OR
GENERAL USE

STEEL

BRONZE, BRASS,
COPPER AND
COMPOSITIONS

WHITE METAL, LEAD,
ZINC, BABBIT AND
ALLOYS

MAGNESIUM
ALUMINUM AND
ALUMINUM ALLOYS

ELECTRICAL INSULATION,
MICA, FIBRE, VULCANITE,
BAKELITE AND OTHER
PLASTICS

WITH GRAIN

CROSS GRAIN

WOOD

CONCRETE

BEDROCK

SOUND
INSULATION

TRANSPARENT
MATERIALS

LIQUIDS

ELECTRIC WINDINGS
AND CABLES

Fig. 8-4 Symbolic section lining

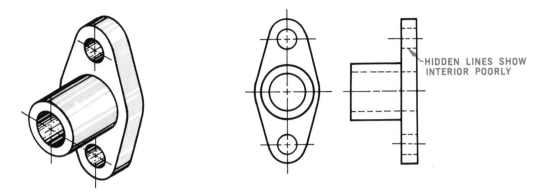

(A) SIDE VIEW NOT SECTIONED

HIDDEN LINES SHOW
INTERIOR POORLY

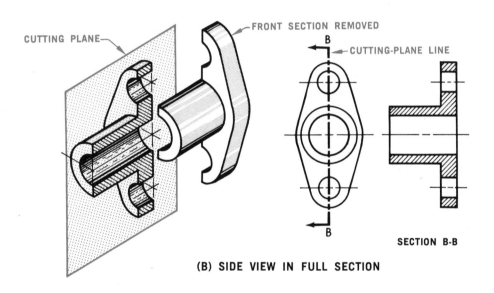

CUTTING PLANE

FRONT SECTION REMOVED

B

CUTTING-PLANE LINE

B

SECTION B-B

(B) SIDE VIEW IN FULL SECTION

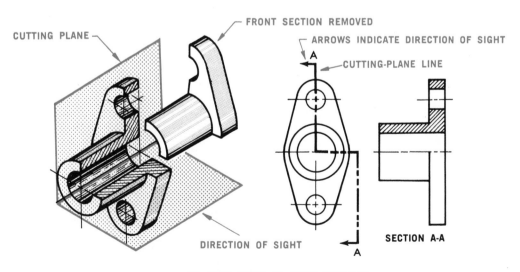

CUTTING PLANE

FRONT SECTION REMOVED

ARROWS INDICATE DIRECTION OF SIGHT

A

CUTTING-PLANE LINE

DIRECTION OF SIGHT

A

SECTION A-A

(C) SIDE VIEW IN HALF SECTION

Fig. 8-5 Full and half sections

half in full view. A normal center line is used on the section view.

The half section drawing should not be used where the dimensioning of internal diameters is required. This will prevent the addition of hidden lines to the portion showing the external features. This type of section is used mostly for assembly drawings where internal and external features are clearly shown and only overall and center-to-center dimensions are required.

COUNTERSINKS, COUNTERBORES, AND SPOT FACES

A *countersunk hole* is a conical depression cut in a piece to receive a countersunk type of flathead screw or rivet as illustrated in figure 8-6. The size is usually shown by a note listing the diameter of the hole first, followed by the diameter of the countersink, the abbreviation CSK, and the angle. A *counterbored hole* is one which has been machined larger to a given depth to receive a fillister, hex-head, or similar type of bolt head. Counterbores are specified by a note giving the diameter of

the hole first, followed by the counterbore diameter, the abbreviation CBORE, and depth of the counterbore. The counterbore and depth may also be indicated by direct dimensioning. A *spot face* is an area where the surface is machined just enough to provide a level seating surface for a bolt head, nut or washer. A spot face is specified by a note listing the diameter of the hole first, followed by the spot face diameter, and the abbreviation SFACE. The depth of the spot face is not usually given.

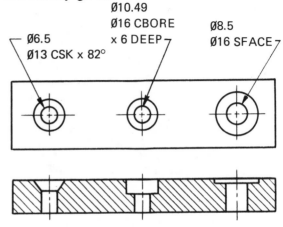

COUNTERSINK COUNTERBORE SPOT FACE

Fig. 8-6 **Dimensioning countersinks, counterbores, and spotfaces**

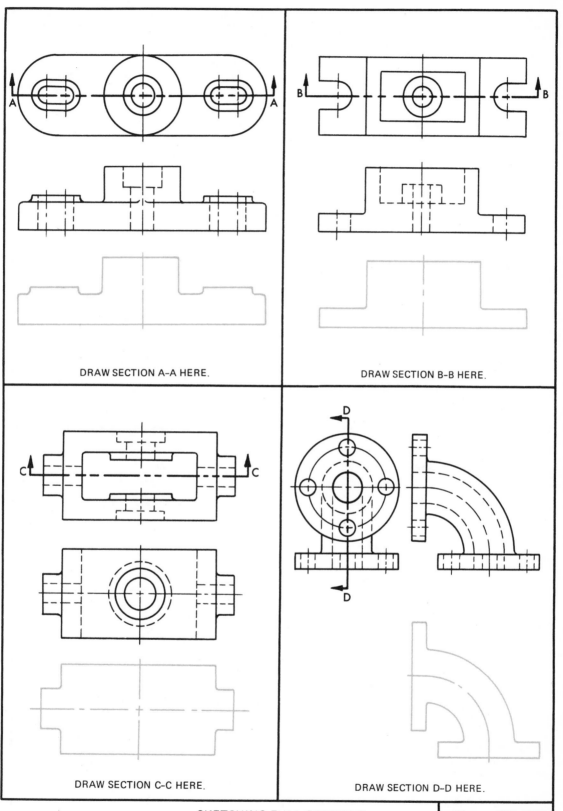

DRAW SECTION A-A HERE.

DRAW SECTION B-B HERE.

DRAW SECTION C-C HERE.

DRAW SECTION D-D HERE.

SKETCHING FULL SECTIONS

A-20

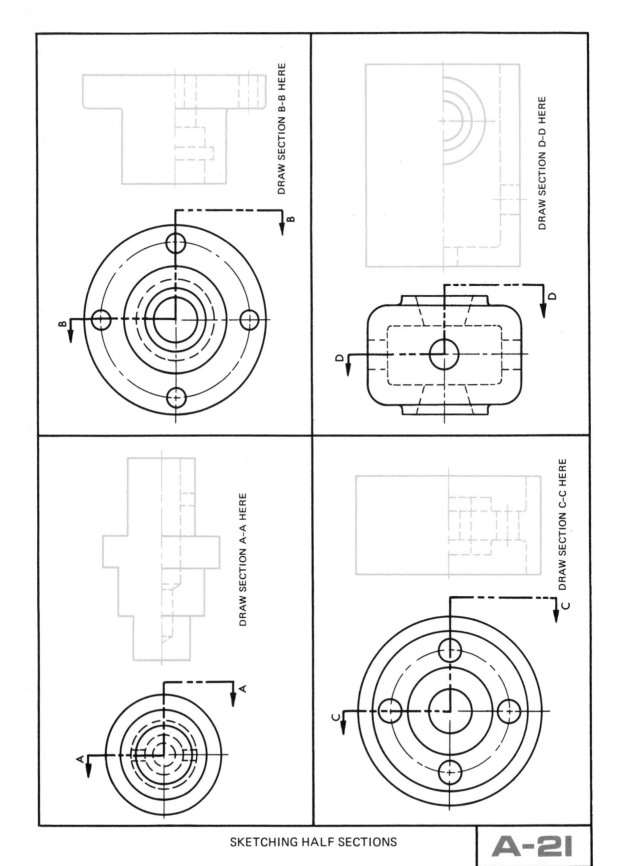

DRAW SECTION B-B HERE

B

B

DRAW SECTION D-D HERE

D

D

DRAW SECTION A-A HERE

A

A

DRAW SECTION C-C HERE

C

C

SKETCHING HALF SECTIONS

A-21

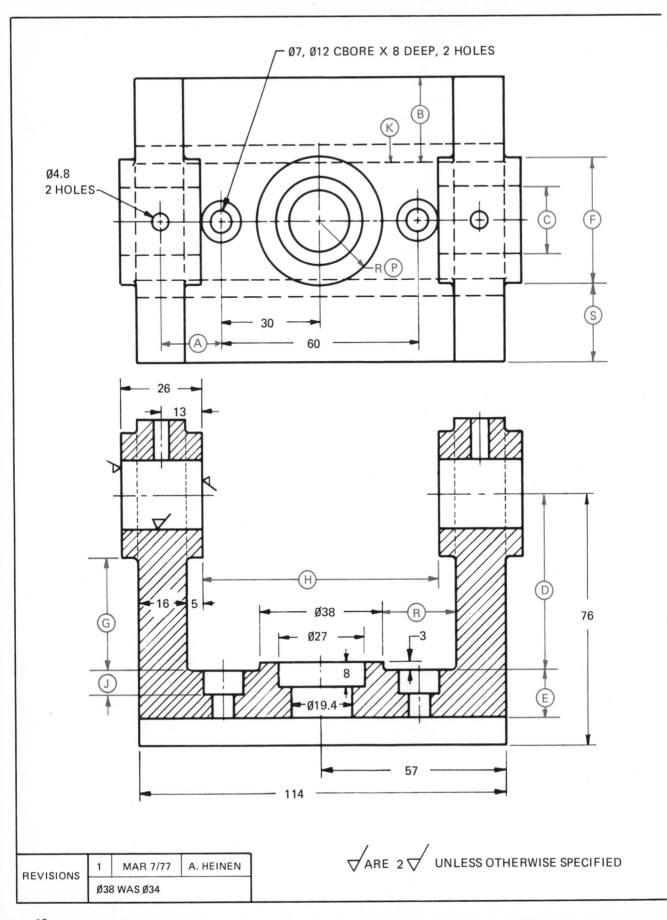

Ø7, Ø12 CBORE X 8 DEEP, 2 HOLES

Ø4.8
2 HOLES

30

60

26

13

16 5

Ø38

Ø27

3

8

Ø19.4

57

114

76

REVISIONS | 1 | MAR 7/77 | A. HEINEN
Ø38 WAS Ø34

▽ARE 2 ▽ UNLESS OTHERWISE SPECIFIED

1. What is the overall width?
2. What is the overall height?
3. What is the center-to-center distance of the Ø4.8 holes?
4. If 2 mm is allowed for each surface requiring finishing, calculate the width of the casting before machining.
5. At what angle to the vertical is the dovetail slot?
6. How many different surfaces require finishing?
7. Which type of lines in the top view represents the dovetail?
8. What was the original size of the Ø38?
9. How wide is the opening in the dovetail?
10. How high is the dovetail?
11. When was the Ø38 dimension altered?
12. How many reference dimensions are shown?
13. Calculate distances (A) to (S).

ANSWERS

1 _____
2 _____
3 _____
4 _____
5 _____
6 _____
7 _____
8 _____
9 _____
10 _____
11 _____
12 _____
13 (A) _____
(B) _____
(C) _____
(D) _____
(E) _____
(F) _____
(G) _____
(H) _____
(J) _____
(K) _____
(L) _____
(M) _____
(N) _____
(P) _____
(Q) _____
(R) _____
(S) _____

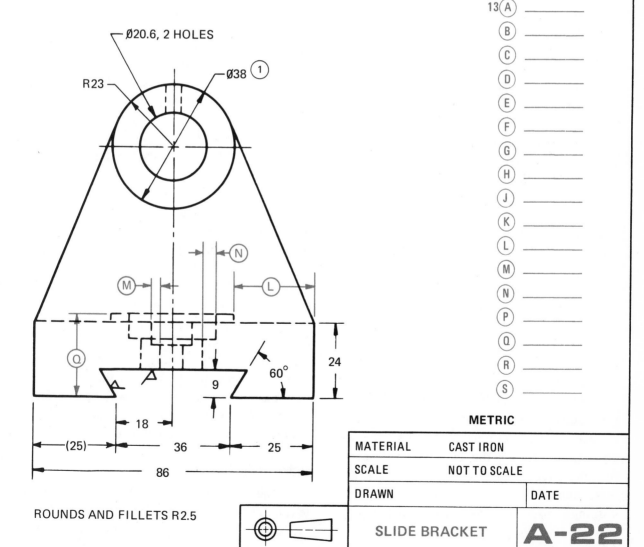

Ø20.6, 2 HOLES
R23
Ø38 (1)
(N)
(M)
(L)
(Q)
60°
24
9
18
(25)
36
25
86

ROUNDS AND FILLETS R2.5

METRIC

MATERIAL	CAST IRON	
SCALE	NOT TO SCALE	
DRAWN		DATE

SLIDE BRACKET

A-22

UNIT

9

SURFACE TEXTURE

The development of modern, high-speed machines has resulted in higher loadings and faster moving parts. To withstand these more severe operating conditions with minimum friction and wear, a particular surface texture is often essential. This requires the designer to accurately describe the needed texture (sometimes called finish) to the persons who are actually making the parts.

Entire machines are rarely designed and manufactured in one plant. They are usually designed in one location, manufactured in another, and perhaps assembled in a third.

All surface finish control begins in the drafting room. The designer is responsible for specifying the correct surface finish for maximum performance and service life at the lowest cost. In selecting the required surface finish for any particular part, the designer bases the decision on past experience with similar parts, field service data, and engineering tests. Many factors including: size and function of the parts, type of loading, speed and direction of movement, operating conditions, physical characteristics of both materials on contact, whether the part is subjected to stress reversals, type and amount of lubricant, contaminants, temperature, etc., influence the designer's choice.

The two principal reasons for surface finish control are friction reduction and the control of wear.

Whenever a lubricating film must be maintained between two moving parts, the surface irregularities must be small enough to prevent penetrating the oil film under even the most severe operating conditions. Bearings, journals, cylinder bores, piston pins, bushings, pad bearings, helical and worm gears, seal surfaces, machine ways, etc. are objects where this condition must be fulfilled.

Surface finish is also important to the wear service of certain pieces subject to dry friction: machine tool bits, threading dies, stamping dies, rolls, clutch plates, brake drums, etc.

Smooth finishes are essential on certain high-precision pieces. In mechanisms such as injectors and high-pressure cylinders, smoothness and lack of waviness are essential to accuracy and pressure retaining ability. Smooth finishes are also used on micrometer anvils, gauges, and gauge blocks, and other items.

Smoothness is often important for the visual appeal of the finished product. For this reason, surface finish is controlled on such articles as rolls, extrusion dies, and precision casting dies.

For gears and other parts, surface finish control may be necessary to insure quiet operation.

In cases where boundary lubrication exists or where surfaces are not compatible, for example, two hard surfaces running together, a certain amount of roughness or character of surface will assist in lubrication.

To meet the requirements for effective control of surface quality under diversified conditions, there is a system for accurately describing the surface.

Surfaces are usually very complex in character. Only the height, width, and direction of surface irregularities will be covered in this section since these are of practical importance in specific applications.

Surface Texture Definitions

The following terms relating to surface texture are illustrated on figure 9-1.

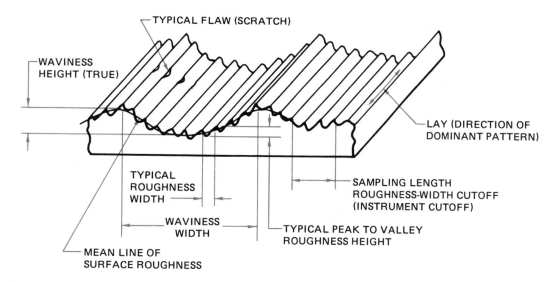

Fig. 9-1 Surface texture characteristics

Micrometre. A micrometre is one millionth of a metre (0.000 001 metre). For written specifications or reference to surface roughness requirements, micrometres may be abbreviated as μm.

Microinch. A microinch is one millionth of an inch (.000 001 inch). For written specifications or reference to surface roughness requirements, microinches may be abbreviated as μin.

Roughness. Roughness consists of the finer irregularities in the surface texture usually including those irregularities which result from the inherent action of the production process. These include traverse feed marks and other irregularities within the limits of the roughness-width cutoff.

Roughness Height. Roughness height is rated as the arithmetical average deviation (AA) expressed in micrometres or microinches measured normal to the center line. ISO and many European countries use the term CLA (center line average) instead of AA. Both mean the same.

Roughness Width. Roughness width is the distance parallel to the nominal surface between successive peaks or ridges which constitute the predominant pattern of the roughness. Roughness width is rated in millimetres.

Roughness-Width Cutoff. The greatest spacing of repetitive surface irregularities to be included in the measurement of average roughness height is the roughness-width cutoff. Roughness-width cutoff is rated in millimetres or inches and must always be greater than the roughness width in order to obtain the total roughness height rating.

Waviness. Waviness is the usually widely spaced component of surface texture and is generally spaced farther apart than the roughness-width cutoff. Waviness may result from machine or work deflections, vibration, chatter, heat treatment or warping strains. Roughness may be considered superimposed on a "wavy" surface.

Lay. The direction of the predominant surface pattern, which is ordinarily determined by the production method used, is the lay. Symbols for the lay are shown in figure 9-2, page 66.

Flaws. Flaws are surface irregularities occuring at one place or at relatively infrequent

SYMBOL	DESIGNATION	EXAMPLE
=	Lay parallel to the line representing the surface to which the symbol is applied	 DIRECTION OF TOOL MARKS
⊥	Lay perpendicular to the line representing the surface to which the symbol is applied	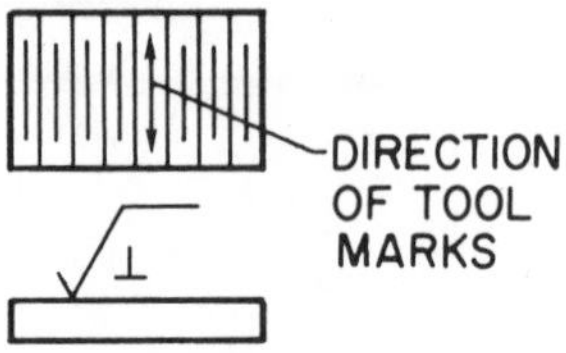 DIRECTION OF TOOL MARKS
X	Lay angular in both directions to line representing the surface to which the symbol is applied	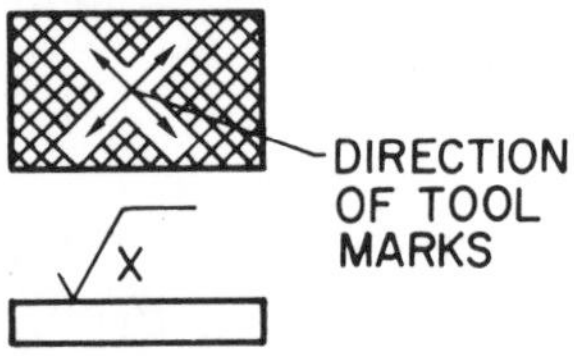 DIRECTION OF TOOL MARKS
M	Lay multidirectional	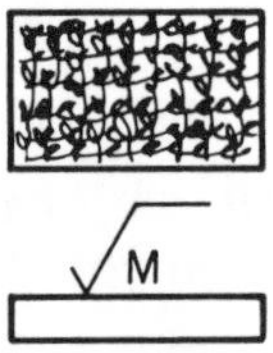
C	Lay approximately circular relative to the center of the surface to which the symbol is applied	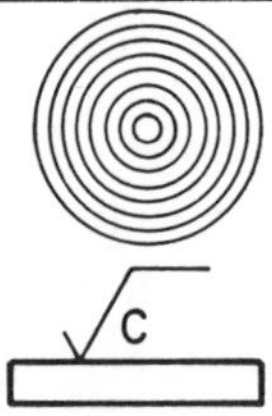
R	Lay approximately radial relative to the center of the surface to which the symbol is applied	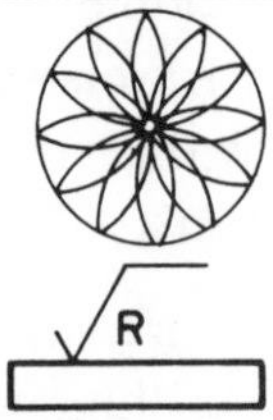
P	Lay nondirectional, pitted, or protuberant	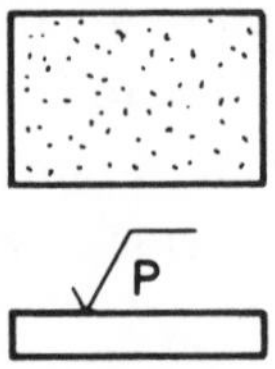

Fig. 9-2 Lay symbols

or widely varying intervals. Flaws include: cracks, blow holes, checks, ridges, scratches, etc. Unless otherwise specified, the effect of flaws is not included in the roughness height measurements.

SURFACE TEXTURE SYMBOL

The surface texture symbol, figure 9-3, denotes surface characteristics on the drawing. Roughness, waviness, and lay are controlled by applying the desired values to the surface texture symbol, figure 9-4 (page 68), or in a general note. The two methods may be used together. The point of the symbol should be on the line indicating the surface, on an extension line from the surface, or on a leader pointing either to the surface or extension line, figure 9-5 (page 69). To be readable from the bottom, the symbol is placed in an upright position when notes or numbers are used. This means the long leg and extension line will be on the right. The symbol applies to the entire surface, unless otherwise specified.

Like dimensions, the symbol for the same surface should not be duplicated on other views. They should be placed on the view with the dimensions showing size or location of the surfaces. Surface texture symbols designate surface texture characteristics and should not be misunderstood as machining symbols (finish marks). For machining symbols, see Unit 7.

SURFACE TEXTURE RATINGS

Roughness Height rating, which is measured in micrometres or microinches, is shown to the left of the long leg of the symbol, figure 9-4. The specification of only one rating indicates the maximum value; any lesser value is acceptable. Specifying two ratings indicates the minimum and maximum values. Anything within that range is acceptable. The maximum value is placed over the minimum.

Waviness Height ratings are indicated in millimetres and positioned as shown in figure 9-4. Any lesser value is acceptable.

Waviness Width ratings are indicated in millimetres positioned as shown in figure 9-4. Any lesser value is acceptable. If the waviness value is a minimum, the abbreviation MIN should be placed after the value.

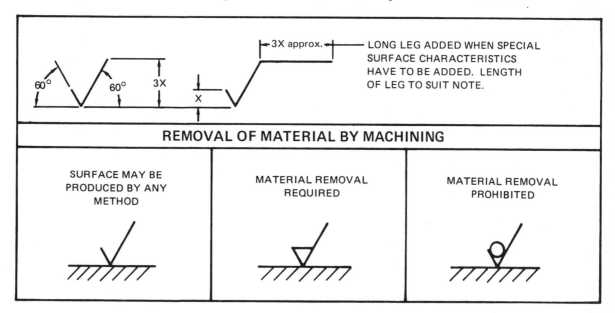

Fig. 9-3 Basic surface texture symbol

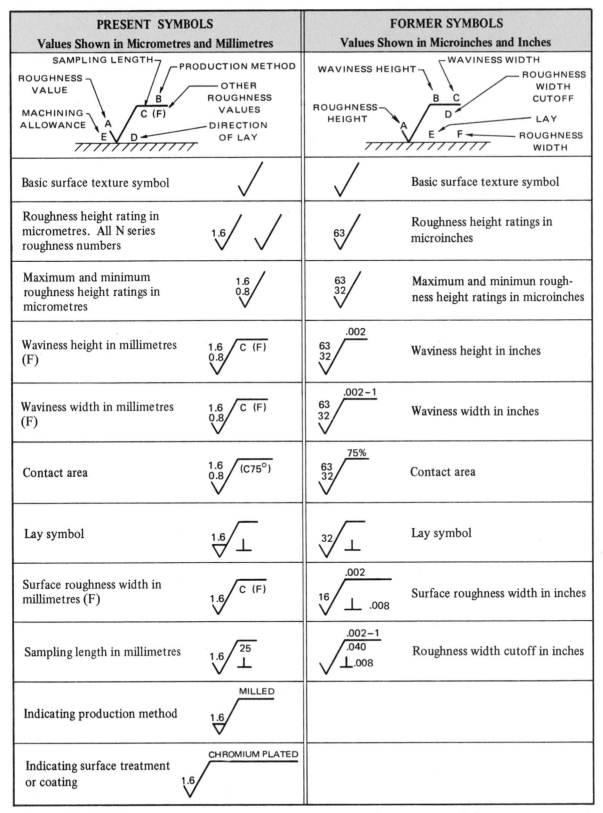

Fig. 9-4 Location of ratings and symbols on surface texture symbol

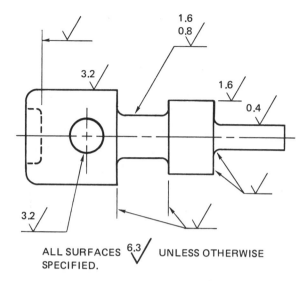

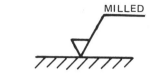

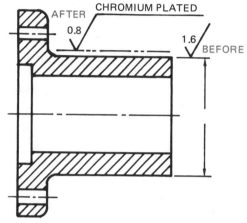

(A) INDICATING PRODUCTION METHOD

ALL SURFACES 6.3 UNLESS OTHERWISE SPECIFIED.

NOTE: VALUES SHOWN ARE IN MICROMETRES.

Fig. 9-5 Application of roughness height and waviness ratings

(B) INDICATING SURFACE TEXTURE BEFORE AND AFTER PLATING

Lay symbols, indicating the directional pattern of the surface texture, are shown in figure 9-2. The symbol is on the right of the long leg of the symbol.

Roughness-Width Cutoff ratings are given in millimetres and located below the horizontal extension. Unless otherwise specified, roughness-width cutoff is 0.8 mm.

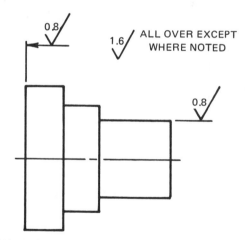

(C) A GENERAL NOTE BESIDE PART

Notes

Usually, a note is used where a given roughness requirement applies to either the whole part or the major portion. Examples are shown in figure 9-6(C).

Fig. 9-6 The use of notes with surface texture symbol

CONTROL REQUIREMENTS

Surface texture control should be specified for surfaces where texture is a functional requirement. For example, most surfaces which have contact with a mating part have a certain texture requirement, especially for roughness. The drawing should reflect the texture necessary for optimum part function without depending upon the variables of machining practices.

Many surfaces do not need a specification of surface texture because the function is unaffected by the surface quality. Such surfaces should not receive surface quality designations because it could unnecessarily increase the product cost.

Figures 9-7, 9-8, and 9-9 show recommended roughness height ratings, the machining methods used to produce them, and the application of the ratings to the surface texture symbol.

ONE- AND TWO-VIEW DRAWINGS

Except for complex objects of irregular shapes, it is seldom necessary to draw more than three views, and for simple parts one- or two-view drawings will often suffice.

In one-view drawings the third dimension may be expressed by a note or by descriptive words or abbreviations, such as *diameter,* or *hexagon across flats,* figure 9-10. Square sections may be indicated by light crossed diagonal lines. The center line through the piece indicates that it is symmetrical. Frequently, the drafter will decide that only two views are necessary to explain the shape of an object fully. One or two views usually show the shape of cylindrical objects adequately.

MULTIPLE-DETAIL DRAWINGS

Details of parts may be shown on separate sheets, or they may be grouped together on one or more large sheets. Often the details of parts are grouped according to the department in which they are made. Metal parts to be fabricated in the machine shop may appear on one detail sheet while parts to be made in the wood shop may be grouped on another.

CHAMFERING

The process of *chamfering,* that is, cutting away the inside or outside corner of an object, is done to facilitate assembly. The recommended method of dimensioning a chamfer is to give an angle and a length, or an angle and a diameter. For angles of 45 degrees only, a note form may be used. This method is permissible only with 45-degree angles because the size may apply to either the longitudinal or radial dimension. Chamfers are never measured along the angular surface.

NECKING

The operation of *necking* or *undercutting* is the cutting of a recess in a diameter. Necking permits two parts to join, figure 9-11. Necking is indicated on a drawing by a note listing the width first and then the diameter. If the radius is shown at the bottom of the undercut, it is assumed that the radius will be equal to half the width unless specified differently and the diameter will apply to the center of the undercut. Where the size of the neck is unimportant, the dimension may be omitted from the drawing.

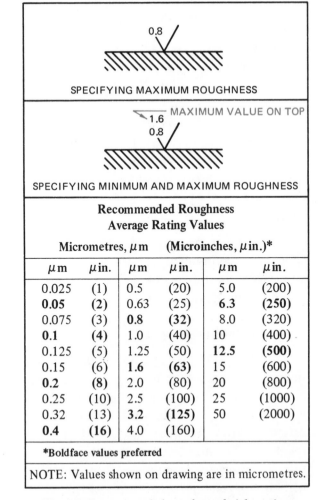

Recommended Roughness Average Rating Values					
Micrometres, μm		(Microinches, μin.)*			
μm	μin.	μm	μin.	μm	μin.
0.025	(1)	0.5	(20)	5.0	(200)
0.05	**(2)**	0.63	(25)	**6.3**	**(250)**
0.075	(3)	**0.8**	**(32)**	8.0	(320)
0.1	**(4)**	1.0	(40)	10	(400)
0.125	(5)	1.25	(50)	**12.5**	**(500)**
0.15	(6)	**1.6**	**(63)**	15	(600)
0.2	**(8)**	2.0	(80)	20	(800)
0.25	(10)	2.5	(100)	25	(1000)
0.32	(13)	**3.2**	**(125)**	50	(2000)
0.4	**(16)**	4.0	(160)		
*Boldface values preferred					
NOTE: Values shown on drawing are in micrometres.					

Fig. 9-7 Recommended roughness height ratings

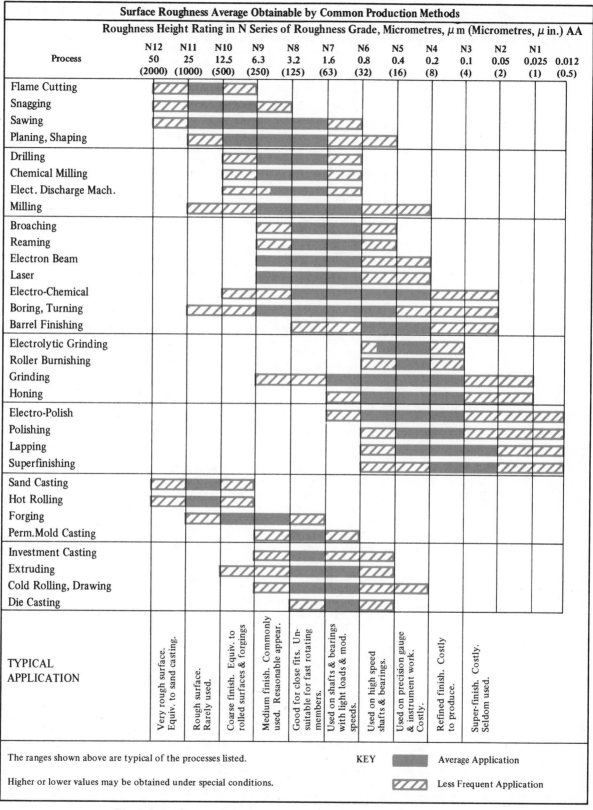

Fig. 9-8 Surface roughness range for common production methods

MICROMETRES AA RATING	MICROINCHES AA RATING	APPLICATION
25.2	1000	Rough, low grade surface resulting from sand casting, torch or saw cutting, chipping, or rough forging. Machine operations are not required because appearance is not objectionable. This surface, rarely specified, is suitable for unmachined clearance areas on rough construction items.
12.5	500	Rough, low grade surface resulting from heavy cuts and coarse feeds in milling turning, shaping, boring, and rough filing, disc grinding and snagging. It is suitable for clearance areas on machinery, jigs, and fixtures. Sand casting or rough forging produces this surface.
6.3	250	Coarse production surfaces, for unimportant clearance and cleanup operations, resulting from coarse surface grind, rough file, disc grind, rapid feeds in turning, milling, shaping, drilling, boring, grinding, etc., where tool marks are not objectionable. The natural surfaces of forgings, permanent mold castings, extrusions, and rolled surfaces also produce this roughness. It can be produced economically and is used on parts where stress requirements, appearance, and conditions of operations and design permit.
3.2	125	The roughest surface recommended for parts subject to loads, vibration, and high stress. It is also permitted for bearing surfaces when motion is slow and loads light or infrequent. It is a medium commercial machine finish produced by relatively high speeds and fine feeds taking light cuts with sharp tools. It may be economically produced on lathes, milling machines, shapers, grinders, etc., or on permanent mold castings, die castings, extrusion, and rolled surfaces.
1.6	63	A good machine finish produced under controlled conditions using relatively high speeds and fine feeds to take light cuts with sharp cutters. It may be specified for close fits and used for all stressed parts, except fast rotating shafts, axles, and parts subject to severe vibration or extreme tension. It is satisfactory for bearing surfaces when motion is slow and loads light or infrequent. It may also be obtained on extrusions, rolled surfaces, die castings and permanent mold castings when rigidly controlled.
0.8	32	A high-grade machine finish requiring close control when produced by lathes, shapers, milling machines, etc., but relatively easy to produce by centerless, cylindrical, or surface grinders. Also, extruding, rolling or die casting may produce a comparable surface when rigidly controlled. This surface may be specified in parts where stress concentration is present. It is used for bearings when motion is not continuous and loads are light. When finer finishes are specified, production costs rise rapidly; therefore, such finishes must be analyzed carefully.
0.4	16	A high quality surface produced by fine cylindrical grinding, emery buffing, coarse honing, or lapping, it is specified where smoothness is of primary importance, such as rapidly rotating shaft bearings, heavily loaded bearing and extreme tension members.
0.2	8	A fine surface produced by honing, lapping, or buffing. It is specified where packings and rings must slide across the direction of the surface grain, maintaining or withstanding pressures, or for interior honed surfaces of hydraulic cylinders. It may also be required in precision gauges and instrument work, or sensitive value surfaces, or on rapidly rotating shafts and on bearings where lubrication is not dependable.
0.1	4	A costly refined surface produced by honing, lapping, and buffing. It is specified only when the design requirements make it mandatory. It is required in instrument work, gauge work, and where packing and rings must slide across the direction of surface grain such as on chrome plated piston rods, etc. where lubrication is not dependable.
0.05 0.025	2 1	Costly refined surfaces produced only by the finest of modern honing, lapping, buffing, and superfinishing equipment. These surfaces may have a satin or highly polished appearance depending on the finishing operation and material. These surfaces are specified only when design requirements make it mandatory. They are specified on fine or sensitive instrument parts or other laboratory items, and certain gauge surfaces, such as precision gauge blocks.

Fig. 9-9 Surface roughness description and application

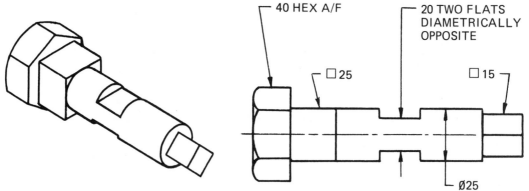

Fig. 9-10 One-view drawing of a turned part

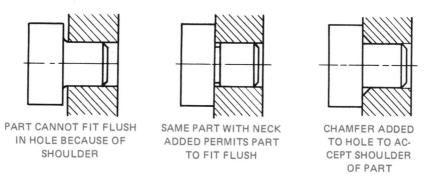

(A) CHAMFER AND NECK APPLICATION

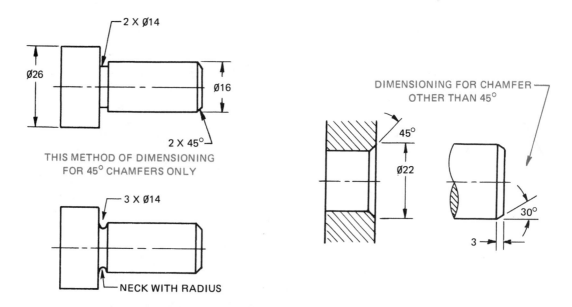

(B) DIMENSIONING CHAMFERS AND NECKS (UNDERCUTS)

Fig. 9-11 Chamfers and necks

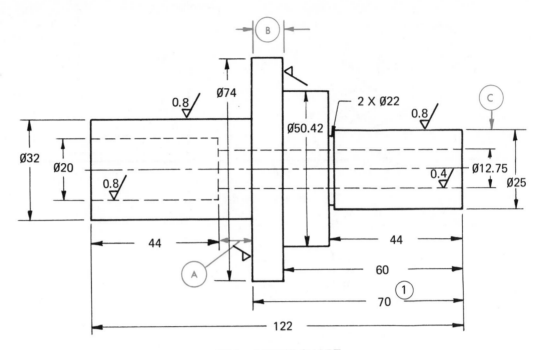

PT 1   LOWER SHAFT
MATERIAL-CRS, 2 REQD   FAO

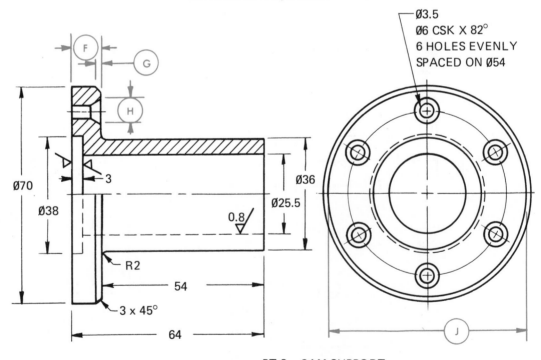

PT 3   CAM SUPPORT
MATERIAL — ALUMINUM, 2 REQD

UNLESS OTHERWISE SPECIFIED –
TOLERANCES ON DIMENSIONS ± 0.5
1.6 ∇ EXCEPT WHERE NOTED

REVISION	1	JAN 12/67	A. HEINEN
	LENGTH WAS 72		

74

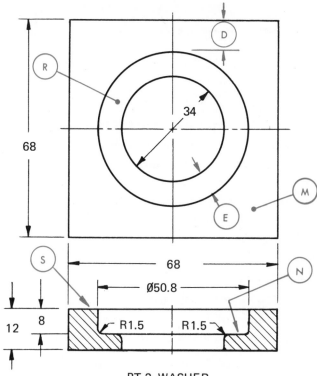

68

68

Ø50.8

12   8

R1.5   R1.5

PT 2 WASHER
MATERIAL — MS. 4 REQD.

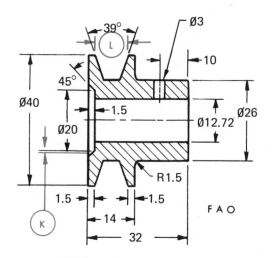

39°

Ø3

10

45°

Ø40

1.5

Ø26

Ø20

Ø12.72

R1.5

1.5   1.5

14

32

FAO

PT 4   V-BELT PULLEY
MATERIAL–CRS, 4 REQD

QUESTIONS

1. Calculate dimensions (A) to (L).

Referring to figure 9-8, what production methods would be suitable to produce the surface texture of

2. Ø12.75 hole in Part 1?
3. Ø32 on Part 1?
4. Ø25.5 hole in Part 3?

Refer to Part 1

5. What is the length of the Ø25.2 portion?
6. What was the original length of the 70 mm dimension?
7. How many hidden circles would be seen if the right end view were drawn?

Refer to Part 2

8. Which surface does (R) represent in the front view?
9. Which surface does (S) represent in the top view?
10. If the part that passes through the washer is Ø32, what is the clearance per side between the two parts?
11. How many sharp edges are broken?

Refer to Part 3

12. How many degrees apart on the Ø54 are the Ø3.5 holes?
13. Is the center line of the countersunk holes in the center of the flange?
14. What operation is performed to allow the head of the mounting screws to rest flush with the flange?
15. What type of section view is shown?
16. What is the amount and degree of chamfer?

Refer to Part 4

17. How deep is the Ø3 hole?
18. How deep is the belt groove?
19. What does FAO mean?
20. What type of section view is shown?

ANSWERS

1 A_____
  B_____
  C_____
  D_____
  E_____
  F_____
  G_____
  H_____
  J_____
  K_____
  L_____
2 _____
3 _____
4 _____
5 _____
6 _____
7 _____
8 _____
9 _____
10 _____
11 _____
12 _____
13 _____
14 _____
15 _____
16 _____
17 _____
18 _____
19 _____
20 _____

METRIC

| SCALE | NOT TO SCALE | |
| DRAWN | | DATE |

HANGER DETAILS   **A-23**

UNIT 10

TOLERANCES AND ALLOWANCES

The history of engineering drawing as a means for the communication of engineering information spans a period of 6 000 years. It seems inconceivable that such an elementary practice as the tolerancing of dimensions, which is taken for granted today, was introduced for the first time about 80 years ago.

Apparently, engineers and workers realized only gradually that *exact* dimensions and shapes could not be attained in the shaping of physical objects. The skilled handicrafters of the past took pride in the ability to work to exact dimensions. This meant that objects were dimensioned more accurately than they could be measured. The use of modern measuring instruments would have shown the deviations from the sizes which were called exact.

It was soon realized that variations in the sizes of parts had always been present, that such variations could be restricted but not avoided, and that slight variation in the size which a part was originally intended to have could be tolerated without impairment of its correct functioning. It became evident that interchangeable parts need not be identical parts, but rather it would be sufficient if the significant sizes which controlled their fits lay between definite limits. Therefore, the problem of interchangeable manufacture developed from the making of parts to a supposedly exact size, to the holding of parts between two limiting sizes, lying so closely together that any intermediate size would be acceptable.

The concept of limits means essentially that a precisely defined *basic* condition (expressed by one numerical value or specification) is replaced by two limiting conditions. Any result lying between these two limits is acceptable. One standard level is replaced by two limiting levels enclosing a zone of acceptability, or tolerance. A workable scheme of interchangeable manufacture that is indispensable to mass production methods has been established.

Definitions

Tolerances. The tolerance of a dimension is the total permissible variation in its size. The tolerance is the difference between the limits of size.

Limits of Size. The limits are the maximum and minimum sizes permitted for a specific dimension.

Allowance. An allowance is the intentional difference in correlated dimensions of mating parts. It is the minimum clearance (positive allowance) or maximum interference (negative allowance) between such parts.

All dimensions on a drawing have tolerances. Some dimensions must be more exact than other dimensions and consequently have smaller tolerances.

When dimensions require a greater accuracy than the general note provides, individual tolerances or limits must be shown for that dimension. For example:

$$64 \pm 0.05 \quad \text{or} \quad \frac{64.05}{63.95}$$
$$\text{or} \quad 63.95 - 64.05$$

Tolerancing Methods

Dimensional tolerances are expressed in either of two ways: *limit dimensioning* or *plus and minus tolerancing.*

Limit Dimensioning. In the limiting dimension method, only the maximum and minimum dimensions are specified, figure 10-1. When placed above one another, the largest dimension is placed on top. When shown with a leader and placed in one line, the smallest size is shown first. A small dash separates the two dimensions.

Plus and Minus Tolerancing. In this method the dimension of the specified size is given first and it is followed by a plus and minus tolerance expression. The tolerance can be bilateral or unilateral, figure 10-2.

A *bilateral tolerance* is a tolerance which is expressed as plus and minus values. These values need not be the same size.

A *unilateral tolerance* is one which applies in one direction from the specified size, so the permissible variation in the other direction is zero. When a unilateral tolerance is specified on the drawing, the plus or minus sign is not shown with the zero.

General tolerance notes greatly simplify the drawing. The following examples illustrate the variety of applications in this system. The values given in the examples are typical only:

EXCEPT WHERE STATED OTHERWISE, TOLERANCES ON DIMENSIONS ± 0.2 mm EXCEPT WHERE STATED OTHERWISE, TOLERANCES ON FINISHED DIMENSIONS TO BE AS FOLLOWS:

DIMENSION (mm)	TOLERANCE (mm)
up to 100	0.1
from 101 to 300	0.2
from 301 to 600	0.5
over 600	1.0

UNLESS OTHERWISE SPECIFIED ±0.2 mm TOLERANCE ON MACHINED DIMENSIONS ±1 mm TOLERANCE ON CAST DIMENSIONS, ANGULAR TOLERANCE ±0.5°

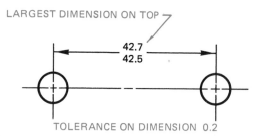

Fig. 10-1 Limit dimensioning

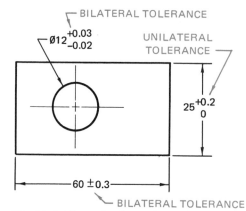

Fig. 10-2 Plus and minus tolerancing

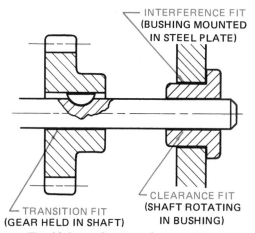

Fig. 10-3 Application of types of fits

Fit

Fit is the general term used to signify the range of tightness or looseness resulting from the application of a specific combination of allowances and tolerances in the design of mating parts. Fits are of three general types: *clearance, interference,* and *transition.* Figures 10-3 and 10-4 (page 78) illustrate the three types of fits.

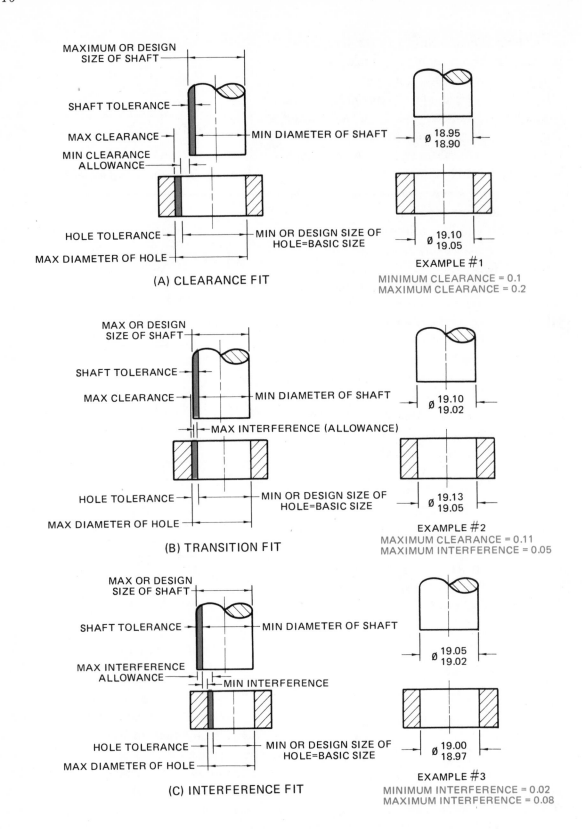

MAXIMUM OR DESIGN SIZE OF SHAFT

SHAFT TOLERANCE

MAX CLEARANCE

MIN CLEARANCE ALLOWANCE

MIN DIAMETER OF SHAFT

Ø 18.95 / 18.90

HOLE TOLERANCE

MAX DIAMETER OF HOLE

MIN OR DESIGN SIZE OF HOLE=BASIC SIZE

Ø 19.10 / 19.05

EXAMPLE #1

(A) CLEARANCE FIT

MINIMUM CLEARANCE = 0.1
MAXIMUM CLEARANCE = 0.2

MAX OR DESIGN SIZE OF SHAFT

SHAFT TOLERANCE

MAX CLEARANCE

MIN DIAMETER OF SHAFT

Ø 19.10 / 19.02

MAX INTERFERENCE (ALLOWANCE)

HOLE TOLERANCE

MAX DIAMETER OF HOLE

MIN OR DESIGN SIZE OF HOLE=BASIC SIZE

Ø 19.13 / 19.05

EXAMPLE #2

MAXIMUM CLEARANCE = 0.11
MAXIMUM INTERFERENCE = 0.05

(B) TRANSITION FIT

MAX OR DESIGN SIZE OF SHAFT

SHAFT TOLERANCE

MIN DIAMETER OF SHAFT

Ø 19.05 / 19.02

MAX INTERFERENCE ALLOWANCE

MIN INTERFERENCE

HOLE TOLERANCE

MAX DIAMETER OF HOLE

MIN OR DESIGN SIZE OF HOLE=BASIC SIZE

Ø 19.00 / 18.97

EXAMPLE #3

MINIMUM INTERFERENCE = 0.02
MAXIMUM INTERFERENCE = 0.08

(C) INTERFERENCE FIT

Fig. 10-4 Types of fits

Clearance Fits. Clearance fits have limits of size prescribed so a clearance always results when mating parts are assembled. Clearance fits are intended for accurate assembly of parts and bearings. The parts can be assembled by hand because the hole is always larger than the shaft.

Interference Fits. Interference fits have limits of size so prescribed that an interference always results when mating parts are assembled. The hole is always smaller than the shaft. Interference fits are for permanent assemblies of parts which require rigidity and alignment, such as dowel pins and bearings in castings. Parts are usually pressed together using an arbor press.

Transition Fits. Transition fits have limits of size indicating that either a clearance or an interference may result when mating parts are assembled. Transition fits are a compromise between clearance and interference fits. They are used for applications where accurate location is important, but either a small amount of clearance or interference is permissible.

INTERSECTION OF UNFINISHED SURFACES

The intersection of unfinished surfaces that are rounded or filleted at the point of theoretical intersection is indicated by a line coinciding with the theoretical point of intersection. The need for this convention is shown by the examples in figure 10-5 (page 80). For a large radius, figure 10-5(C), no line is drawn.

Members such as ribs and arms that blend into other features end in curves called *runouts*.

REFERENCES AND SOURCE MATERIALS

1. F.L. Spalding, "The Development of Standards for Dimensioning and Tolerancing," *Graphic Science,* 8 No. 2, (1966).
2. Extracted from *American National Standard Drafting Practices: Dimensioning and Tolerancing* (ANSI Y14.5 — 1973), with the permission of the publisher, The American Society of Mechanical Engineers, 345 East 47th Street, New York, N.Y. 10017.

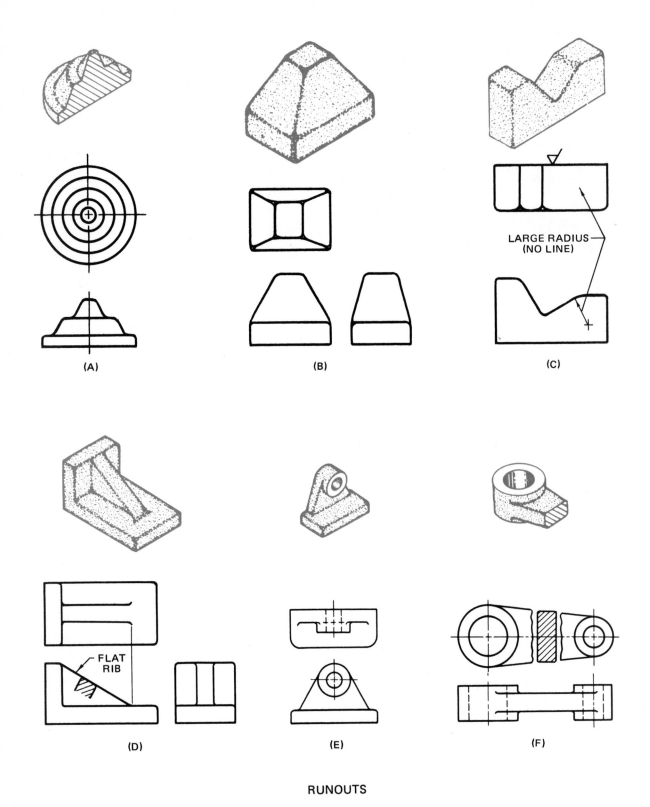

RUNOUTS

Fig. 10-5 Rounded and filleted intersections

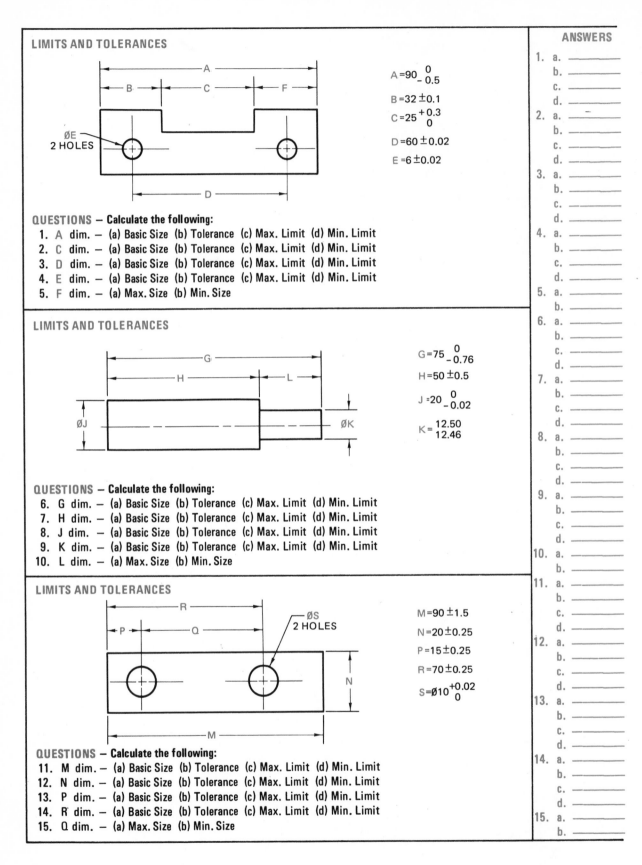

LIMITS AND TOLERANCES

A =90 $^{0}_{-0.5}$

B =32 ±0.1

C =25 $^{+0.3}_{0}$

D =60 ±0.02

E =6 ±0.02

ØE
2 HOLES

QUESTIONS – Calculate the following:
1. A dim. – (a) Basic Size (b) Tolerance (c) Max. Limit (d) Min. Limit
2. C dim. – (a) Basic Size (b) Tolerance (c) Max. Limit (d) Min. Limit
3. D dim. – (a) Basic Size (b) Tolerance (c) Max. Limit (d) Min. Limit
4. E dim. – (a) Basic Size (b) Tolerance (c) Max. Limit (d) Min. Limit
5. F dim. – (a) Max. Size (b) Min. Size

LIMITS AND TOLERANCES

G =75 $^{0}_{-0.76}$

H =50 ±0.5

J =20 $^{0}_{-0.02}$

K = $\frac{12.50}{12.46}$

ØJ   ØK

QUESTIONS – Calculate the following:
6. G dim. – (a) Basic Size (b) Tolerance (c) Max. Limit (d) Min. Limit
7. H dim. – (a) Basic Size (b) Tolerance (c) Max. Limit (d) Min. Limit
8. J dim. – (a) Basic Size (b) Tolerance (c) Max. Limit (d) Min. Limit
9. K dim. – (a) Basic Size (b) Tolerance (c) Max. Limit (d) Min. Limit
10. L dim. – (a) Max. Size (b) Min. Size

LIMITS AND TOLERANCES

M =90 ±1.5

N =20 ±0.25

P =15 ±0.25

R =70 ±0.25

S =Ø10 $^{+0.02}_{0}$

ØS
2 HOLES

QUESTIONS – Calculate the following:
11. M dim. – (a) Basic Size (b) Tolerance (c) Max. Limit (d) Min. Limit
12. N dim. – (a) Basic Size (b) Tolerance (c) Max. Limit (d) Min. Limit
13. P dim. – (a) Basic Size (b) Tolerance (c) Max. Limit (d) Min. Limit
14. R dim. – (a) Basic Size (b) Tolerance (c) Max. Limit (d) Min. Limit
15. Q dim. – (a) Max. Size (b) Min. Size

ANSWERS

1. a. b. c. d.
2. a. b. c. d.
3. a. b. c. d.
4. a. b. c. d.
5. a. b.
6. a. b. c. d.
7. a. b. c. d.
8. a. b. c. d.
9. a. b. c. d.
10. a. b.
11. a. b. c. d.
12. a. b. c. d.
13. a. b. c. d.
14. a. b. c. d.
15. a. b.

CLEARANCE FITS

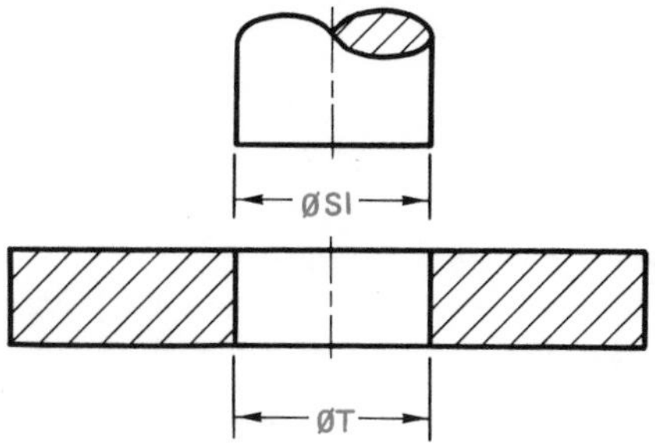

$SI = \emptyset\,^{24.075}_{24.060}$

$T = \emptyset\,^{25.025}_{25.000}$

QUESTIONS
16. What is the tolerance on the shaft (SI)?
17. What is the tolerance on the hole (T)?
18. What is the minimum clearance between parts?
19. What is the maximum clearance between parts?

INTERFERENCE FITS

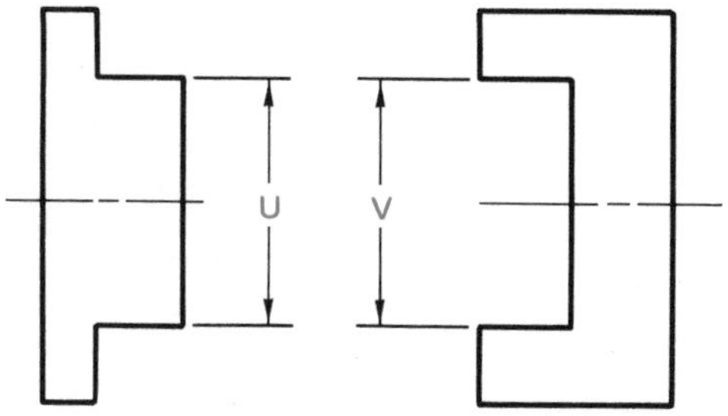

$U = \,^{40.12}_{40.09}$

$V = \,^{40.05}_{40.00}$

QUESTIONS
20. What is the tolerance on the part (U)?
21. What is the tolerance on the slot (V)?
22. What is the minimum interference between the parts?
23. What is the maximum interference between the parts?

CLEARANCE AND INTERFERENCE FITS

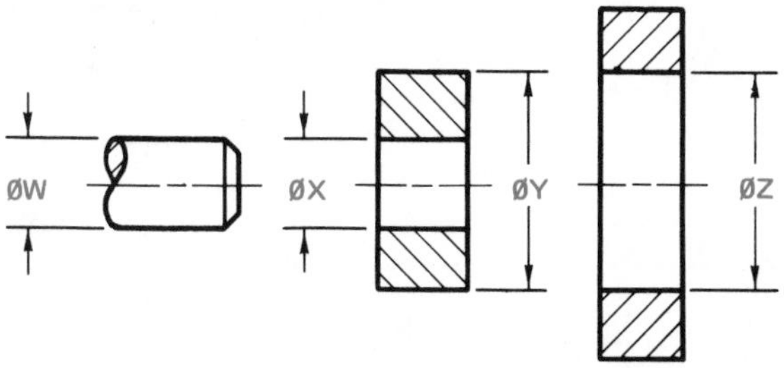

$X = \emptyset18 \pm 0.02$

$Z = \emptyset\,^{30.05}_{30.00}$

QUESTIONS
24. Dimension shaft (W) to have (a) a tolerance of 0.5 and (b) a minimum clearance of 0.2 .
25. Dimension bushing (Y) to have (a) a tolerance of 0.08 and (b) a maximum interference of 0.2 .

NOTE: Only the dimensions pertaining to the questions are given on the drawings.

ASSIGNMENT
Calculate the limits, tolerance, fits, etc. of the parts as outlined in the questions following the individual parts.

A-24

ANSWERS	
16.	
17.	
18.	
19.	
20.	
21.	
22.	
23.	
24. a.	
b.	
25. a.	
b.	

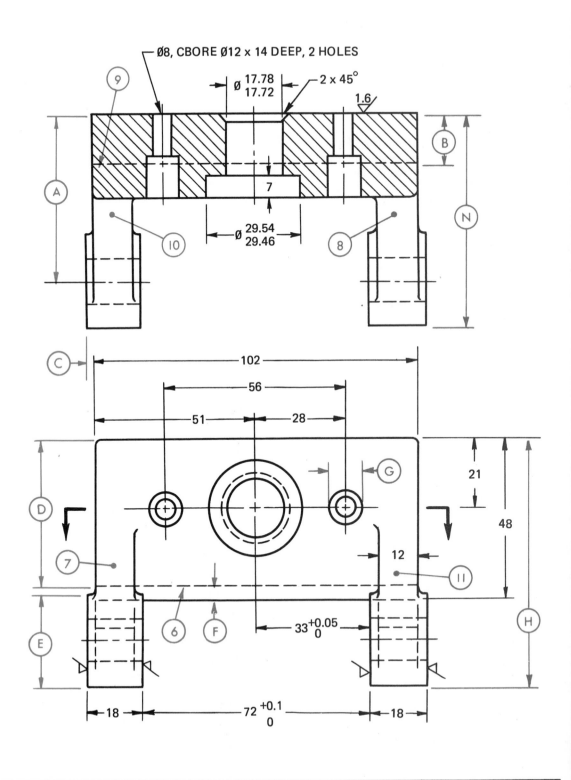

QUESTIONS

1. How many surfaces are to be finished?

2. Except where noted otherwise, what is the tolerance on all dimensions?

3. What is the tolerance on the Ø12.70 – 12.76 holes?

4. What are the limit dimensions for the 40.64 dimension shown on the side view?

5. What are the limit dimensions for the 26 dimension shown on the side view?

6. What is the maximum distance permissible between the centers of the Ø8 hole?

7. Express the Ø12.70 – 12.76 holes as a plus-and-minus tolerance dimension.

8. How many surfaces require a $\overset{6.3}{\nabla}$ finish?

9. What are the limit dimensions for the 7 dimension shown on the top view?

10. Locate surfaces ④ on the top view.

11. How many surface ⑤ s are there?

12. Locate line ③ in the top view.

13. Locate line ⑥ in the side view.

14. Which surfaces in the front view indicate line ④ in the side view?

15. Calculate distances Ⓐ to Ⓝ .

ANSWERS

1. _____
2. _____
3. _____
4. _____
5. _____
6. _____
7. _____
8. _____
9. _____
10. _____
11. _____
12. _____
13. _____
14. _____

15. Ⓐ _____
Ⓑ _____
Ⓒ _____
Ⓓ _____
Ⓔ _____
Ⓕ _____
Ⓖ _____
Ⓗ _____
Ⓘ _____
Ⓙ _____
Ⓚ _____
Ⓛ _____
Ⓜ _____
Ⓝ _____

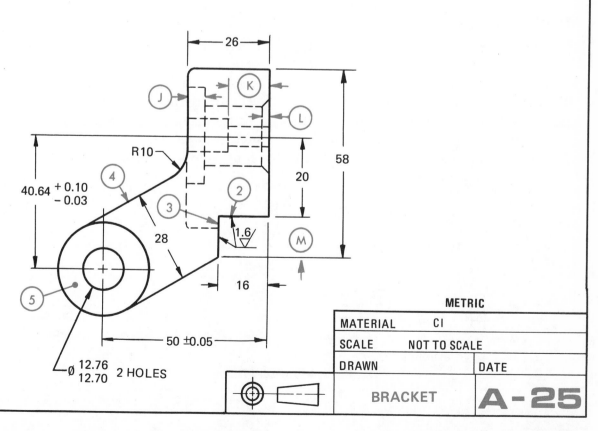

METRIC		
MATERIAL	CI	
SCALE	NOT TO SCALE	
DRAWN		DATE
BRACKET		A-25

UNIT 11

FASTENERS

Fastening devices are vital to most phases of industry. They are used in assembling manufactured products, the machines and devices used in the manufacturing processes, and in the construction of all types of buildings.

There are two basic types of fastening: *semi-permanent* and *removable.* Rivets are semi-permanent fastenings; bolts, screws, studs, nuts, pins, and keys are removable fastenings. With the progress of industry, fastening devices have become standardized. A thorough knowledge of the design and graphic representation of the common fasteners is essential for interpreting engineering drawings, figure 11-1.

THREADED FASTENERS

Thread Representation

True representation of a screw thread is seldom provided on working drawings because of the time involved and the drawing cost. Three types of conventions are generally used for screw thread representation: pictorial, schematic, and simplified representation. Simplified representation is used whenever it will clearly indicate the requirements. Schematic and pictorial representations require more drafting time, but they are sometimes used to avoid confusion with other parallel lines or to more clearly portray particular aspects of threads. ANSI and CSA thread representation vary slightly, figures 11-2 and 11-3.

Thread Standards

With the progress and growth of industry, there is a growing need for uniform, interchangeable, threaded fasteners. Aside from the threaded forms previously mentioned, the pitch of the thread and the major diameters are factors affecting standards.

METRIC THREADS

Metric threads are grouped into diameter-pitch combinations differentiated by the pitch applied to specified diameters. The *pitch* for metric threads is the distance between corresponding points on adjacent teeth. In addition to a coarse and fine pitch series, a series of constant pitches is available.

For each of the two main thread elements — pitch diameter and crest diameter — there are numerous tolerance grades. The number of the tolerance grade reflects the tolerance

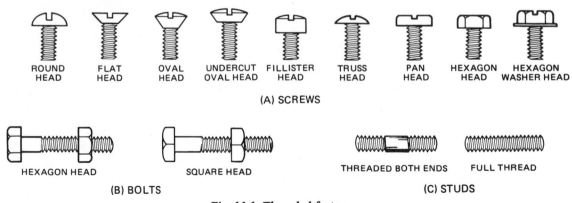

ROUND HEAD   FLAT HEAD   OVAL HEAD   UNDERCUT OVAL HEAD   FILLISTER HEAD   TRUSS HEAD   PAN HEAD   HEXAGON HEAD   HEXAGON WASHER HEAD

(A) SCREWS

HEXAGON HEAD   SQUARE HEAD   THREADED BOTH ENDS   FULL THREAD

(B) BOLTS   (C) STUDS

Fig. 11-1 Threaded fasteners

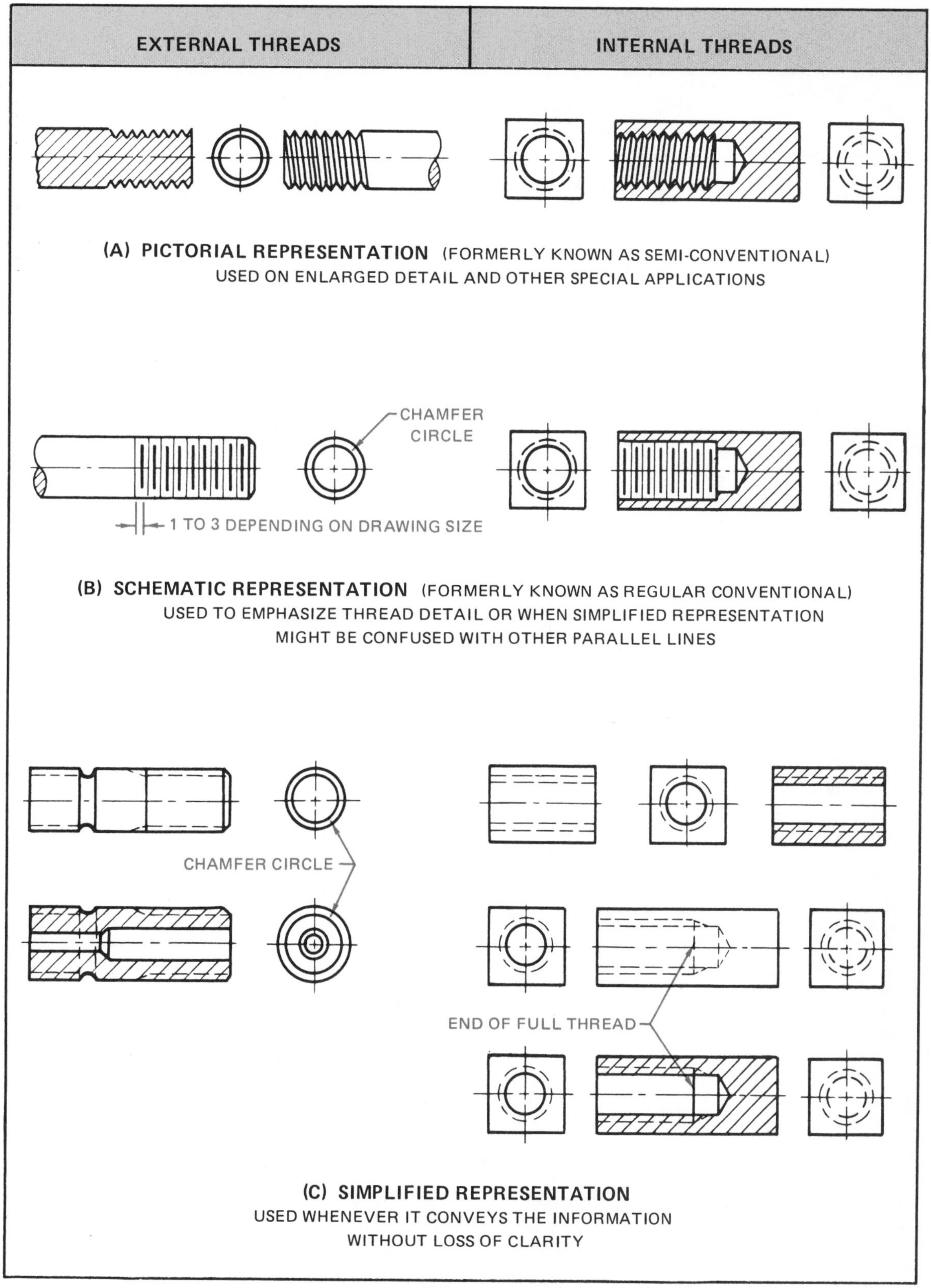

EXTERNAL THREADS	INTERNAL THREADS

(A) PICTORIAL REPRESENTATION (FORMERLY KNOWN AS SEMI-CONVENTIONAL)
USED ON ENLARGED DETAIL AND OTHER SPECIAL APPLICATIONS

CHAMFER CIRCLE

1 TO 3 DEPENDING ON DRAWING SIZE

(B) SCHEMATIC REPRESENTATION (FORMERLY KNOWN AS REGULAR CONVENTIONAL)
USED TO EMPHASIZE THREAD DETAIL OR WHEN SIMPLIFIED REPRESENTATION
MIGHT BE CONFUSED WITH OTHER PARALLEL LINES

CHAMFER CIRCLE

END OF FULL THREAD

(C) SIMPLIFIED REPRESENTATION
USED WHENEVER IT CONVEYS THE INFORMATION
WITHOUT LOSS OF CLARITY

Fig. 11-2 Standard thread conventions (ANSI)

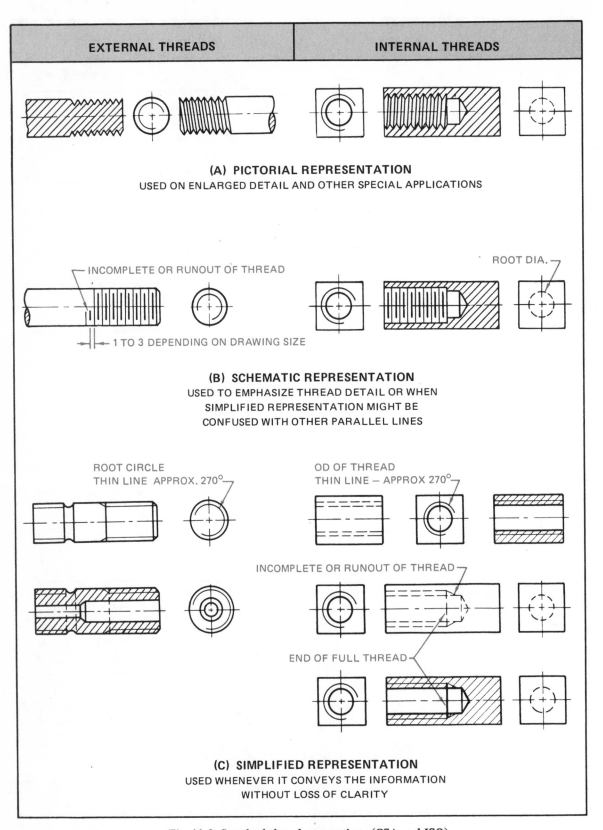

EXTERNAL THREADS | **INTERNAL THREADS**

(A) PICTORIAL REPRESENTATION
USED ON ENLARGED DETAIL AND OTHER SPECIAL APPLICATIONS

— INCOMPLETE OR RUNOUT OF THREAD

ROOT DIA.

1 TO 3 DEPENDING ON DRAWING SIZE

(B) SCHEMATIC REPRESENTATION
USED TO EMPHASIZE THREAD DETAIL OR WHEN
SIMPLIFIED REPRESENTATION MIGHT BE
CONFUSED WITH OTHER PARALLEL LINES

ROOT CIRCLE
THIN LINE APPROX. 270°

OD OF THREAD
THIN LINE — APPROX 270°

INCOMPLETE OR RUNOUT OF THREAD

END OF FULL THREAD

(C) SIMPLIFIED REPRESENTATION
USED WHENEVER IT CONVEYS THE INFORMATION
WITHOUT LOSS OF CLARITY

Fig. 11-3 Standard thread conventions (CSA and ISO)

size. For example: Grade 4 tolerances are smaller than Grade 6 tolerances; Grade 8 tolerances are larger than Grade 6 tolerances.

In each case, Grade 6 tolerances should be used for medium quality length of engagement applications. The tolerance grades below Grade 6 are intended for applications involving fine quality and/or short lengths of engagement. Tolerance grades above Grade 6 are intended for coarse quality and/or long periods of engagement.

In addition to the tolerance grade, positional tolerance is required. The positional tolerance defines the maximum-material limits of the pitch and crest diameters of the external and internal threads and indicates their relationship to the basic profile.

In conformance with current coating (or plating) thickness requirements and the demand for ease of assembly, a series of tolerance positions reflecting the application of varying amounts of allowance has been established:

For External Threads:

Tolerance position "e" (large allowance)
Tolerance psoition "g" (small allowance)
Tolerance position "h" (no allowance)

For Internal Threads:

Tolerance position "G" (small allowance)
Tolerance position "H" (no allowance)

Thread Designation

ISO metric screw threads are defined by nominal size (basic major diameter) and pitch, both expressed in millimetres. An "M" specifying an ISO metric screw thread precedes the nominal size and an "X" separates the nominal size from the pitch. For the coarse thread series, the pitch is shown only when the dimension for the length of thread is required. When specifying the length of thread, an "X" is used to separate the length of thread from the rest of the designations. For external threads, the length of thread may be given as a dimension on the drawing.

For example, a 10 mm diameter, 1.25 pitch, fine thread series is expressed as M10 X 1.25. A 10 mm diameter, 1.5 pitch,

coarse thread series is expressed as M10; the pitch is not shown unless the length of thread is required. If the latter thread was 25 mm long and this information was required on the drawing, the thread callout would be M 10 X 1.5 X 25.

In addition to the basic designation, complete designation for an ISO metric screw thread includes a tolerance class identification. A dash separates the tolerance class designation from the basic designation and includes the symbol for the pitch diameter tolerance followed immediately by the symbol for crest diameter tolerance. Each of these symbols consists of a numeral indicating the grade tolerance followed by a letter indicating the tolerance position (a capital letter for external threads and lower case letter for internal threads). Where the pitch and crest diameter symbols are identical, the symbol is necessary only once. Figure 11-4 illustrates the labeling of metric threads.

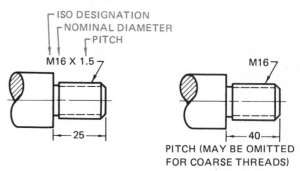

(A) BASIC THREAD CALLOUT

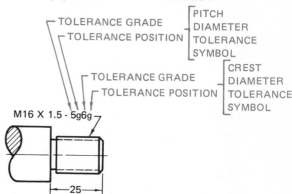

(B) ADDITIONAL THREAD CALLOUT

Fig. 11-4 Specifications for metric threads

INCH THREADS

Until 1976, nearly all threaded assemblies in North America were designed using inch-sized threads. In this system the pitch is equal to the distance between corresponding points and adjacent threads and is expressed as:

$$\frac{1}{\text{Number of Threads per inch}}$$

The number of threads per inch is set for different diameters in a thread "series." For the Unified National system there are the *coarse thread series* and the *fine thread series.*

There is also an *extra-fine thread series, UNEF,* for use where a small pitch is wanted, such as on thin-walled tubing. For special work and diameters larger than those specified in the coarse and fine series, the Unified Thread system has three series that allow the same number of threads per inch regardless of the diameter. These are the 8-thread series, the 12-thread series, and the 16-thread series.

Three classes of external thread (Classes 1A, 2A, and 3A) and three classes of internal thread (Classes 1B, 2B, and 3B) are provided. These classes differ from each other in the amounts of the allowances and tolerances provided.

The general characteristics and uses of the classes are:

Classes 1A and 1B. These classes produce the loosest fit, that is, the most play in assembly. They are useful for work where ease of assembly and disassembly is essential, such as for some automotive work and for stove bolts and other rough bolts and nuts.

Classes 2A and 2B. These classes are designed for the ordinary good grade of commercial products including machine screws and fasteners and most interchangeable parts.

Classes 3A and 3B. These classes are intended for exceptionally high-grade commercial products needing a particularly close or snug fit. Classes 3A and 3B are used only when

the high cost of precision tools and machines is warranted.

Thread Designation

Thread designation for both external and internal inch threads is expressed in this order: diameter (nominal or major diameter), number of threads per inch, thread form and series, and class of fit, figure 11-5. Figure 11-6 shows a comparison of metric threads and inch threads.

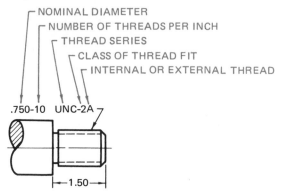

Fig. 11-5 Specifications for inch threads

INCH THREADS	METRIC THREADS
#10 (.190)	M5
1/4 (.250)	M6
5/16 (.312)	M8
3/8 (.375)	M10
7/16 (.438)	M12
1/2 (.500)	
9/16 (.562)	M14*
5/8 (.625)	M16
3/4 (.750)	M18*
	M20
7/8 (.875)	M22*
	M24
1 (1.000)	
	M27*
1 1/8 (1.125)	M30
1 1/4 (1.250)	
	M33*
1 3/8 (1.375)	M36
1 1/2 (1.500)	
	M39*

*NON PREFERRED SIZES

Fig. 11-6 Size comparison of inch and metric threads

RIGHT- AND LEFT-HANDED THREADS

Unless designated differently, threads are assumed to be right hand. A bolt being threaded into a tapped hole would be turned in a right-hand (clockwise) direction. For some special uses, such as turnbuckles, left-hand threads are required. When such a thread is necessary, the letters "LH" are added after the thread designation.

KNURLING

Knurling is the machining of a surface to create uniform depressions. Knurling permits a better grip. Knurling is shown on drawings as either a straight or diamond pattern. The pitch of the knurl may be specified. It is unnecessary to hatch in the whole area to be knurled if enough is shown to clearly indicate the pattern. Knurls are specified on the drawing by a note calling for the type and pitch. The length and diameter of the knurl are shown as dimensions, figure 11-7.

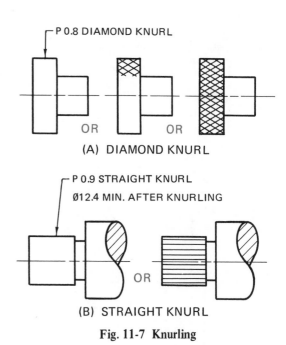

P 0.8 DIAMOND KNURL

OR   OR

(A) DIAMOND KNURL

P 0.9 STRAIGHT KNURL
Ø12.4 MIN. AFTER KNURLING

OR

(B) STRAIGHT KNURL

Fig. 11-7 Knurling

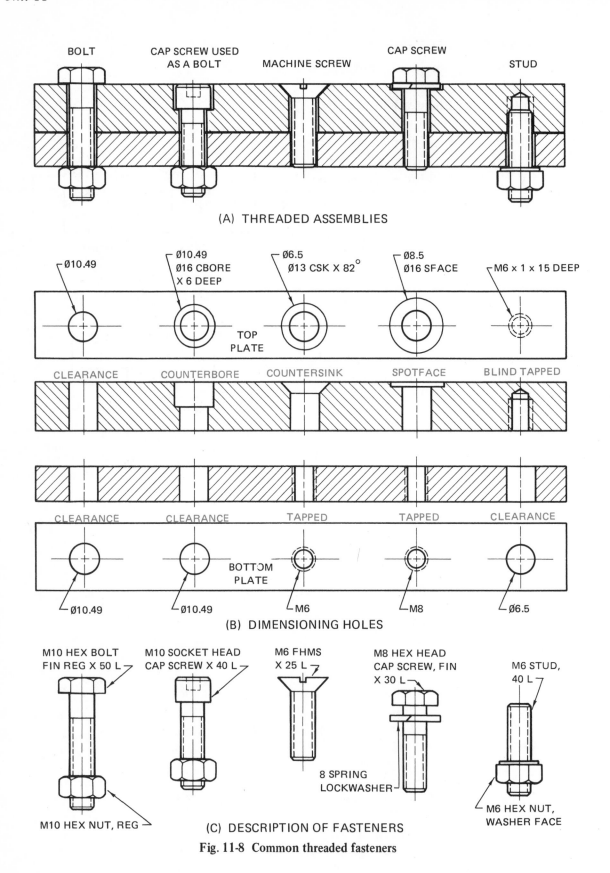

(A) THREADED ASSEMBLIES

(B) DIMENSIONING HOLES

(C) DESCRIPTION OF FASTENERS

Fig. 11-8 Common threaded fasteners

QUESTIONS

1. Calculate distances (A) to (J).

2. How many threaded holes or shafts are shown?

3. How many chamfers are shown?

4. How many necks are shown?

5. How many surfaces require finishing?

Refer to Part 1

6. What are the limit sizes for the Ø16 dimension?

7. What operation provides better gripping when turning the clamping nut?

8. What is the tap drill size?

9. Is the thread right hand or left hand?

10. What is the depth of the threads?

Refer to Part 2

11. What is the length of thread?

12. Is the thread pitch fine or coarse?

13. What heat treatment does the part undergo?

14. How many threads are there in a 10 mm length?

Refer to Part 3

15. What is the maximum permissible length of the part?

16. What size thread is cut on the outside of the piece?

17. What is the distance between the last thread and the flange?

18. What is the tolerance on the center-to-center distance between the tapped holes?

19. What is the smallest diameter to which the hole through the stuffing box can be made?

20. What are the limits for the Ø50 dimension?

NOTE: EXCEPT WHERE NOTED —
TOLERANCE ON DIMENSIONS ±0.5
TOLERANCE ON ANGLES ±0.5 °
1.6⟍ EXCEPT WHERE NOTED

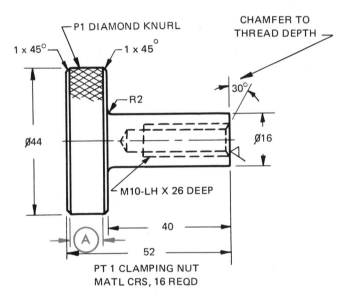

PT 1 CLAMPING NUT
MATL CRS, 16 REQD

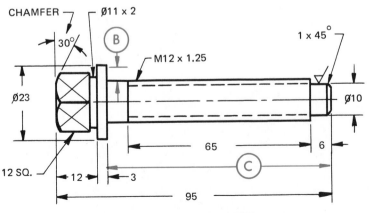

PT 2 TOOL POST SCREW
MATERIAL CRS CASE HARDEN, 8 REQD

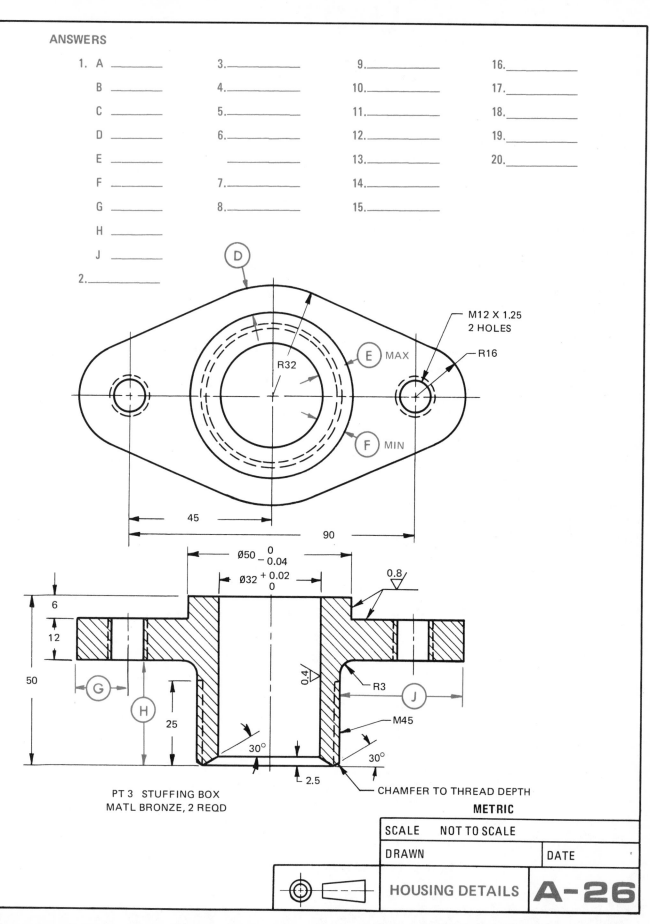

1. A _____   3. _____   9. _____   16. _____

B _____   4. _____   10. _____   17. _____

C _____   5. _____   11. _____   18. _____

D _____   6. _____   12. _____   19. _____

E _____   _____   13. _____   20. _____

F _____   7. _____   14. _____

G _____   8. _____   15. _____

H _____

J _____

2. _____

D

M12 X 1.25
2 HOLES

R16

E MAX

R32

F MIN

45

90

Ø50 $^{0}_{-0.04}$

Ø32 $^{+0.02}_{0}$

0.8

6

12

50

G

H

25

0.4

R3

J

M45

30°

2.5

30°

CHAMFER TO THREAD DEPTH

PT 3   STUFFING BOX
MATL BRONZE, 2 REQD

METRIC

SCALE	NOT TO SCALE	
DRAWN		DATE
	HOUSING DETAILS	**A-26**

95

REVOLVED AND REMOVED SECTIONS

Revolved and removed sections are used to show the cross-sectional shape of ribs, spokes, or arms, when the shape is not obvious in the regular views. End views are often not needed when a revolved section is used.

REVOLVED SECTIONS

For a *revolved section* a center line is drawn through the shape on the plane to be described, the part is imagined to be rotated 90 degrees, and the view that would be seen when rotated is superimposed on the view. If the revolved section does not interfere with the view on which it is revolved, then the view is not broken unless it would facilitate clearer dimensioning. When the revolved section interferes or passes through lines on the view on which it is revolved, the view is usually broken. Often the break is used to shorten the length of the object. When superimposed on the view, the outline of the revolved section is a thin continuous line, figure 12-1.

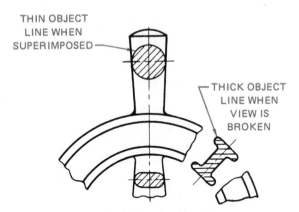

THIN OBJECT LINE WHEN SUPERIMPOSED

THICK OBJECT LINE WHEN VIEW IS BROKEN

Fig. 12-1 Revolved section

REMOVED SECTIONS

The *removed section* differs from the revolved section in that the section is removed to an open area on the drawing instead of being drawn directly on the view, figures 12-2, 12-3, and 12-4. Frequently, the removed section is drawn to an enlarged scale for clarification and easier dimensioning.

Removed sections of symmetrical parts are placed on the extension of the center line where possible.

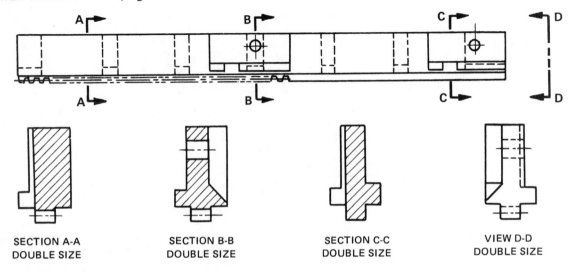

SECTION A-A
DOUBLE SIZE

SECTION B-B
DOUBLE SIZE

SECTION C-C
DOUBLE SIZE

VIEW D-D
DOUBLE SIZE

Fig. 12-2 Removed sections and removed view

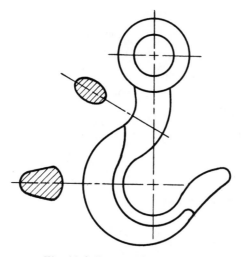

Fig. 12-3 Removed section of crane hook

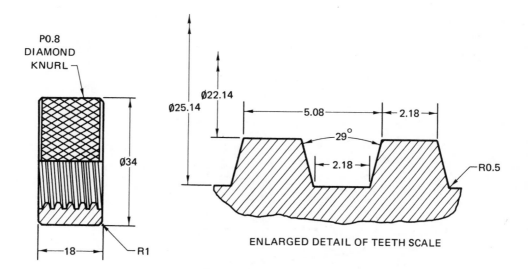

ENLARGED DETAIL OF TEETH SCALE

Fig. 12-4 Removed section of nut

QUESTIONS

1. Calculate distances (A) to (K) .

2. How many surfaces require finishing?

3. What tolerance is permitted on dimensions which do not specify a certain tolerance?

4. What type of section view is used on part 1?

5. What type of section view is used on part 2?

6. How many holes are there?

Refer to Part 1

7. What is the maximum center-to-center distance between the holes?

8. Which surface finish is required?

9. Express the largest hole in plus or minus dimensioning.

10. What is the maximum permissible wall thickness at the larger hole?

11. What would be the minimum permissible wall thickness at the smaller hole?

Refer to Part 2

12. Name two machines that could produce the type of finish required for the slot.

13. What is the nominal size machine screw used in the counterbored holes?

14. What type of machine screw would be used?

15. What is the distance between the center of the counterbored holes and the center of the slot?

EXCEPT WHERE STATED OTHERWISE:
TOLERANCES ARE ±0.2
ROUNDS AND FILLETS R2
MACHINE FINISH 0.8

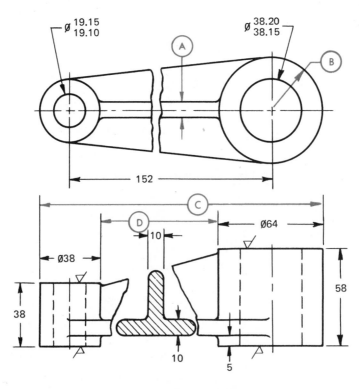

PT I SHAFT SUPPORT
MATL − CI

ANSWERS

1. A ______	J ______	7. ______	13. ______
B ______	K ______	8. ______	14. ______
C ______	2. ______	9. ______	15. ______
D ______	3. ______	10. ______	
E ______	4. ______	11. ______	
F ______	5. ______	12. ______	
G ______	6. ______	______	
H ______			

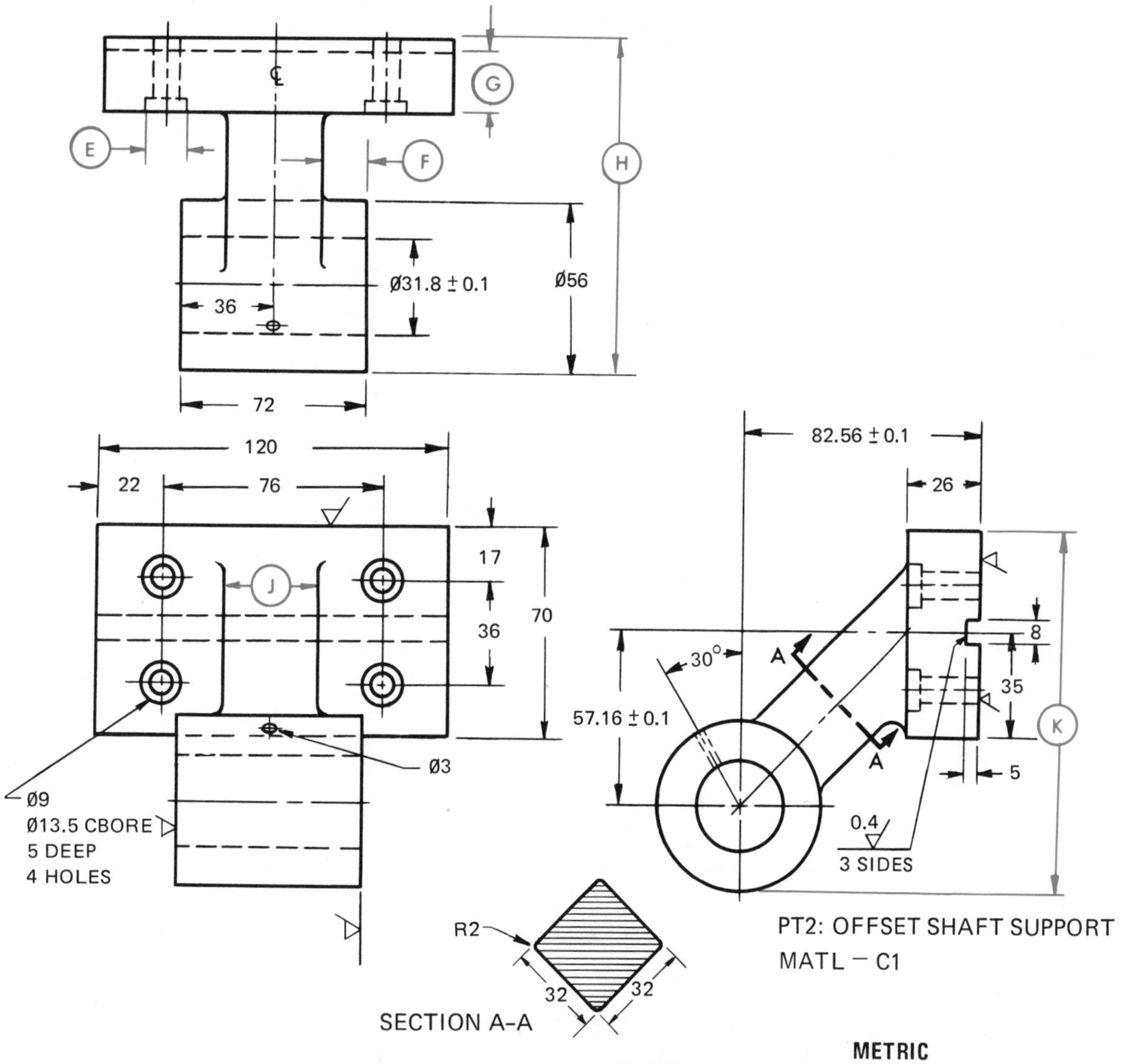

G
E
F
H
Ø31.8 ± 0.1
Ø56
36
72
120
22
76
17
36
70
J
Ø3
Ø9
Ø13.5 CBORE
5 DEEP
4 HOLES
82.56 ± 0.1
26
8
35
5
30°
A
A
57.16 ± 0.1
K
0.4
3 SIDES
R2
32
32
SECTION A-A
PT2: OFFSET SHAFT SUPPORT
MATL — C1

METRIC

SCALE NOT TO SCALE		
DRAWN		DATE
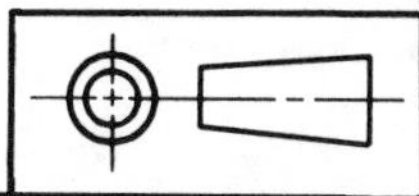	SHAFT SUPPORTS	A-27

UNIT
13

KEYS

A *key* is a piece of metal lying partly in a groove in the shaft, called a *keyseat,* and extending into another groove, called a *keyway,* in the hub, figure 13-1. The key is used to secure gears, pulleys, cranks, handles, and similar machine parts to shafts, so that the motion of the part is transmitted to the shaft or the motion of the shaft to the part, without slippage. The key is also a safety feature. Because of its size, when overloading occurs, the key slows or breaks before the part or shaft breaks.

Common key types are square, flat and Woodruff. Tables in the appendix give standard square and flat key sizes recommended for various shaft diameters and the necessary dimensions for Woodruff Keys.

The Woodruff key is semicircular and fits into a semicircular keyseat in the shaft and a rectangular keyway in the hub. Woodruff keys are currently only available in inch sizes and are identified by a four-digit key number. The first two numbers give the key width in thirty-seconds of an inch. The last two numbers give the nominal diameter in eighths of an

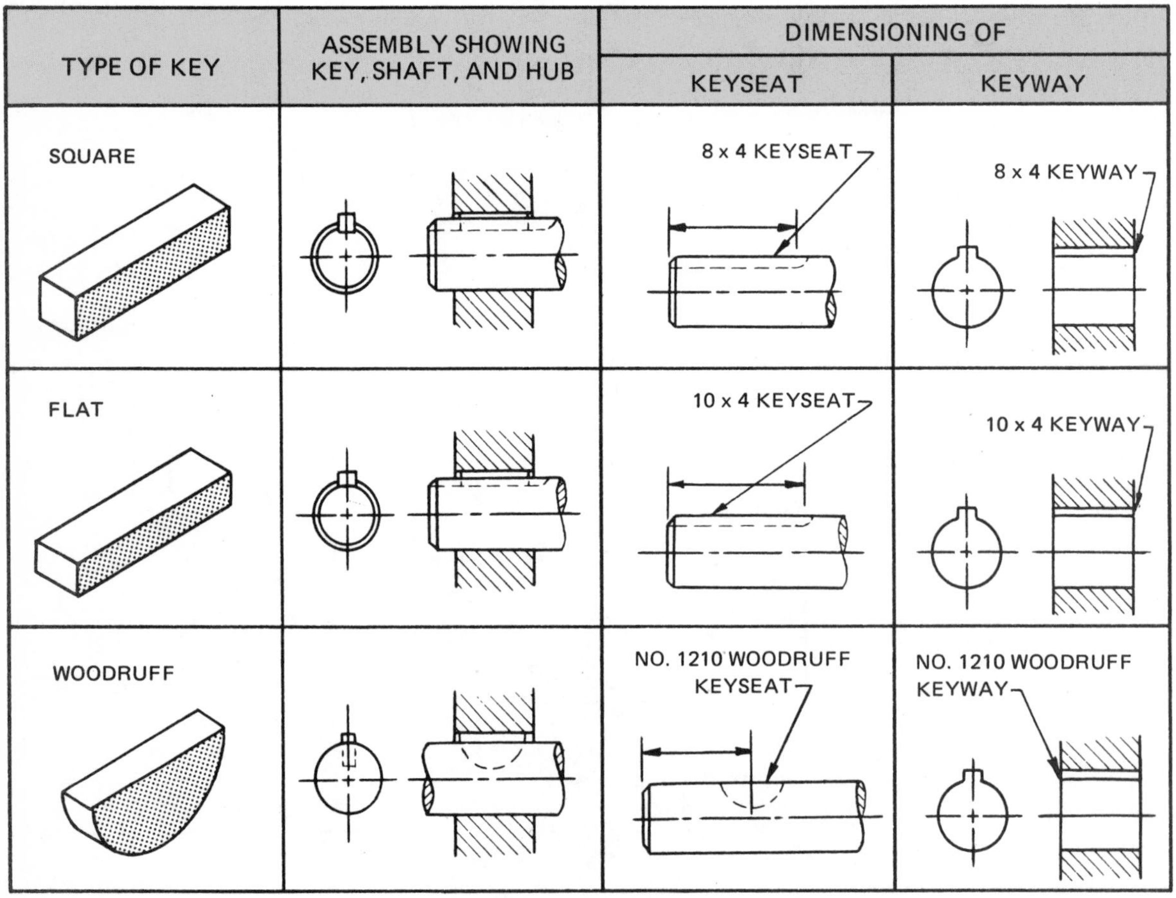

TYPE OF KEY	ASSEMBLY SHOWING KEY, SHAFT, AND HUB	DIMENSIONING OF	
		KEYSEAT	KEYWAY
SQUARE		8 x 4 KEYSEAT	8 x 4 KEYWAY
FLAT		10 x 4 KEYSEAT	10 x 4 KEYWAY
WOODRUFF		NO. 1210 WOODRUFF KEYSEAT	NO. 1210 WOODRUFF KEYWAY

Fig. 13-1 Common keys

inch. For example, a No. 1210 Woodruff key indicates a key 12/32 inch wide by 10/8 (1 1/4) inches in diameter.

Dimensioning of Keyways and Keyseats

All dimensions of keyways and keyseats for square and flat keys, except the length of the flat portion of the keyseat which is given by a direct dimension on the drawing, are shown on the drawing by a note specifying first the width and then the depth, figure 13-1. This type of dimensioning is the standard method used for unit production where the machinist is expected to fit the key into the keyway and keyseat.

For interchangeable assembly and mass production purposes, keyway and keyseat dimensions are given in limit dimensions to assure proper fits and are located from the opposite side of the hole or shaft, figure 13-2.

SETSCREWS

Setscrews are used as semipermanent fasteners to hold a collar, sheave, or gear on a shaft against rotational or translational forces. In contrast to most fastening devices, the setscrew is essentially a compression device. Forces developed by the screw point during tightening produce a strong clamping action that resists relative motion between assembled parts. The basic problem in setscrew

selection is finding the best combination of setscrew form, size, and point style providing the required holding power.

Setscrews are categorized by their forms and the desired point style, figure 13-3. Each of the standardized setscrew forms is available in six point styles. Selection of a specific form or point is influenced by functional, as well as other considerations.

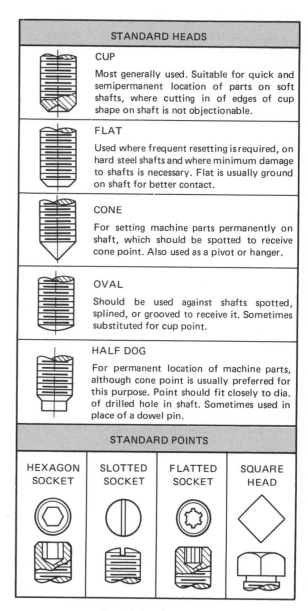

Fig. 13-3 Setscrews

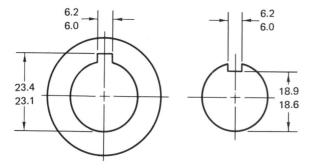

Fig. 13-2 Dimensioning keyways and keyseats for interchangeable assembly

The selection of the type of driver and thus, the setscrew form, is usually determined by factors other than tightening. Despite higher tightening ability, the protrusion of the square head is a major disadvantage. Compactness, weight saving, safety, and appearance may dictate the use of flush-seating socket or slotted headless forms.

The conventional approach to setscrew selection is usually based on a rule-of-thumb procedure: the setscrew diameter should be roughly equal to one-half the shaft diameter. This rule of thumb often gives satisfactory results, but it has a limited range of usefulness. When a setscrew and key are used together, the screw diameter should be equal to the width of the key.

FLATS

A *flat* is a slight depression usually cut on a shaft to serve as a surface on which the end of a setscrew can rest when holding an object in place, figure 13-4.

BOSSES AND PADS

A *boss* is a relatively small cylindrical projection above the surface of an object. A pad is a slight, noncircular projection above the surface of an object, figure 13-5. Bosses and pads are used to provide additional clearance or strength in the area where they are used.

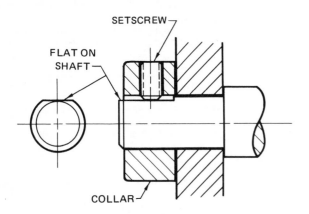

Fig. 13-4 Flat and setscrew application

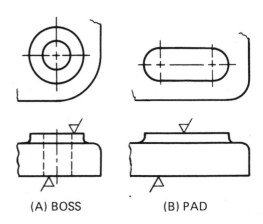

Fig. 13-5 Bosses and pads

QUESTIONS

Refer to Part 1

1. What was the original length of the shaft?

2. What symbol is used to indicate that the 50 dimension is not to scale?

3. At how many places are threads being cut?

4. Specify for any left-hand threads the thread diameter and the pitch.

5. What is the pitch for the M24 thread?

6. What is the entire length of that portion of the shaft which includes the (A) M22 thread, (B) M30 thread, (C) M24 thread?

7. What distance is there between the last thread and the shoulder on the Ø22 shaft?

8. How many dimensions have been changed?

9. What type of section view is used?

10. What is the largest size to which the Ø30 shaft can be turned?

11. Which scale was used on this drawing?

Refer to Part 2

12. How many holes are there?

13. What are the overall width and depth dimensions of the base?

14. How many surfaces are to be finished?

15. What is the diameter of the bosses on the base?

16. How wide is the pad on the upright shaft?

17. What is the depth of the keyway?

18. How far does the horizontal hole overlap the vertical hole? (Use maximum sizes of holes and minimum center-to-center distances.)

19. How much material was added when the change to the bosses was made?

20. Which scale was used on this drawing?

21. What is the maximum permissible center-to-center distance of the two large holes?

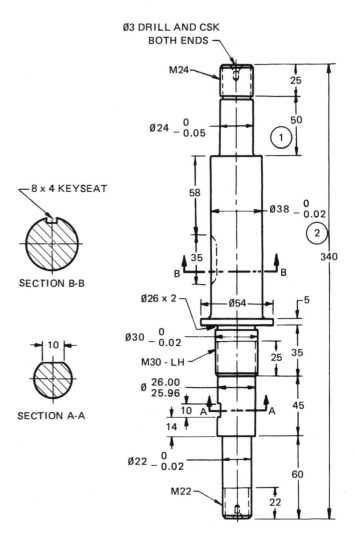

Ø3 DRILL AND CSK BOTH ENDS

8 x 4 KEYSEAT

SECTION B-B

SECTION A-A

CHAMFER STARTING END OF ALL THREADS
45° TO THREAD DEPTH
NOTE: ALL FILLETS R30

PT 1 SPINDLE SHAFT, SCALE 1:2, MATL — CRS
2 REQD

REVISIONS	1	12/1/76	R.H.	2	12/8/76	C.J.	3	12/15/76	G.H.
		50 WAS 52			340 WAS 345			20 WAS 18	

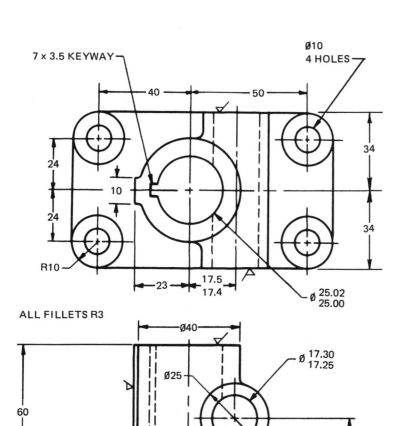

7 x 3.5 KEYWAY

Ø10
4 HOLES

40

50

34

24

10

24

34

R10

23

17.5
17.4

Ø 25.02
25.00

ALL FILLETS R3

Ø40

Ø 17.30
17.25

Ø25

60

3

20

16

33

PT 2 COLUMN BRACKET, SCALE 1:1
MATL – WROUGHT IRON 4 REQD

UNLESS OTHERWISE SPECIFIED
TOLERANCE ON DIMENSIONS ± 0.5

METRIC

SCALE	AS SHOWN	
DRAWN		DATE
RACK DETAILS		A-28

UNIT
14

PRIMARY AUXILIARY VIEWS

Many objects have surfaces that are perpendicular to only one plane of projection. These surfaces are referred to as inclined sloped surfaces. In the remaining two orthographic views such surfaces appear to be foreshortened and their true shape is not shown, figure 14-1. When an inclined surface has important characteristics that should be shown clearly and without distortion, an auxiliary view is used to completely explain the shape of the object.

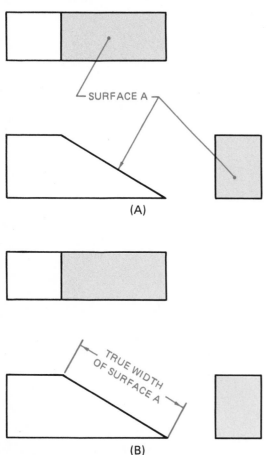

(A)

(B)

SURFACE A

TRUE WIDTH OF SURFACE A

Fig. 14-1 (A) Surface A is perpendicular only to the front plane of projection. (B) Therefore, only the front view shows the true width of surface A, but no view shows the true shape.

For example, figure 14-2 clearly shows why an auxiliary view is required. The circular

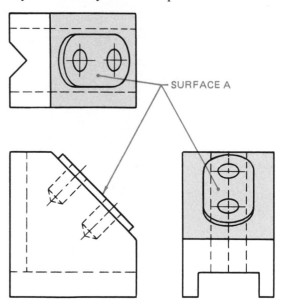

SURFACE A

(A) REGULAR VIEWS DO NOT SHOW TRUE FEATURES OF SURFACE A.

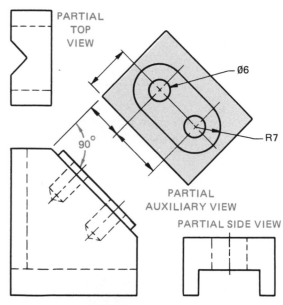

PARTIAL TOP VIEW

Ø6

R7

90°

PARTIAL AUXILIARY VIEW

PARTIAL SIDE VIEW

(B) AUXILIARY VIEW ADDED TO SHOW TRUE FEATURES OF SURFACE A.

Fig. 14-2 The need for auxiliary views

features on the sloped surface on the front view cannot be seen in their true shape on either the top or side views. The auxiliary view is the only view which shows the actual shape of these features. Note that only the sloped surface details are shown. Background detail is often omitted on auxiliary views and regular views to simplify the drawing and

avoid confusion. A break line is used to signify the break in an incomplete view. The break line is not required if only the exact surface is drawn for either an auxiliary view or a partial regular view. Dimensions for the detail on the inclined face are placed on the auxiliary view, showing the detail in its true shape.

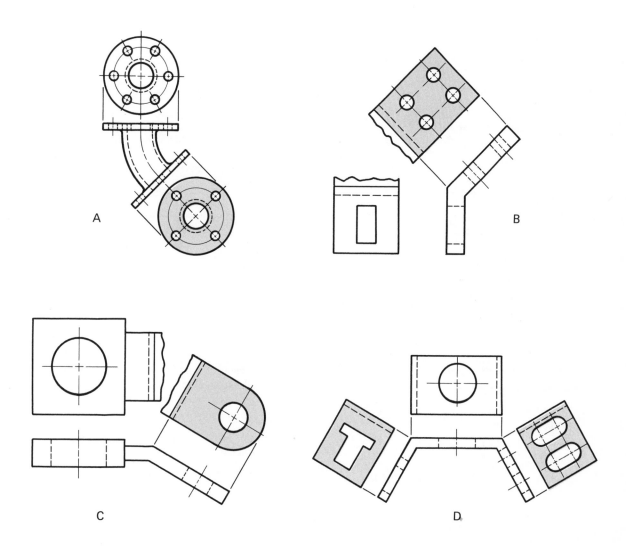

NOTE: CONVENTIONAL BREAK OF PROJECTED SURFACE
ONLY NEED BE SHOWN ON PARTIAL VIEWS

Fig. 14-3 Examples of auxiliary-view drawings

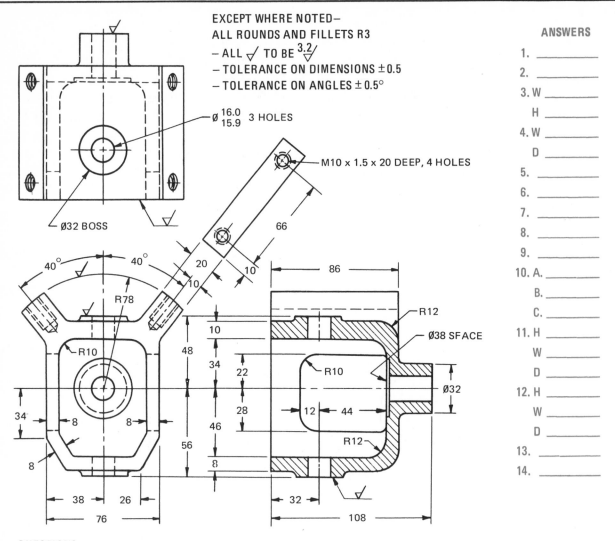

EXCEPT WHERE NOTED—
ALL ROUNDS AND FILLETS R3
– ALL ✓ TO BE 3.2✓
– TOLERANCE ON DIMENSIONS ±0.5
– TOLERANCE ON ANGLES ±0.5°

Ø 16.0 / 15.9 3 HOLES

M10 x 1.5 x 20 DEEP, 4 HOLES

Ø32 BOSS

66

40°   40°

R78

20

10   10

R10

48

10

34

22

R10

28

12   44

46

R12

56

8

34

8   8

R12

8

38   26

32

76

86

R12

Ø38 SFACE

Ø32

108

QUESTIONS

1. What is the diameter of the bosses?

2. What is the tolerance on the holes in the bosses?

3. What are the width and height of the cutouts in the sides of the box?

4. What are the width and depth of the legs of the box?

5. How many degrees (nominal) are there between the legs of the box?

6. What is the maximum surface roughness in micrometers permitted on the machined surfaces?

7. How many surfaces are machined?

8. What would be the inside diameter of the mating part that this box fits into?

9. If the mating part is 8 mm thick, and a flat and lock washer are used, what size cap screws would be used to fasten the parts together? (See table 12, Appendix.)

10. What is the thickness of (A) the side walls of the box; (B) the top of the box; (C) the bottom of the box? Disregard the bosses.

11. Give overall inside dimensions of the box.

12. What are the overall outside dimensions of the box? (Do not include legs or bosses.)

13. Of what material is the box cast?

14. How many screws are required to fasten the box to the mating part?

METRIC	
MATERIAL	WROUGHT IRON
SCALE	NOT TO SCALE
DRAWN	DATE

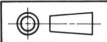

A-29

SECONDARY AUXILIARY VIEWS

As mentioned earlier, auxiliary views show the true lengths of lines and the true shapes of surfaces which cannot be described in the ordinary views.

A primary auxiliary view is drawn by projecting lines from a regular view where the inclined surface appears as an edge.

The auxiliary view in figure 15-1(A), which shows the true projection and true shape of surface X, is called a *primary auxiliary view* because it is projected directly from the regular front view.

Some surfaces are inclined so that they are not perpendicular to any of the three viewing planes. In this case, they appear as a surface in all three views, but never in their true shape. These are referred to as *oblique surfaces,* figure 15-2 (page 110). Because the oblique surface is not perpendicular to the viewing planes, it cannot be parallel to them. Consequently, the surface appears foreshortened in all three views. If a true view is required for this surface, two auxiliary views, a primary and a *secondary view,* must be drawn.

The chamfered corners **Z** have threaded holes on the oblique surfaces. To show the true shape and location of the threaded holes **M** a second auxiliary view must be shown, as

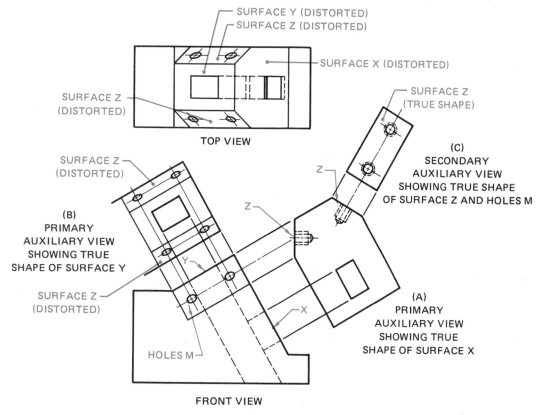

TOP VIEW

SURFACE Y (DISTORTED)
SURFACE Z (DISTORTED)
SURFACE X (DISTORTED)
SURFACE Z (DISTORTED)

SURFACE Z (TRUE SHAPE)

(C) SECONDARY AUXILIARY VIEW SHOWING TRUE SHAPE OF SURFACE Z AND HOLES M

(B) PRIMARY AUXILIARY VIEW SHOWING TRUE SHAPE OF SURFACE Y

SURFACE Z (DISTORTED)

SURFACE Z (DISTORTED)

HOLES M

(A) PRIMARY AUXILIARY VIEW SHOWING TRUE SHAPE OF SURFACE X

FRONT VIEW

NOTE: MANY HIDDEN LINES ARE OMITTED FOR CLARITY

Fig. 15-1 Primary and secondary auxiliary views

at **(C)**. This auxiliary view is projected from the first or primary auxiliary view, and is known as a *secondary auxiliary view*. The view at **(B)** is a primary auxiliary view because it is projected from one of the regular views. Notice that the side view is not drawn since the auxiliary views provide the information usually shown on this view.

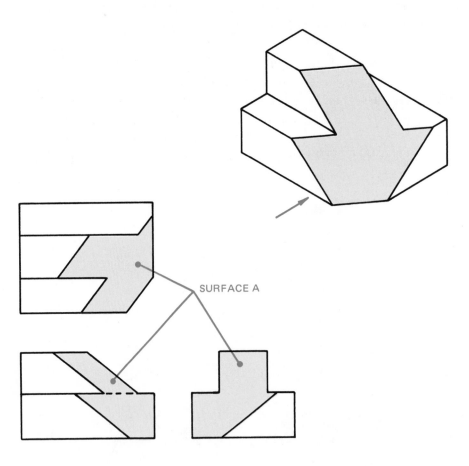

SURFACE A

Fig. 15-2 Surface A is not perpendicular to any of the three viewing planes and is therefore considered oblique.

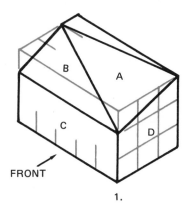

1.

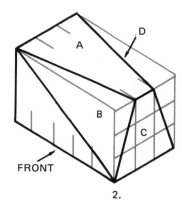

2.

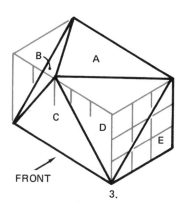

3.

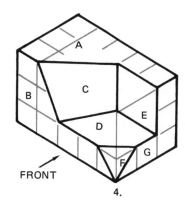

4.

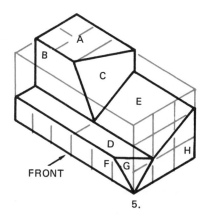

5.

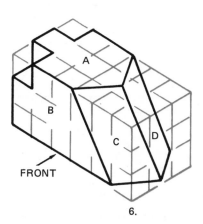

6.

USING THE SAME NUMBER OF SQUARES, DRAW THE FRONT,
TOP, AND SIDE VIEWS. IDENTIFY THE OBLIQUE SURFACES ON
THE THREE VIEWS WITH THE APPROPRIATE LETTERS.

A-30

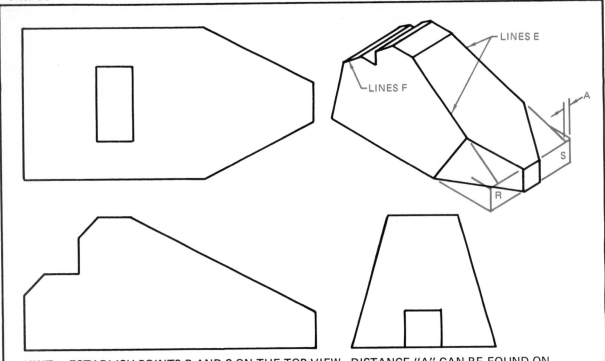

HINT — ESTABLISH POINTS R AND S ON THE TOP VIEW. DISTANCE "A" CAN BE FOUND ON THE SIDE VIEW. DRAW LINES "E" ON THE TOP VIEW. ESTABLISH LINES "F" THE SAME WAY. IDENTIFY THE OBLIQUE SURFACES.

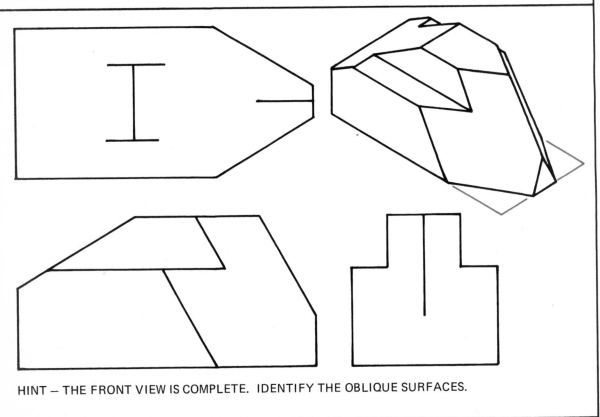

HINT — THE FRONT VIEW IS COMPLETE. IDENTIFY THE OBLIQUE SURFACES.

COMPLETE THE MISSING VIEWS.

A-31

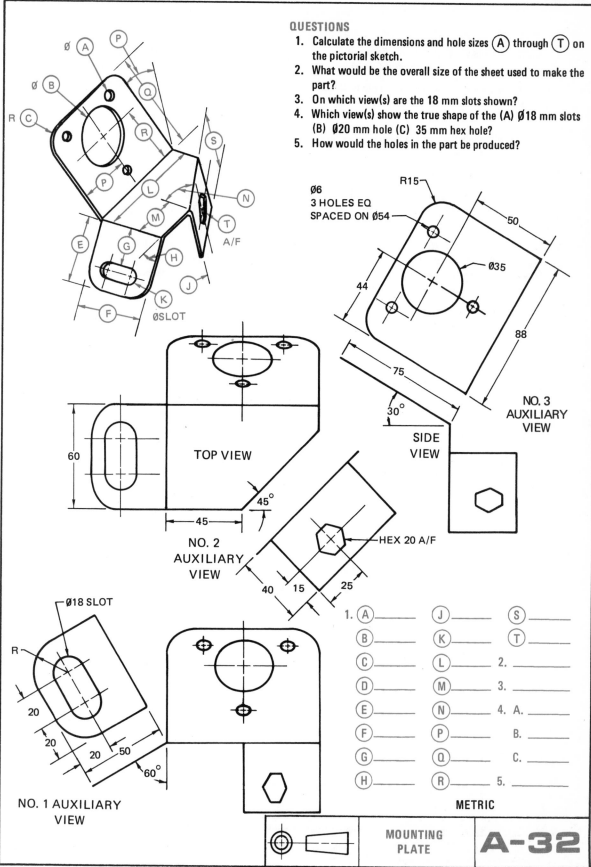

QUESTIONS

1. Calculate the dimensions and hole sizes (A) through (T) on the pictorial sketch.
2. What would be the overall size of the sheet used to make the part?
3. On which view(s) are the 18 mm slots shown?
4. Which view(s) show the true shape of the (A) Ø18 mm slots (B) Ø20 mm hole (C) 35 mm hex hole?
5. How would the holes in the part be produced?

Ø6
3 HOLES EQ
SPACED ON Ø54

R15

50

Ø35

44

88

75

30°

NO. 3
AUXILIARY
VIEW

SIDE
VIEW

TOP VIEW

60

45

45°

NO. 2
AUXILIARY
VIEW

HEX 20 A/F

40   15   25

Ø18 SLOT

R

20

20

20   50

60°

NO. 1 AUXILIARY
VIEW

1. (A)_____   (J)_____   (S)_____
   (B)_____   (K)_____   (T)_____
   (C)_____   (L)_____   2. _____
   (D)_____   (M)_____   3. _____
   (E)_____   (N)_____   4. A. _____
   (F)_____   (P)_____   B. _____
   (G)_____   (Q)_____   C. _____
   (H)_____   (R)_____   5. _____

METRIC

MOUNTING
PLATE

A-32

113

1. Which other view(s) show the true height of the 12 mm dimension shown in the front view?

2. Which other view(s) show the true width of the 15 mm dimension shown in the primary auxiliary view?

3. What view(s) show the true (A) height, (B) depth, (C) width of the part?

4. How many surfaces require finishing?

5. List the surface(s) or feature(s) which are shown in their true shape or size in the primary auxiliary view but are distorted in all other views.

6. List the surface(s) or feature(s) which are shown in their true shape or size in the secondary auxiliary view but are distorted in all other views.

7. What would be the thickness of the base ㉖ before machining?

8. If the hexagon bar support was fastened to another member, —

   (Refer to the Appendix when necessary.)

   (A) How many cap screws would be used?

   (B) What would be the cap screw size?

   (C) If the supporting member had tapped holes which were 20 mm deep and flat washers were used under the cap screw heads; what would be the cap screw length?

   (D) If the cap screws were of the fine thread series, how would you call out these cap screws?

   (E) What would be the I.D. and O.D. of the flat washers used under these cap screws?

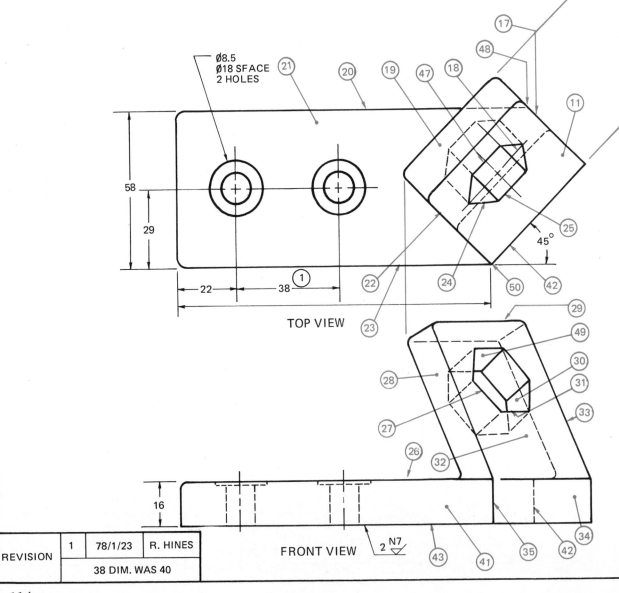

Ø8.5
Ø18 SFACE
2 HOLES

TOP VIEW

FRONT VIEW

REVISION | 1 | 78/1/23 | R. HINES
38 DIM. WAS 40

114

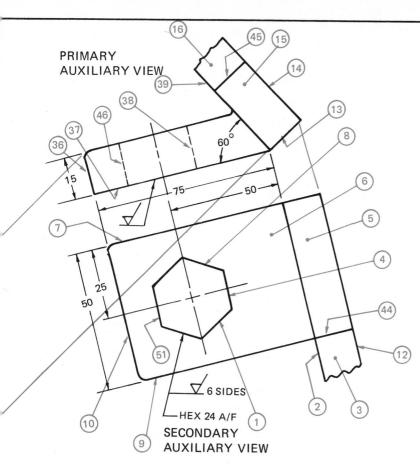

PRIMARY
AUXILIARY VIEW

60°

15

75   50

25
50

6 SIDES

HEX 24 A/F

SECONDARY
AUXILIARY VIEW

NOTE: UNLESS OTHERWISE SPECIFIED
– TOLERANCE ON DIMENSIONS ± 0.5
– TOLERANCE ON ANGLES ± 0.5°
ALL SURFACES SHOWN ▽ HAVE AN
N7 FINISH.

ANSWERS

1. _____

2. _____

3. A. _____

   B. _____

   C. _____

4. _____

5. _____

   _____

6. _____

7. _____

8. A. _____

   B. _____

   C. _____

   D. _____

   E. _____

ASSIGNMENT

In the table below, a certain feature is shown in several views. Fill in the number representing the same feature in the other views.

Front View	Top View	Primary Auxiliary View	Secondary Auxiliary View
		—	1
	23	—	
26			
		38	
		—	9
—	20		—
33			
	—	14	
			5
	19		
		37	
		—	8
35		—	
42			—
	47		

METRIC

MATERIAL	CI
SCALE	1:2
DRAWN	DATE
HEXAGON BAR SUPPORT	A-33

UNIT 16

FIRST-ANGLE PROJECTION

In Europe and many other countries, first-angle, instead of third-angle orthographic projection, is used on engineering drawings. Except for their location on the drawing, the views are identical. A comparison of these projections follows:

Position of Views

Third-Angle or American Projection. In reference to the front view (as seen from direction F) the other views in figure 16-1 are arranged as follows:

- The view from above is placed above.
- The view from below is placed underneath.
- The view from the left is placed on the left.
- The view from the right is placed on the right.
- The view from the rear may be placed on the left or right according to convenience.

The third-angle ISO projection symbol is shown on the drawing.

First-Angle or European Projection. In reference to the front view (as seen from direction F) the other views in figure 16-2 are arranged as follows:

- The view from above is placed underneath.
- The view from below is placed above.
- The view from the left is placed on the right.
- The view from the right is placed on the left.
- The view from the rear may be placed on the left or right according to convenience.

The first-angle ISO projection symbol is shown on the drawing.

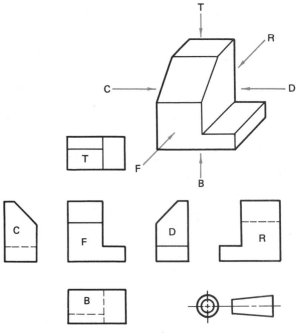

Fig. 16-1 Third-angle projection

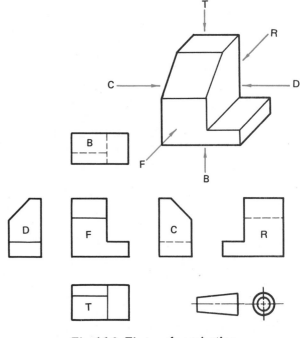

Fig. 16-2 First-angle projection

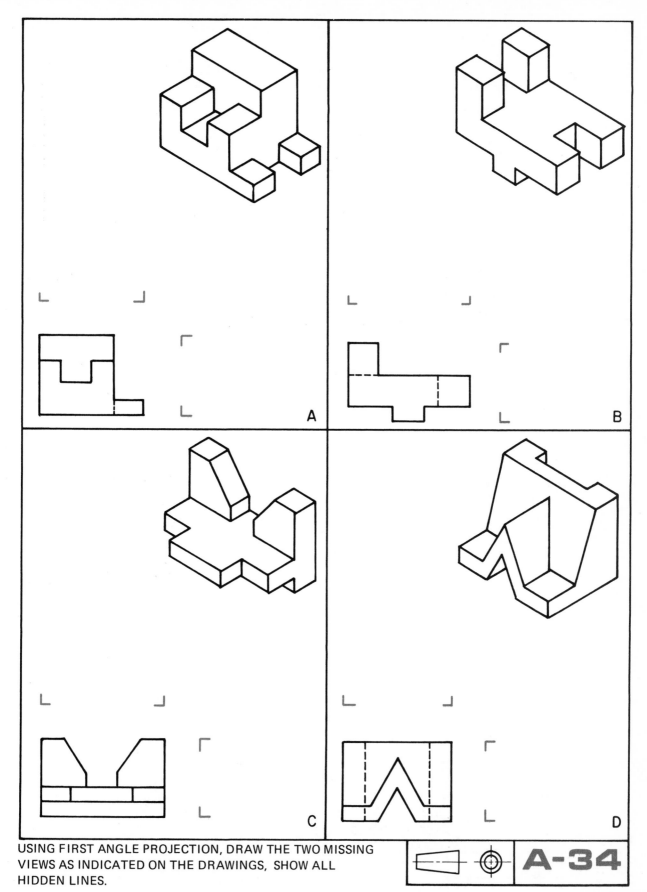

USING FIRST ANGLE PROJECTION, DRAW THE TWO MISSING
VIEWS AS INDICATED ON THE DRAWINGS, SHOW ALL
HIDDEN LINES.

A-34

QUESTIONS

1. Calculate dimensions (A) to (H) .

2. How many surfaces have a $3\overset{\text{N7}}{\nabla}$ finish?

3. What type of projection is used on this drawing?

4. What are the limit dimensions of the Ø140?

5. What is the tolerance on the Ø28 hole?

6. Which size of tap drill would be used for the M24 hole? See Appendix.

7. What is the roughness height of the N7 finish in micrometres.

8. How long is the Ø28 hole?

9. Which numbers in (A) the front view, (B) view A, indicate the size of radius Y?

10. What would be the overall (A) width, (B) height, and (C) depth of the bracket before machining?

11. Locate surface (T) in (A) the front view, (B) view A.

12. Locate surface (M) in view A.

13. Locate surface (R) in the side view.

14. Locate the Ø28 hole in (A) the side view, (B) view A.

15. Locate surface 11 in the front view.

16. Locate surface 8 in the side view.

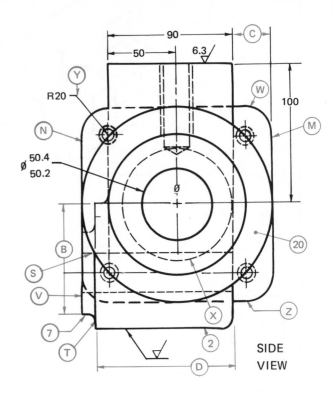

SIDE VIEW

SKETCH SECTION **A-A** IN THE SPACE PROVIDED.

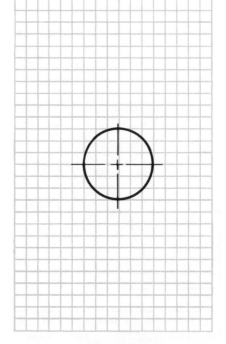

SECTION A-A
10 mm SQUARES

REVISIONS	1			

118

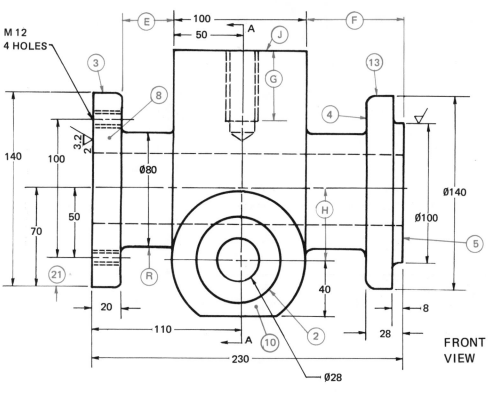

M 12
4 HOLES

③

⑧

3.2
2

Ø80

Ⓔ 100
Ⓐ 50

E 100 50 A

J

G

⑬

④

⑤

Ø140
Ø100

H

140
100

70
50

㉑

R

20

110

A

⑩

230

Ø28

②

40

8
28

**FRONT
VIEW**

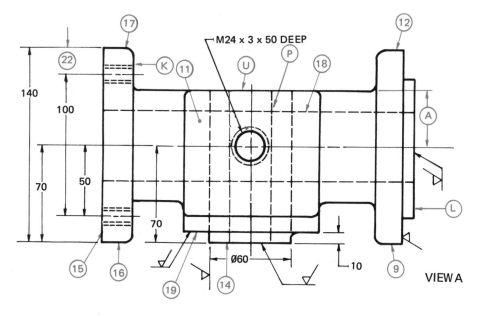

⑰
㉒

K ⑪

M24 x 3 x 50 DEEP

U

P

⑱

⑫

140
100

A

70
50

70

Ø60

10

⑨

L

⑮ ⑯

⑲ ⑭

VIEW A

NOTE: EXCEPT WHERE NOTED
– TOLERANCES ON DIMENSIONS ± 0.5
– ▽ TO BE 3 N7 ▽
– ROUNDS AND FILLETS R3.

METRIC

ANSWERS

1. Ⓐ _____
   Ⓑ _____
   Ⓒ _____
   Ⓓ _____
   Ⓔ _____
   Ⓕ _____
   Ⓖ _____
   Ⓗ _____

2. _____
3. _____
4. _____
5. _____
6. _____
7. _____
8. _____
9. A. _____
   B. _____
10. A. _____
    B. _____
    C. _____
11. A. _____
    B. _____
12. _____
13. _____
14. A. _____
    B. _____
15. _____
16. _____

MATERIAL	WROUGHT IRON
SCALE	
DRAWN	DATE

FLANGED OIL
FEED BRACKET

A-35

ARROWLESS DIMENSIONING

To avoid having many dimensions extending away from the object, arrowless, or ordinate, dimensioning may be used, as shown in figure 17-1. In this system, the "zero" lines are used as reference lines and each of the dimensions shown without arrowheads indicates the distance from the zero line. There is never more than one zero line in each direction.

Arrowless dimensioning is particularly useful when such features are produced on a general-purpose machine, such as a jig borer, a tape-controlled drill, or a turret-type press.

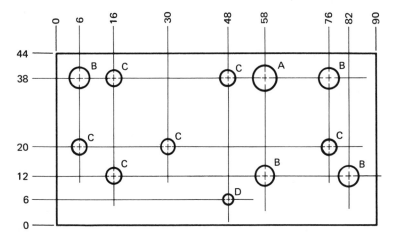

HOLE SYMBOL	HOLE Ø
A	6
B	5
C	4
D	3

Fig. 17-1 Arrowless (ordinate) dimensioning

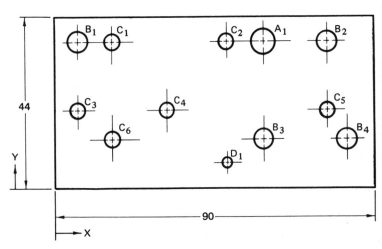

Hole Symbol	Ø Hole	Location	
		X	Y
A_1	6	58	38
B_1	5	6	38
B_2	5	76	38
B_3	5	58	12
B_4	5	82	12
C_1	4	16	38
C_2	4	48	38
C_3	4	6	20
C_4	4	30	20
C_5	4	76	20
C_6	4	16	12
D_1	3	48	6

Fig. 17-2 Tabular dimensioning

TABULAR DIMENSIONING

When there are many holes or repetitive features, such as in a chassis or a printed circuit board, and where the multitude of center lines would make a drawing difficult to read, tabular dimensioning is recommended. In this system each hole or feature is assigned a letter or a letter and numeral subscript. The feature dimensions and the feature location along the **X** and **Y** axes are given in a table as shown in figure 17-2. The numbering and lettering of the features are usually from left to right and from bottom to top.

REFERENCES AND SOURCE MATERIALS

1. *Dimensioning and Tolerancing,* ANSI Y14.5 (1973).

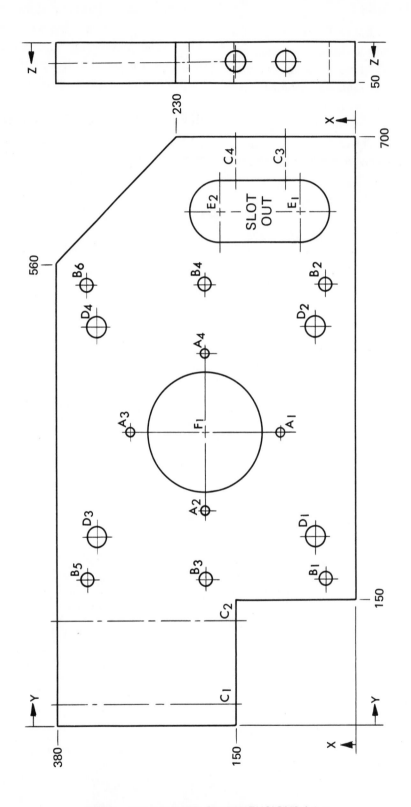

NOTE: TOLERANCE ON DIMENSIONS ± 0.5

122

1. What is the width of the part?

2. What is the height of the part?

3. What are the width and height of the chamfer on the corner?

4. What is the width of slot E?

5. What is the height of slot E?

6. How much wood is left between slot E and the right-hand edge of the part?

7. What are the center distances between holes

    (A) B_5 and B_6 (D) D_3 and D_4 (F) C_1 and C_2

    (B) D_2 and D_4 (E) A_1 and A_3 (G) B_3 and A_4 ?

    (C) B_1 and B_5

8. How much wood is left between holes

    (A) A_1 and E_1 (C) C_1 and C_2

    (B) A_2 and B_3 (D) B_3 and B_4 ?

1. _____

2. _____

3. W _____

   H _____

4. _____

5. _____

6. _____

7. A. _____

   B. _____

   C. _____

   D. _____

   E. _____

   F. _____

   G. _____

8. A. _____

   B. _____

   C. _____

   D. _____

HOLE SYMBOL	HOLE DIA.	LOCATION X ↑	LOCATION Y →	HOLE SYMBOL	HOLE DIA.	LOCATION X ↑	LOCATION Y →	LOCATION Z ←
A_1	10	94	356	C_1	20		24	25
A_2	10	190	260	C_2	20		126	25
A_3	10	286	356	C_3	20	90		25
A_4	10	190	452	C_4	20	150		25
B_1	16	38	176	D_1	24	50	230	
B_2	16	38	536	D_2	24	50	482	
B_3	16	190	176	D_3	24	330	230	
B_4	16	190	536	D_4	24	330	482	
B_5	16	342	176	E_1	76	70	622	
B_6	16	342	536	E_2	76	170	622	
				F_1	144	190	356	

MATERIAL	DRY MAPLE
SCALE	NOT TO SCALE
DRAWN	DATE

METRIC

SUPPORT BRACKET

A-36

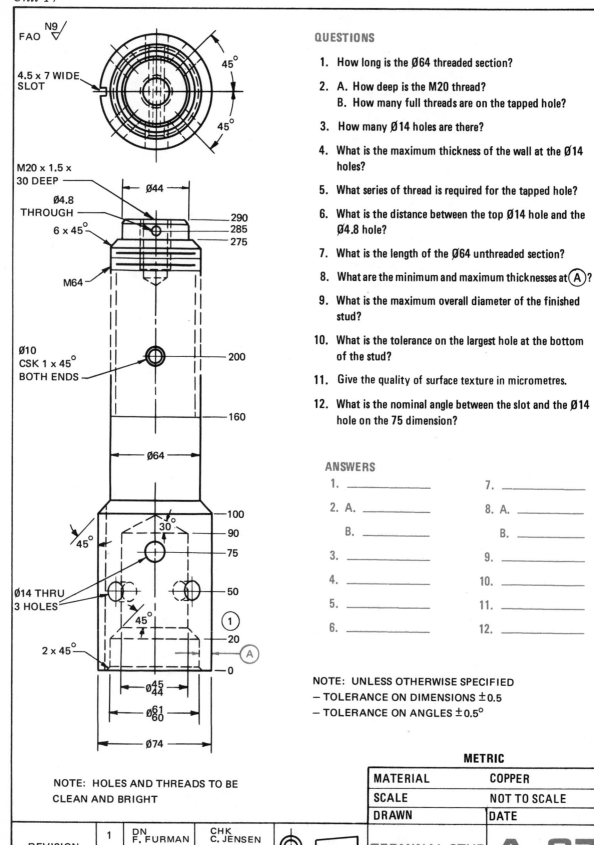

FAO $\overset{N9}{\nabla}$

4.5 x 7 WIDE SLOT

M20 x 1.5 x 30 DEEP

Ø4.8 THROUGH

6 x 45°

M64

Ø44

290
285
275

Ø10
CSK 1 x 45°
BOTH ENDS

200

160

Ø64

100
90
75
50
20
0

30°

45°

Ø14 THRU
3 HOLES

45°

2 x 45°

$\overset{45}{\underset{44}{Ø}}$

$\overset{61}{\underset{60}{Ø}}$

Ø74

1

A

QUESTIONS

1. How long is the Ø64 threaded section?

2. A. How deep is the M20 thread?
   B. How many full threads are on the tapped hole?

3. How many Ø14 holes are there?

4. What is the maximum thickness of the wall at the Ø14 holes?

5. What series of thread is required for the tapped hole?

6. What is the distance between the top Ø14 hole and the Ø4.8 hole?

7. What is the length of the Ø64 unthreaded section?

8. What are the minimum and maximum thicknesses at Ⓐ?

9. What is the maximum overall diameter of the finished stud?

10. What is the tolerance on the largest hole at the bottom of the stud?

11. Give the quality of surface texture in micrometres.

12. What is the nominal angle between the slot and the Ø14 hole on the 75 dimension?

ANSWERS

1. _____   7. _____

2. A. _____   8. A. _____

   B. _____   B. _____

3. _____   9. _____

4. _____   10. _____

5. _____   11. _____

6. _____   12. _____

NOTE: UNLESS OTHERWISE SPECIFIED
– TOLERANCE ON DIMENSIONS ±0.5
– TOLERANCE ON ANGLES ±0.5°

NOTE: HOLES AND THREADS TO BE
CLEAN AND BRIGHT

METRIC	
MATERIAL	COPPER
SCALE	NOT TO SCALE
DRAWN	DATE

REVISION	1	DN F. FURMAN	CHK C. JENSEN		TERMINAL STUD	A-37
	DIMENSION WAS 24		8/6/78			

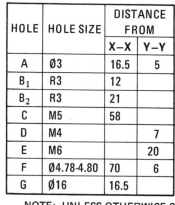

HOLE	HOLE SIZE	DISTANCE FROM	
		X–X	Y–Y
A	Ø3	16.5	5
B₁	R3	12	
B₂	R3	21	
C	M5	58	
D	M4		7
E	M6		20
F	Ø4.78-4.80	70	6
G	Ø16	16.5	

NOTE: UNLESS OTHERWISE SPECIFIED
TOLERANCE ON DIMENSIONS ±0.5

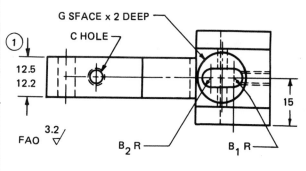

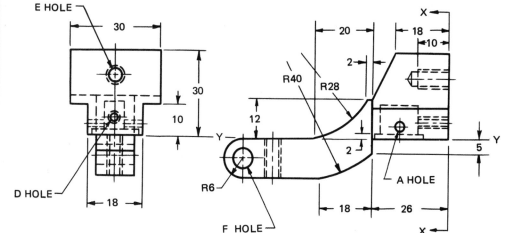

ANSWERS

1. _____
2. _____
3. _____
4. _____
5. _____
6. _____
7. _____
8. _____
9. _____
10. _____
11. _____
12. _____
13. _____
14. A. _____
    B. _____
15. _____

QUESTIONS

1. What is the overall width?

2. What is the overall height?

3. What is the distance from **X-X** to the center of hole **F**?

4. What is the distance from **Y-Y** to the center of hole **E**?

5. What is the horizontal distance between the center lines of **A** and **F** holes?

6. What is the horizontal distance between the center lines of **C** hole and **B₁** radius?

7. How many full threads does **E** hole have? (See Appendix.)

8. How deep is the spotface?

9. What is the tolerance of **F** hole?

10. What type of projection is used on this drawing?

11. How far apart are the **D** and **E** holes?

12. What is the tolerance on the thickness of the material at **F** hole?

13. What is the length of the **C** hole?

14. What is the distance from line **X-X** to the termination of both ends of the 28 radius?

15. What does FAO mean?

METRIC

MATERIAL	ALUMINUM
SCALE	NOT TO SCALE
DRAWN	DATE

REVISION	1	DN 6/8/78 F. NEWMAN	CH A. HEINEN		CONTACTOR	A-38
		DIMENSION WAS 12.4/12.7				

125

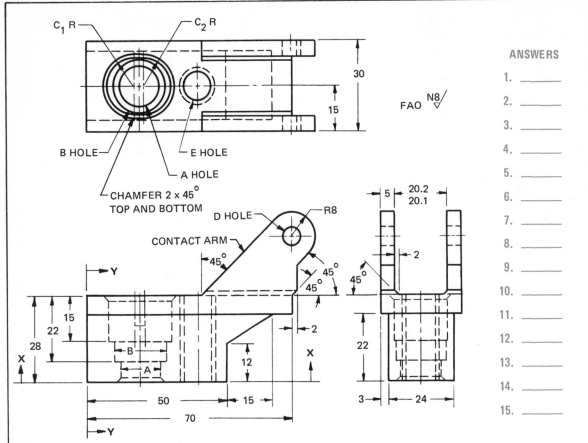

FAO $\overset{N8}{\triangledown}$

CHAMFER 2 x 45°
TOP AND BOTTOM

QUESTIONS

1. What is the overall width?

2. What is the overall height?

3. Give the chamfer angle.

4. What is the distance from **Y-Y** to **E** hole?

5. What is the distance from **X-X** to **D** hole?

6. How many complete threads does the tapped hole have?

7. Which thread series is the tapped hole?

8. What is the surface finish in micrometres?

9. How deep is the **C** counterbore from the top of the surface?

10. How wide is the **C** counterbore?

11. Give the distance between the $\mathcal{C}_L$ of C_1 and C_2 radii.

12. What is the nominal thickness of the contact arm at **D** hole?

13. What is the tolerance on the distance between the contact arms?

14. What is the tolerance on **D** hole?

15. What is the center distance between **A** and **E** holes?

NOTE: UNLESS OTHERWISE SPECIFIED
– TOLERANCE ON DIMENSIONS ±0.5
– TOLERANCE ON ANGLES ±0.5°

HOLE SYMBOL	HOLE SIZE	LOCATION	
		X–X	Y–Y
A	Ø13.5		18
B	Ø17		18
C_1	R9		16
C_2	R9		20
D	Ø6.5 - 6.6	50	70
E	M12 x 1.25		38

MATERIAL	MANGANESE BRONZE
SCALE	NOT TO SCALE
DRAWN	DATE

METRIC

CONTACT ARM

A-39

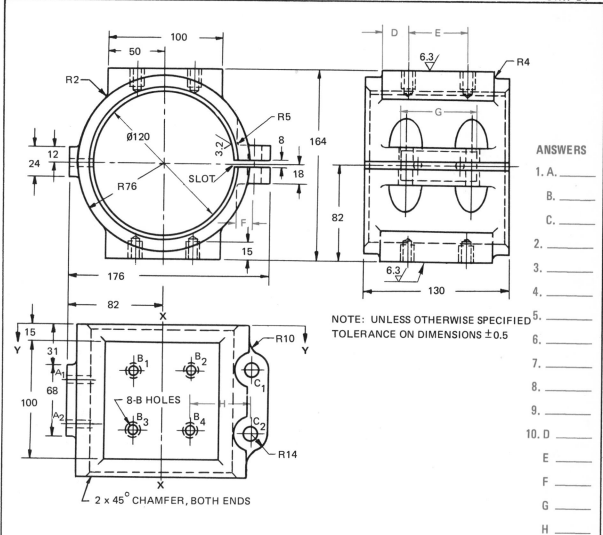

NOTE: UNLESS OTHERWISE SPECIFIED
TOLERANCE ON DIMENSIONS ±0.5

1. A. _____
   B. _____
   C. _____
2. _____
3. _____
4. _____
5. _____
6. _____
7. _____
8. _____
9. _____
10. D _____
    E _____
    F _____
    G _____
    H _____

QUESTIONS

1. What is the center-to-center distance between the following:
   A. the A_1 and A_2 holes
   B. the B_1 and B_2 holes
   C. the C_1 and C_2 holes
2. What finish is required on the 100 x 100 face?
3. How far apart are the two surfaces of the slot?
4. How many surfaces require finishing?
5. What is the depth of the A holes?
6. What is the length of the threading in the B holes?
7. What size bolts would be used in the C holes if 1 mm clearance is used? ?
8. How many chamfers are called for?
9. How many tapped holes are there?
10. Calculate distances D, E, F, G, and H.

HOLE	HOLE SIZE	LOCATION	
		X–X	Y–Y
A_1	M5		46
A_2	M5	25	84
B_1	M12	25	40
B_2	M12	25	40
B_3	M12	25	90
B_4	M12	25	90
C_1	Ø13	80	38
C_2	Ø13	80	92

MATERIAL	COPPER
SCALE	NOT TO SCALE
DRAWN	DATE

METRIC

TERMINAL BLOCK

A-40

UNIT 18

PIPING

Until one hundred years ago, water was the only important fluid conveyed from place to place through pipe. Today nearly every conceivable fluid is handled in pipe during its production, processing, transportation, or utilization. During the age of atomic energy and rocket power liquid metals, sodium, and nitrogen have been added to the list of more common fluids such as oil, water, and acids being transported through pipe. Many gases are also being stored and delivered through piping systems.

Pipe is also used for hydraulic and pneumatic mechanisms and used extensively for the controls of machinery and other equipment. Piping is also used as a structural element for columns and handrails.

The nominal size of pipe is given in inches, but the inside diameter, outside diameter, and wall thickness are given in millimetres.

Kinds of Pipe

Steel and Wrought Iron Pipe. This pipe carries water, steam, oil, and gas and is commonly used under high temperatures and pressures. Standard steel or cast iron pipe is specified by the nominal diameter, which is always less than the actual inner diameter (ID) of the pipe. Until recently, this pipe was available in only three masses — standard, extra strong, and double extra strong, figure 18-1. In order to use common fittings with these different pipe weights, the outside diameter (OD) of each of the different pipes remained the same. The extra metal was added to the ID to increase the wall thickness of the extra strong and double extra strong pipe.

The demand for a greater variety of pipe for use under increased pressure and temperature led to the introduction of ten different pipe masses, each designated by a schedule number. Standard pipe is now called schedule 40 pipe. Extra strong pipe is schedule 80.

Cast Iron Pipe. This is often installed underground to carry water, gas, and sewage.

Seamless Brass and Copper Pipe. These pipes are used extensively in plumbing because of their ability to withstand corrosion.

Copper Tubing. This is used in plumbing and heating and where vibration and misalignment are factors, such as in automotive, hydraulic, and pneumatic design.

Plastic Pipe. This pipe or tubing, because of its resistance to corrosion and chemicals, is often used in the chemical industry. It is easily installed. However, it is not recommended where heat or pressure is a factor.

Pipe Joints and Fittings

Parts joined to pipe are called *fittings*. They may be used to change size or direction and to join or provide branch connections. There are three general classes of fittings: *screwed,* *welded,* and *flanged.* Other methods such as soldering, brazing, and gluing are used for cast iron pipe, copper, and plastic tubing.

(A) STANDARD SCHEDULE 40

(B) EXTRA STRONG SCHEDULE 80

(C) DOUBLE EXTRA STRONG

Fig. 18-1 Comparison of wall thicknesses

Pipe fittings are specified by the nominal pipe size, the name of the fitting, and the material. Some fittings, for example, tees, crosses, and elbows, connect different sizes of pipe. These are called reducing fittings. Their nominal pipe sizes must be specified. The largest opening of the through run is given first, followed by the opposite end and the outlet. Figure 18-2 illustrates the method of designating sizes of reducing fittings.

Screwed Fittings. Screwed fittings are generally used on small pipe design of 65 mm or less.

There are two types of American Standard Pipe Thread: tapered and straight. The tapered thread is more common. Straight threads are used for special applications which are listed in the ANSI Handbook.

Tapered threads are designated on drawings as NPT (National Pipe Thread) or whichever standard is used and may be drawn either with or without the taper, figure 18-3. When drawn in tapered form, the taper is exaggerated. Straight pipe threads are designated on drawings as NPTS and standard thread symbols are used. Pipe threads are assumed to be tapered unless specified otherwise.

Welded Fittings. Welded fittings are used where connections will be permanent and on high pressure and temperature lines. The ends of the pipe and pipe fittings are usually bevelled to accommodate the weld.

Flanged Fittings. Flanged joint fittings provide a quick way to disassemble pipe. Flanges are attached to the pipe ends by welding, screwing, or lapping.

Valves

Valves are used in piping systems to stop or regulate the flow of fluids and gases. A few of the more common types are described below.

Gate Valves. These are used to control the flow of liquids. The wedge, or gate, lifts to

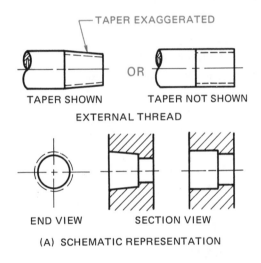

(A) SCHEMATIC REPRESENTATION

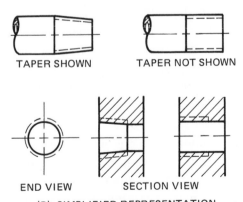

(B) SIMPLIFIED REPRESENTATION

Fig. 18-3 Pipe thread conventions

NOTE: NOMINAL PIPE SIZES IN INCHES

Fig. 18-2 Order of specifying the openings of reducing fittings

allow full, unobstructed flow and lowers to stop the flow. They are generally used where operation is infrequent and are not intended for throttling or close control.

Globe Valves. These are used to control the flow of liquids or gases. The design of the glove valve requires two changes in the direction of flow, slightly reducing the pressure in the system. The globe valve is recommended for the control of air, steam, gas, and other compressibles where instantaneous on-and-off operation is essential.

Check Valves. Check valves permit flow in one direction, but check all reverse flow. They are operated by pressure and velocity of line flow alone and have no external means of operation.

PIPING DRAWINGS

Piping drawings show the size and location of pipes, fittings, and valves. Because of the detail required to accurately describe these items, there is a set of symbols to indicate them on drawings.

There are two types of piping drawings in use, *single-line* and *double-line* drawings, figure 18-4. Double-line drawings take longer to draw and are therefore not recommended for production drawings. They are, however, suitable for catalogs and other applications where the appearance is more important than the extra drafting time.

Single-Line Drawings

Single-line drawings, also known as simplified representation, of pipe lines provide substantial savings without loss of clarity or reduction of comprehensiveness of information. Therefore, the simplified method is used whenever possible.

Single-line piping drawing uses a single line to show the arrangement of the pipe and

fittings. The center line of the pipe, regardless of pipe size, is drawn as a thick line to which symbols are added. The size of the symbol is left to the discretion of the drafter. When pipe lines carry different liquids, such as cold or hot water, a coded line symbol is often used.

Drawing Projection

Two methods of projection are used, orthographic and isometric, figure 18-5 (page 133). Orthographic projection is recommended for the representation of single pipes which are either straight or bent in one plane only. However, this method is also used for more complicated pipings.

Isometric projection is recommended for all pipes bent in more than one plane and for assembly and layout work because the finished drawing is easier to understand.

Crossings. The crossing of pipes without connections is usually drawn without interrupting the line representing the hidden line, figure 18-6(A), page 134. But, when it is desirable to indicate that one pipe must pass behind the other, the line representing the pipe furthest away from the viewer will be shown with a break or interruption where the other pipe passes in front of it.

Connections. Permanent connections or junctions, whether made by welding or other processes, are indicated on the drawing by a heavy dot, figure 18-6(F). A general note or specification may indicate the process used.

Detachable connections or junctions are indicated by a single thick line. Specifications, a general note, or the bill of material will indicate the type of fitting, for example, flanges, union, or coupling. The specifications will also indicate whether the fittings are flanged, threaded, or welded.

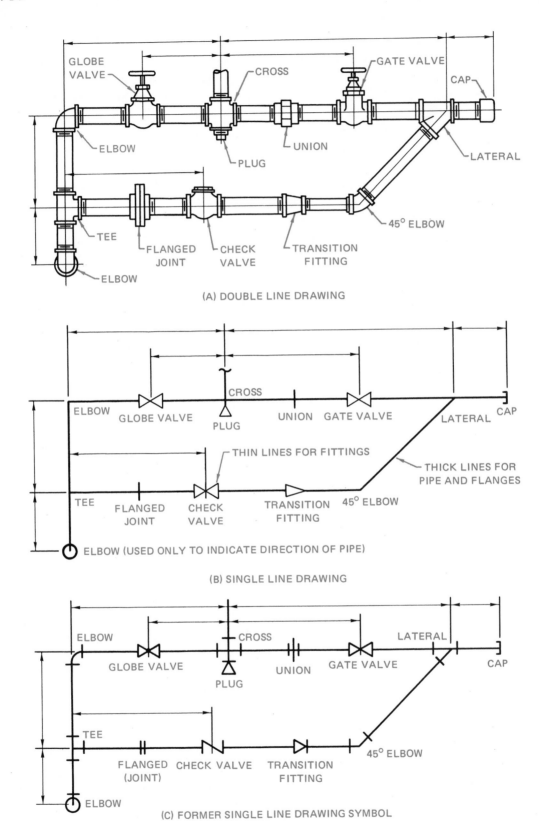

Fig. 18-4 Piping drawing symbols

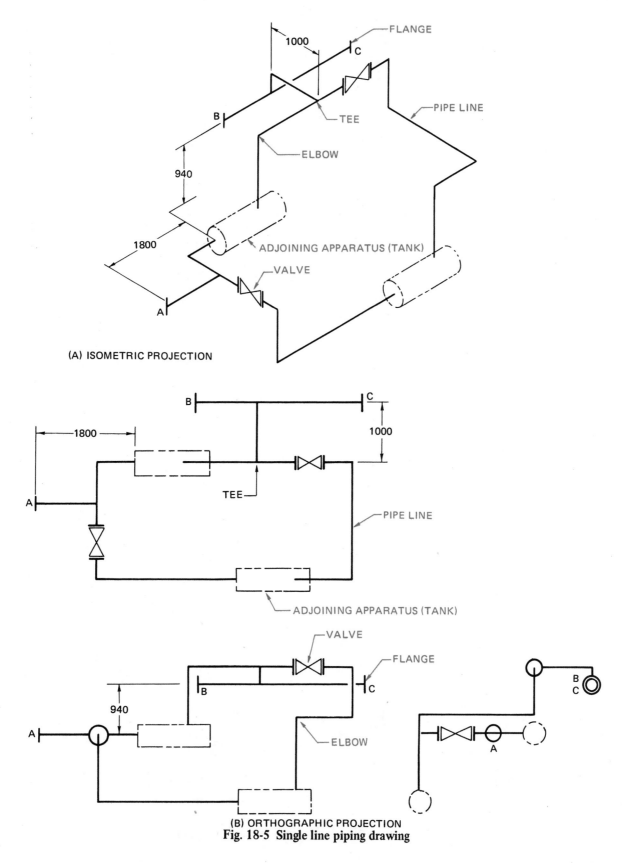

(A) ISOMETRIC PROJECTION

(B) ORTHOGRAPHIC PROJECTION
Fig. 18-5 Single line piping drawing

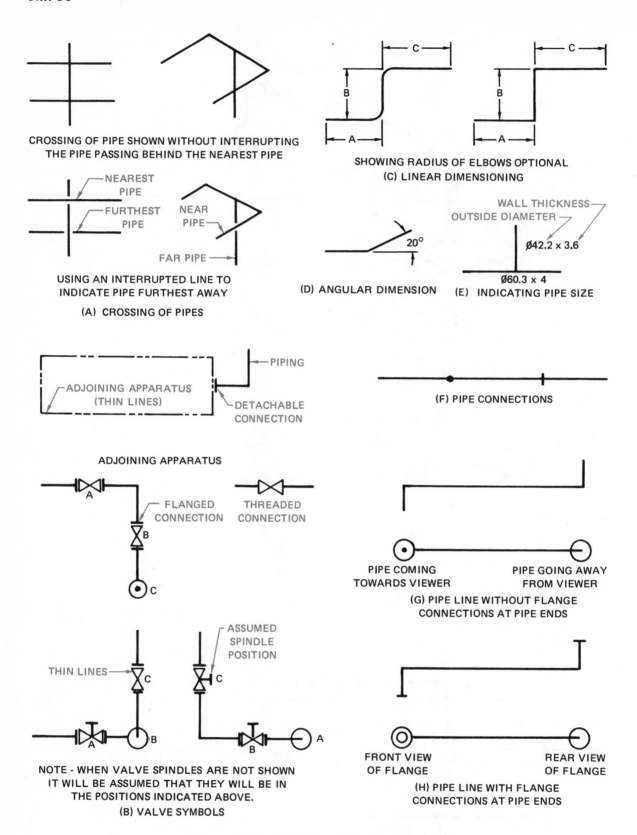

CROSSING OF PIPE SHOWN WITHOUT INTERRUPTING
THE PIPE PASSING BEHIND THE NEAREST PIPE

NEAREST PIPE

FURTHEST PIPE

NEAR PIPE

FAR PIPE

USING AN INTERRUPTED LINE TO
INDICATE PIPE FURTHEST AWAY

(A) CROSSING OF PIPES

SHOWING RADIUS OF ELBOWS OPTIONAL
(C) LINEAR DIMENSIONING

20°

(D) ANGULAR DIMENSION

WALL THICKNESS
OUTSIDE DIAMETER

Ø42.2 × 3.6

Ø60.3 × 4

(E) INDICATING PIPE SIZE

PIPING

ADJOINING APPARATUS
(THIN LINES)

DETACHABLE
CONNECTION

ADJOINING APPARATUS

(F) PIPE CONNECTIONS

FLANGED
CONNECTION

THREADED
CONNECTION

PIPE COMING
TOWARDS VIEWER

PIPE GOING AWAY
FROM VIEWER

(G) PIPE LINE WITHOUT FLANGE
CONNECTIONS AT PIPE ENDS

ASSUMED
SPINDLE
POSITION

THIN LINES

NOTE - WHEN VALVE SPINDLES ARE NOT SHOWN
IT WILL BE ASSUMED THAT THEY WILL BE IN
THE POSITIONS INDICATED ABOVE.

(B) VALVE SYMBOLS

FRONT VIEW
OF FLANGE

REAR VIEW
OF FLANGE

(H) PIPE LINE WITH FLANGE
CONNECTIONS AT PIPE ENDS

Fig. 18-6 Single line piping drawing symbols and dimensions

REPRESENTING FITTINGS

If specific symbols are not standardized, fittings such as tees, elbows, crosses, etc., are not specially drawn, but are represented, like pipe, by a continuous line. The circular symbol for a tee or elbow may be used when necessary to indicate whether the piping is viewed from the front or back, as shown in figure 18-6(H). Elbows on isometric drawings may be shown without the radius. However, if this method is used, the direction change of the piping must be shown clearly.

Adjoining Apparatus. If needed, adjoining apparatus, such as tanks, machinery, etc., not belonging to the piping itself, are shown by outlining them with a thin phantom line.

Dimensioning

- Dimensions for pipe and pipe fittings are always given from center-to-center of pipe and to the outer face of the pipe end or flange, figure 18-6(C).

- Individual pipe lengths are usually cut to suit by the pipe fitter. However, the total length of pipe required is usually called for in the bill of material.

- Pipe and fitting sizes and general notes are placed on the drawing beside the part concerned or, where space is restricted, with a leader.

- A Bill of Material is usually provided with the drawing.

- Pipes with bends are dimensioned from vertex to vertex.

- Radii and angles of bends are indicated as shown in figure 18-7(D). Whenever possible, the smaller of the supplementary angles is specified.

- The outer diameter and wall thickness of the pipe are indicated on the line representing the pipe, or in the Bill of Material, general note, or specifications, figure 18-6(E).

Orthographic Piping Symbols

Pipe Symbols. If flanges are not attached to the ends of the pipe lines when drawn in orthographic projection, pipe line symbols indicating the direction of the pipe are required. If the pipe line direction is toward the front (or viewer), it is shown by two concentric circles, the smaller one of which is a large solid dot, figure 18-6(G). If the pipe line direction is toward the back (or away from the viewer), it will be shown by one solid circle. No extra lines are required on the other views.

Flange Symbols. Irrespective of their type and sizes, flanges are to be represented by:
- two concentric circles for the front view,
- one circle for the rear view,
- a short stroke for the side view,

while using lines of equal thickness as chosen for the representation of pipes, figure 18-6(H).

Valve Symbols. Symbols representing valves are drawn with continuous thin lines (not thick lines as for piping and flanges). The valve spindles should only be shown if it is necessary to define their positions. It will be assumed that unless otherwise indicated, the valve spindle is in the position shown in figure 18-6(B).

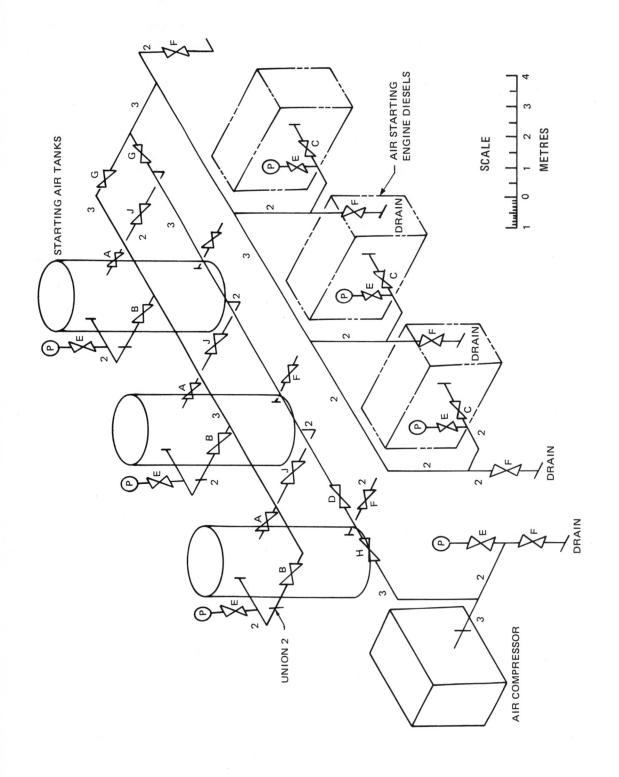

STARTING AIR TANKS

AIR STARTING
ENGINE DIESELS

SCALE
METRES

AIR COMPRESSOR

UNION 2

DRAIN

DRAIN

DRAIN

DRAIN

Safety valves are provided for the compressor and the air storage tanks. Check valves are installed on the air storage tank feed lines and the compressor discharge lines to prevent accidental discharge of the tanks.

Piping is arranged so that the compressor will either fill the storage tanks and/or pump directly to the engines. Any of the three storgae tanks may be used for starting, and pressure gauges indicate their readiness. The engines are fitted with quick-opening valves to admit air quickly at full pressure and shut it off at the instant rotation is obtained. A bronze globe valve is installed to permit complete shut-down of the engine for repairs, and regulation of the flow of air. Drains are provided at low points to remove condensate from the air storage tanks, lines, and engine feeds.

Globe valves are recommended throughout this hookup except on the main shutoff lines where gate valves are used because of infrequent operation.

ASSIGNMENT:

1. Complete the Bill of Material calling for all valves and fittings.

QUESTIONS (Use scale provided where necessary):

1. What size pipe is used for the main feed line?
2. Disregarding the length of the valves and fittings, what is the total approximate length of
   (a) the 3-inch pipe?   (b) the 2-inch pipe?
3. What is the approximate center line to center line spacing of the diesel engines?
4. How many cm^3 of air is contained in each of the starting air tanks?
5. Why are diesel engines used instead of public power supply?
6. What is the purpose of the air compressor?
7. Why is a gate valve used instead of a globe valve in the main line shutoff?
8. State the uses of the following parts.
   (a) Valve C   (c) Gauge P   (e) Valve D
   (b) Valve G   (d) Valve F   (f) Valve H

ANSWERS

1. _____	7. _____	d. _____
2. a. _____	_____	_____
b. _____	8. a. _____	e. _____
3. _____		_____
4. _____	b. _____	f. _____
5. _____	_____	_____
	c. _____	
6. _____	_____	
_____		

NOTE
ALL PIPING M.I.
ALL DIMENSIONS IN METRES
EXCEPT FOR PIPE SIZES WHICH
ARE GIVEN IN INCHES

Code	Valve	Service
A	Bronze Globe	Air Storage Tank Feed Lines
B	Bronze Globe	Air Storage Tank Discharge Lines
C	Bronze Globe	Diesel Engine Shutoff Control
D	Bronze Globe	Air Compressor Discharge
E	Bronze Globe	Pressure Gauge Shutoff
F	Bronze Globe	Drain Valves
G	Spindle Gates	Main Line Shutoff
H	Bronze Check	Air Compressor Check
J	Bronze Check	Air Storage Tank Feed Lines
P	Pressure Gauge	Discharge or Feed Lines

QTY	ITEM	MAT'L	DESCRIPTION	PT NO
				14
				13
				12
				11
				10
				9
				8
				7
				6
				5
				4
				3
				2
				1

METRIC	SCALE	
	DRAWN	DATE
ISOMETRIC PROJECTION	ENGINE STARTING AIR SYSTEM	A-41

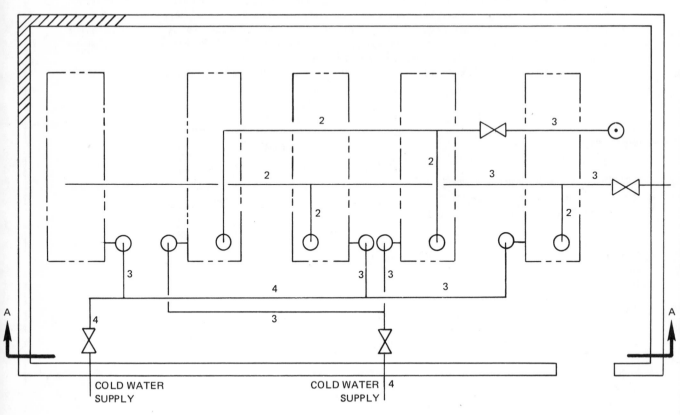

PLAN VIEW OF BOILER ROOM

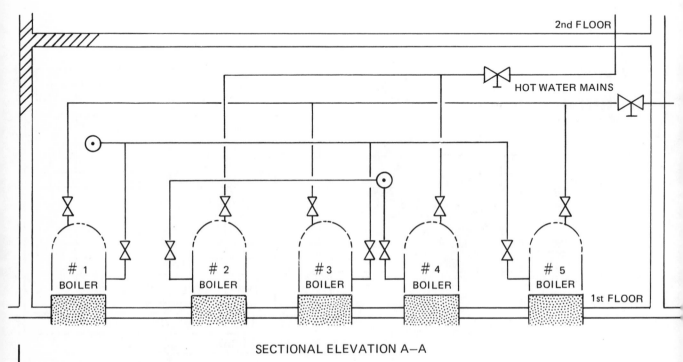

SECTIONAL ELEVATION A—A

1. Using the scale provided, complete the isometric view of the Boiler Room piping shown in orthographic projection on the opposite page. It is not necessary to show the walls and floors.

2. On a separate sheet, make up a Bill of Material calling for all valves, fittings, and pipe. (Give the approximate length of each size of pipe.)

3. From the Bill of Material, add part numbers to the isometric drawing for the valves and fittings only.

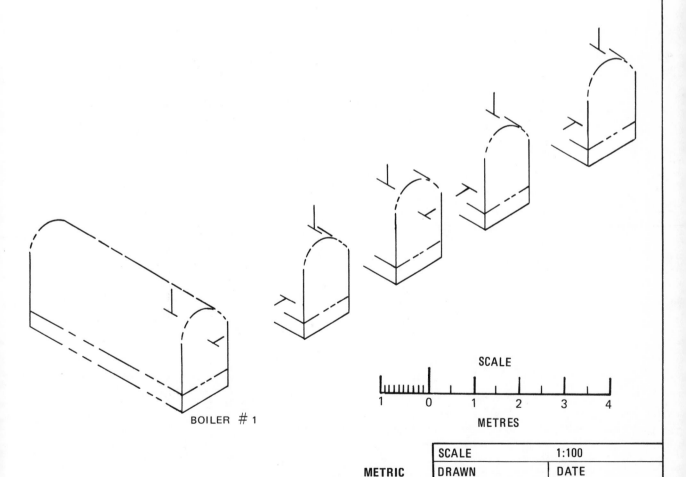

BOILER #1

SCALE

1   0   1   2   3   4

METRES

SCALE	1:100
DRAWN	DATE

METRIC

BOILER ROOM   **A-42**

UNIT
19

TUBING

Metal tubing has many uses. It is used in agricultural machines, airplane parts, automotive parts, conveyors and machine parts. Solid and flexible nonmetallic tubing, such as fiber tubing, has many uses in the electrical industry. For example, nonmetallic tubing is used in insulators and support tubes.

Dimensioning Tubes

Although tubes have three size dimensions (OD, ID, and wall thickness), only two of these dimensions are specified. The dimensions specified would be: (1) OD and ID; (2) OD and wall thickness; or (3) ID and wall thickness.

Simplified drafting methods are often used in the detailing of pipes and tubes. The drafter must decide which method is the most efficient and timesaving. Factors such as productions methods and machines and the craftsperson in the shop's ability to interpret the drafter's drawings will influence the choice of method. Figure 19-1 gives two simplified methods for detailing the same tube.

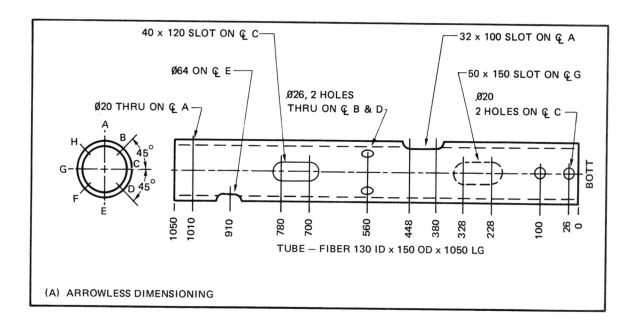

40 × 120 SLOT ON ℄ C

32 × 100 SLOT ON ℄ A

Ø64 ON ℄ E

50 × 150 SLOT ON ℄ G

Ø20 THRU ON ℄ A

Ø26, 2 HOLES
THRU ON ℄ B & D

Ø20
2 HOLES ON ℄ C

1050 1010 910 780 700 560 448 380 328 228 100 26 0

BOTT

TUBE — FIBER 130 ID × 150 OD × 1050 LG

(A) ARROWLESS DIMENSIONING

ANGLE
LOCATION

0°

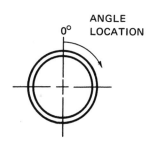

PT. NO.	ITEM	BILL OF MATERIAL	QTY.
		MATERIAL	
1	SUPPORT	TUBE — FIBER 130 ID × 150 OD × 1050 LG 90° — 26 AND 100 — 2 HOLES Ø20 ZERO — 1010 — 1 HOLE Ø20 — THRU 45° — 560 — 1 HOLE Ø26 — THRU 135° — 560 — 1 HOLE Ø26 — THRU ZERO — 380 AND 448 — 1 SLOT 32 WIDE 90° — 700 AND 780 — 1 SLOT 40 WIDE 270° — 228 AND 328 — 1 SLOT 50 WIDE 180° — 910 — 1 HOLE Ø64	1

NOTE: ORDER OF DIMENSIONS ON MATERIAL LIST:

1. ANGLE INDICATES DEGREES CLOCKWISE FROM CENTER LINE SHOWN.

2. DIMENSIONS FROM BOTTOM OF PART TO CENTER LINE OF HOLE, HOLES OR SLOTS.

3. SIZE OF HOLE OR HOLES FOR SLOTS.

NOTE: ORDER OF DIMENSIONS SHOWN WHEN DONE ON AUTO-MATIC DRILL. THIS ORDER CALLS FOR SMALLEST HOLE FIRST THEN INCREASING TO LARGEST HOLE. IF DONE BY OPERATOR, AN ALTERNATIVE ORDER OF LISTING HOLES MAY BE GIVEN, I.E., ALL HOLES DONE ON 0 DEGREES FOLLOWED BY ALL HOLES DONE ON 45 DEGREES, ETC.

METRIC

(B) TABULAR DIMENSIONING SHOWN IN BILL OF MATERIAL

Fig. 19-1 Dimensioning for pipes and tubes

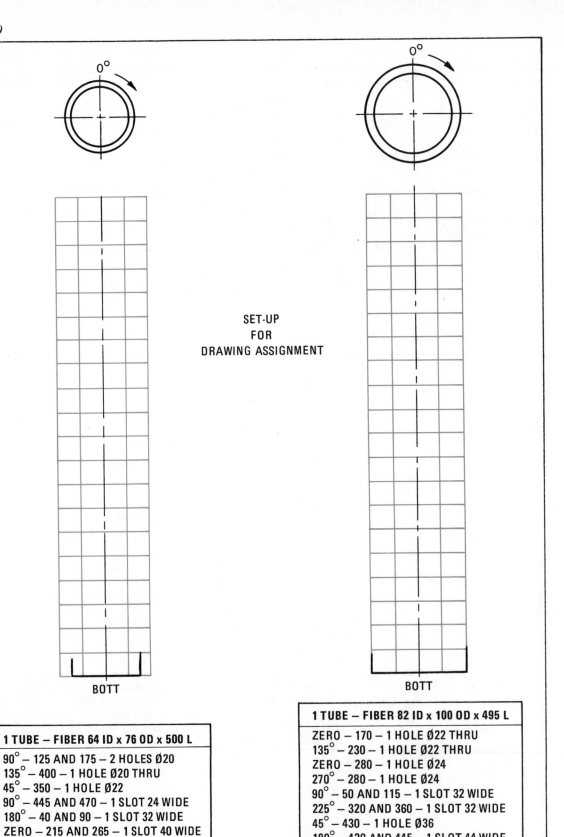

SET-UP
FOR
DRAWING ASSIGNMENT

0°

0°

BOTT

BOTT

1 TUBE – FIBER 82 ID x 100 OD x 495 L

ZERO – 170 – 1 HOLE Ø22 THRU
135° – 230 – 1 HOLE Ø22 THRU
ZERO – 280 – 1 HOLE Ø24
270° – 280 – 1 HOLE Ø24
90° – 50 AND 115 – 1 SLOT 32 WIDE
225° – 320 AND 360 – 1 SLOT 32 WIDE
45° – 430 – 1 HOLE Ø36
180° – 420 AND 445 – 1 SLOT 44 WIDE

1 TUBE – FIBER 64 ID x 76 OD x 500 L

90° – 125 AND 175 – 2 HOLES Ø20
135° – 400 – 1 HOLE Ø20 THRU
45° – 350 – 1 HOLE Ø22
90° – 445 AND 470 – 1 SLOT 24 WIDE
180° – 40 AND 90 – 1 SLOT 32 WIDE
ZERO – 215 AND 265 – 1 SLOT 40 WIDE

**MAKE DRAWINGS OF THE TUBES DESCRIBED IN THE TABLES COMPLETE WITH
ARROWLESS DIMENSIONING, SCALE 1:4 (25 mm SQUARES.)**

A-43

ASSIGNMENT: IN THE SPACES PROVIDED, DETAIL THE FIBER TUBES SHOWN BELOW IN TABULAR DIMENSIONING FORM, LISTING THE DETAILS (STARTING FROM THE BOTTOM OF THE TUBE) ACCORDING TO THE COLUMN HEADINGS. NOTE — ALL HOLES ON ONE SIDE ONLY UNLESS SPECIFIED BY THE WORD "THRU."

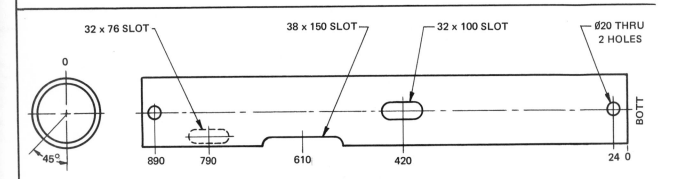

MATERIAL — HD FIBER 98 ID x 7 WALL x 915 LG

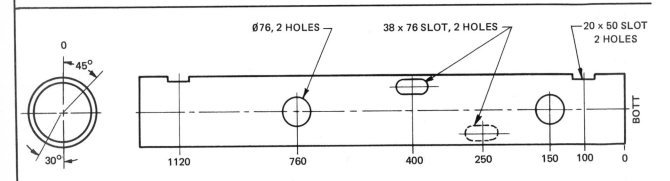

MATERIAL — KRAFT PAPER 150 OD x 8 WALL x 1220 LG

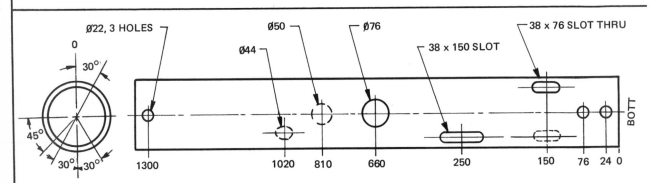

MATERIAL — HD FIBER 108 ID x 10 WALL x 1350 LG

COLUMN I	COLUMN 2
DETAIL THE TUBES BY CALLING FOR THE SMALLEST HOLE FIRST, THEN INCREASING TO THE LARGEST HOLE. 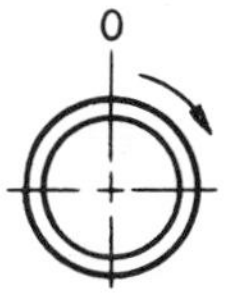 VIEW FROM BOTTOM	DETAIL THE TUBES BY TUBE ROTATION, I.E., ALL HOLES ON 0°, FOLLOWED BY ALL HOLES ON 45°, ETC. 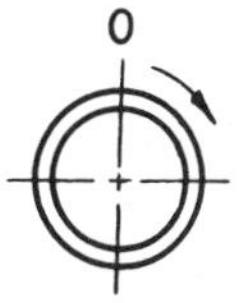 VIEW FROM BOTTOM

NOTE: UNLESS OTHERWISE SPECIFIED

– TOLERANCE ON DIMENSIONS ± 0.5

– TOLERANCE ON ANGLES ± 0.5°

METRIC	SCALE	NOT TO SCALE
	DRAWN	DATE
	TUBE SUPPORTS	A-44

UNIT

20

STEEL SPECIFICATIONS

Carbon steels are the workhorses of product design. They account for over 90 percent of total steel production. More carbon steels are used in product manufacturing than all other metals combined. Far more research is going into carbon steel metallurgy and manufacturing technology than into all other steel mill products. Various technical societies and trade associations have issued the specifications covering the composition of metals. They serve as a selection guide and are a way for the buyer to conveniently specify certain known and recognized requirements. The main technical societies and trade associations are:

AISI – American Iron and Steel Institute

CSA – Canadian Standards Association

SAE – Society of Automotive Engineers

SAE and AISI Systems of Steel Identification

The specifications for steel bar are based on a number code indicating the composition of each type of steel covered. They include both plain-carbon and alloy steels. The code is a 4-number system. The first two figures indicate the alloy series and the last two figures the carbon content in hundredths of a percent, figure 20-1. Therefore, the figure XX15 indicates 0.15 of 1 percent carbon.

For example, AISI 4830 is a molybdenum-nickel steel containing 0.2 – 0.3 percent molybdenum, 3.25 – 3.75 percent nickel and 0.3 percent carbon. In addition to the four number designation the suffix "H" is used for steels produced to specify hardenability limits, and the prefix "E" indicates a steel made by the basic electric-furnace method.

Originally, the second figure indicated the percentage of the major alloying element present. This was true of many of the alloy steels. However, this had to be varied in order to classify all the steels that became available.

Alloying materials are added to steel to improve properties such as strength, hardness, machinability, corrosion, electrical conductivity, and ease of forming. Figure 20-2 lists many types of steel and their uses.

Effect of Alloys on Steel

Carbon. Increasing the carbon content increases the tensile strength and hardness.

Sulphur. When sulphur content is over 0.06 percent the metal tends toward *red shortness* (brittleness in steel when it is red-hot). Free cutting steel, for threading and screw machine work, is obtained by increasing sulphur content to from about 0.075 to 0.1 percent.

Phosphorous. Phosphorous produces brittleness and general cold shortness. It also strengthens low carbon steel, increases resistance to corrosion, and improves machinability.

Manganese. Manganese is added during the making of steel to prevent red shortness and increase hardenability.

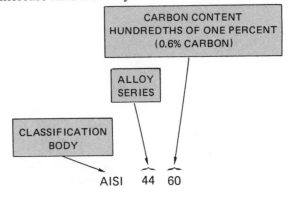

Fig. 20-1 Steel designation system

Type of Steel		Number Symbol	Principle Properties	Common Uses
Carbon Steels				
– Nonresulfurized (Basic Carbon Steel)		10XX		Chains, Rivets, Shafts, Pressed Steel Products
– Plain Carbon		10XX		
– Low Carbon Steel (0.06% to 0.2% Carbon)		1006 to 1020	Toughness and Less Strength	
– Medium Carbon Steel (0.2% to 0.5% Carbon)		1020 to 1050	Toughness and Strength	Gears, Axles, Machine Parts, Forgings, Bolts and Nuts
– High Carbon Steel (Over 0.5% Carbon)		1050 and over	Less Toughness and Greater Hardness	Saws, Drills, Knives, Razors, Finishing Tools, Music Wire
– Resulfurized (Free Cutting)		11XX	Improves Machinability	Threads, Splines, Machined Parts
– Phosphorized		12XX	Increases Strength and Hardness but Reduces Ductility	
– Manganese Steels		13XX	Improves Surface Finish	
Molybdenum Steels				
0.15 – 0.30 Mo		40XX		
0.08 – 0.35 Mo	0.4 – 1.1 Cr	41XX		
1.65 – 2.00 Ni	0.4 – 09 Cr	43XX	High Strength	Axles, Forgings, Gears, Cams, Mechanism Parts
0.2 – 0.3 Mo				
0.45 – 0.60 Mo		44XX		
0.7 – 2.0 Ni	0.15 – 0.30 Mo	46XX		
0.9 – 1.2 Ni	0.35 – 0.55 Cr	47XX		
0.15 – 0.40 Mo				
3.25 – 3.75	0.2 – 0.3 Mo	48XX		
Chromium Steels				
0.3 – 0.5 Cr		50XX	Hardness, Great Strength and Toughness	Gears, Shafts, Bearings Springs, Connecting Rods
0.70 – 1.15 Cr		51XX		
1.00 C	0.90 – 1.15 Cr	E51100		
1.00 C	0.90 – 1.15 Cr	E52100		
Chromium Vanadium Steels				Punches and Dies, Piston Rods, Gears, Axles
0.5 – 1.1 Cr	0.10 – 0.15 V	61XX	Hardness and Strength	
Nickel – Chromium – Molybdenum Steels				
0.4 – 0.7 Ni 0.4 – 0.6 Cr 0.15 – 0.25 Mo		86XX	Rust Resistance, Hardness and Strength	Food Containers, Surgical Equipment
0.4 – 0.7 Ni 0.4 – 0.6 Cr 0.2 – 0.3 Mo		87XX		
0.4 – 0.7 Ni 0.4 – 0.6 Cr 0.3 – 0.4 Mo		88XX		
Silicon Steels			Springiness and Elasticity	
1.8 – 2.2 Si		92XX		Springs

Fig. 20-2 Designations, uses, and properties of steel

Silicon. Silicon is used as a general purpose deoxidizer. It strengthens low-alloy steels and increases hardenability.

Copper. Copper is used to increase atmospheric corrosion resistance.

Nickel. Nickel strengthens and toughens ferrite and pearlite steels.

Molybdenum. Molybdenum increases hardenability and coarsening temperature.

Chromium. Chromium increases hardenability, corrosion resistance, oxidation, and abrasion.

Vanadium. Vanadium elevates coarsening temperatures, increases hardenability, and is a strong deoxidizer.

Boron. Boron increases hardenability of lower carbon steels and has better machinability than standard alloy steels.

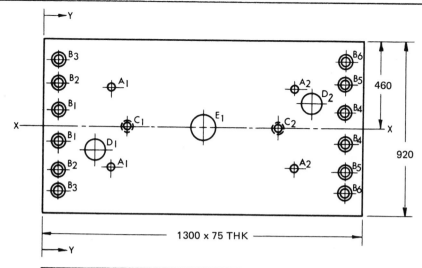

1. _____
2. _____
3. _____
4. _____
5. _____
6. _____
7. _____
8. _____
9. _____
10. _____
11. _____
12. _____
13. _____
14. _____
15. _____

HOLE	HOLE SIZE	LOCATION	
		X	Y
A_1	Ø16, 2 HOLES	200	280
A_2		200	1020
B_1	Ø32, Ø64 CBORE 30 DEEP, NEAR SIDE ONLY, 12 HOLES	76	50
B_2		228	50
B_3		342	50
B_4		76	1250
B_5		228	1250
B_6		342	1250
C_1	M20, 2 HOLES		342
C_2			958
D_1	Ø78, 2 HOLES	126	210
D_2		126	1090
E_1	Ø100.0–100.2, 1 HOLE		650

NOTE: UNLESS OTHERWISE SPECIFIED, TOLERANCE ON DIMENSIONS ± 0.5.

QUESTIONS

1. What is the thickness of the material?

2. What type of steel is specified?

3. What indicates that the drawing is not to scale?

4. How many counterbored holes are there?

5. What is the diameter of the CBORE?

6. How many Ø78 holes are there?

7. Which letter specifies the Ø78 holes?

8. How many tapped holes of different sizes are specified?

9. What is the total number of tapped holes?

10. What is the thread pitch of the C holes?

11. What is the tolerance given for the E hole?

12. What is the high limit of the E hole?

13. What would be the maximum limit for Ø78 hole if a ± 0.2 leeway is permitted?

14. Is the E hole on the center of the part?

15. How far apart are holes C_1 and C_2?

METRIC

MATERIAL	SAE 1020
SCALE	NOT TO SCALE
DRAWN	DATE

CROSSBAR   **A-45**

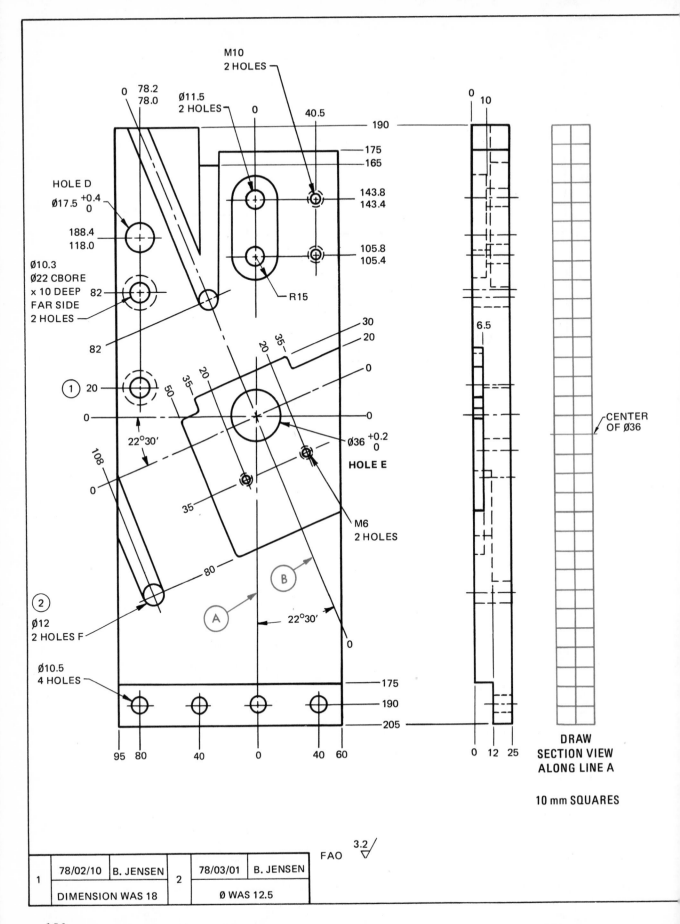

M10
2 HOLES

Ø11.5
2 HOLES

0   78.2
           78.0

0   40.5
190
175
165

HOLE D
Ø17.5 +0.4
        0   143.8
                                        143.4
188.4
118.0
                                        105.8
                                        105.4
Ø10.3
Ø22 CBORE
x 10 DEEP   82
FAR SIDE   R15
2 HOLES

82

                                        30
                                        20

① 20   35
                            20
                            0

0   0
                    50
                    35
                    20

22°30'   Ø36 +0.2
                                 0
                            HOLE E
108

0   35

                            M6
                            2 HOLES

②
                    80
Ø12
2 HOLES F   Ⓐ   Ⓑ   22°30'

                                        0
Ø10.5
4 HOLES   175
                                        190
                                        205

95 80   40   0   40 60

0   10
6.5

CENTER
OF Ø36

0 12 25

DRAW
SECTION VIEW
ALONG LINE A

10 mm SQUARES

FAO   3.2 ▽

1	78/02/10	B. JENSEN	2	78/03/01	B. JENSEN
	DIMENSION WAS 18			Ø WAS 12.5	

150

1. What are the overall (A) width, (B) height, (C) depth of the parts?

2. What is the width of the slots that are cut out from the F holes?

3. What is their depth?

4. What class of surface texture is required?

5. What is the depth of the recess adjacent to the (A) E hole, (B) Ø11.5 holes?

6. How many degrees are there between line B and the horizontal?

7. What was the size of the F holes before they were changed?

8. What tolerance is required on the E hole?

9. What is the low tolerance limit on the E hole?

10. What tolerance is required on the D hole?

11. What is the high limit on the D hole?

12. Determine the maximum vertical distance from D hole to the four Ø10.5 holes located at the bottom of the part?

13. Determine the (A) maximum, (B) minimum, distance between the two M10 holes.

14. Determine the maximum horizontal distance between D hole and the M10 tapped holes.

15. What are the width, height, and depth of the cutout at the bottom of the part?

16. How deep are (A) the Ø11.5 holes, (B) the Ø10.3 holes?

17. How deep are the M6 holes?

18. How many full threads do the M6 tapped holes have?

19. What are the main alloys in the material?

NOTE: ALL RADII R2 UNLESS OTHERWISE SPECIFIED.

DIMENSIONS ARE TAKEN FROM PLANES DESIGNATED O-O, AND ARE PARALLEL TO THESE PLANES.

UNLESS OTHERWISE STATED:
± 0.5 ON DIMENSIONS
. ± 0.5° ON ANGLES.

CENTER OF Ø36

DRAW
SECTION VIEW
ALONG LINE B

1. A. _____
   B. _____
   C. _____
2. _____
3. _____
4. _____
5. A. _____
   B. _____
6. _____
7. _____
8. _____
9. _____
10. _____
11. _____
12. _____
13. A. _____
    B. _____
14. _____
15. W _____
    H _____
    D _____
16. A. _____
    B. _____
17. _____
18. _____
19. _____
_____

METRIC	
MATERIAL	SAE 4020
SCALE	NOT TO SCALE
DRAWN	DATE

OIL CHUTE **A-46**

UNIT 21

CASTINGS

Irregular or odd-shaped parts which are difficult to make from metal plate or bar stock may be cast to the desired shape. Casting processes for metals can be classified by either the type of mold or pattern or the pressure or force used to fill the mold. Conventional sand, shell, and plaster molds utilize a durable pattern, but the mold is used only once. Permanent molds and die-casting dies are machined in metal or graphite sections and are employed for a large number of castings. Investment casting and the relatively new full mold process involve both an expendable mold and pattern.

Sand Mold Casting

The most widely used casting process for metals uses a permanent pattern of metal or wood that shapes the mold cavity when loose molding material is compacted around the pattern. This material consists of a relatively fine sand plus a binder that serves as the adhesive.

Figures 21-1 and 21-2 show a typical sand mold, with the various provisions for pouring the molten metal and compensating for contraction of the solidifying metal, and a sand core for forming a cavity in the casting. Sand molds are prepared in flasks which consist of two or more sections: bottom (drag), top (cope), and intermediate sections (checks) when required.

The cope and drag are equipped with pins and lugs to insure the alignment of the flask. Molten metal is poured into the sprue and connecting runners provide flow channels for the metal to enter the mold cavity through gates. Riser cavities are located over the highest section of the casting.

The gating system, besides providing a way for the molten metal to enter the mold, functions as a venting system for the removal of gases from the mold and acts as a riser to furnish liquid metal to the casting during solidification.

In producing sand molds, a metal or wooden pattern must first be made. The pattern is slightly larger in every dimension than the part to be cast to allow for shrinkage when the casting cools. This is known as *shrinkage allowance,* and the patternmaker allows for it by using a shrink rule for each of the cast metals. Since shrinkage and draft are taken care of by the patternmaker, they are of no concern to the drafter.

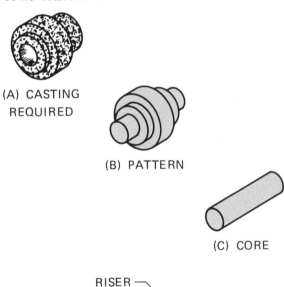

(A) CASTING REQUIRED

(B) PATTERN

(C) CORE

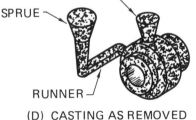

(D) CASTING AS REMOVED FROM MOLD

Fig. 21-1 Sand casting parts

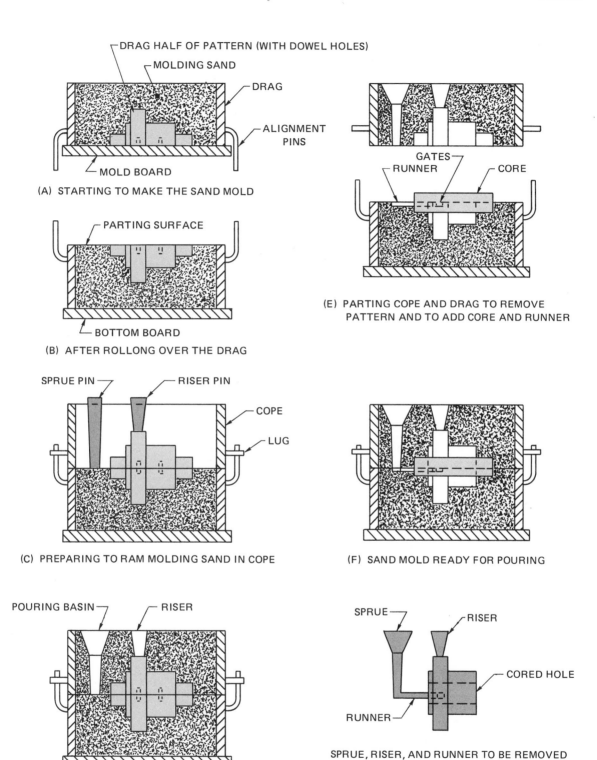

(A) STARTING TO MAKE THE SAND MOLD

(B) AFTER ROLLONG OVER THE DRAG

(C) PREPARING TO RAM MOLDING SAND IN COPE

(D) REMOVING RISER AND GATE SPRUE PINS AND ADDING POURING BASIN

(E) PARTING COPE AND DRAG TO REMOVE PATTERN AND TO ADD CORE AND RUNNER

(F) SAND MOLD READY FOR POURING

(G) CASTING AS REMOVED FROM THE MOLD

Fig. 21-2 Sequence in preparing a sand casting

Additional metal, known as machining or finish allowance, must be provided on the casting where a surface is to be finished. Depending on the material being cast, from 2 to 4 mm is usually allowed on small castings for each surface requiring finishing.

When casting a hole or recess in a casting, a core is often used. A core is a mixture of sand and a bonding agent that is baked and hardened to the desired shape of the cavity in the casting, plus an allowance to support the core in the sand mold. In addition to the shape of the casting desired, the pattern must be designed to produce areas in the mold cavity to locate and hold the core. The core must be solidly supported in the mold, permitting only that part of the core that corresponds to the shape of the cavity in the casting to project into the mold.

Preparation of Sand Molds. The drag portion of the flask is first prepared in an upside-down position with the pins pointing down. The drag half of the pattern is placed in position on the mold board and a light coating of parting compound is used as a release agent. The molding sand is then rammed or pressed into the drag flask. A bottom board is placed on the drag, the whole unit is rolled over, and the mold board is removed.

The cope half of the pattern is placed over the drag half and the cope portion of the flask is placed in position over the pins. A light coating of parting compound is sprinkled throughout. Next the sprue pin and riser pin, tapered for easy removal, are located, and the molding sand is rammed into the cope flask. The sprue pin and riser pin are then removed. A pouring basin may or may not be formed at the top of the gate sprue.

Now the cope is lifted carefully from the drag and the pattern is exposed. The runner and gate, passageways for the molten metal into the mold cavity, are formed in the drag sand. The pattern is removed and the core is placed in position. The cope is then put back on the drag.

The molten metal is poured into the pouring basin and runs down the sprue to a runner and through the gate and into the mold cavity. When the mold cavity is filled, the metal will begin to fill the sprue and the riser. Once the sprue and riser have been filled, the pouring should stop.

When the metal has hardened, the sand is broken and the casting removed. The excess metal, gates, and risers, are removed and then remelted.

Full Mold Casting

The characteristic feature of the *full mold process* is the use of gasifiable patterns made of foamed plastic. These are not extracted from the mold, but are vaporized by the molten metal.

The full mold process is suitable for individual castings, and for small series of up to five castings. The full mold process is very economical, and it reduces the delivery time required for prototypes, articles urgently needed for repair jobs, and individual large machine parts.

REFERENCES AND SOURCE MATERIALS

1. J.F. Wallace, "Casting" *Machine Design,* 37, No. 21 (1965).

1. Which line in the top view represents surface (1) ?
2. Locate surface (A) in the left-side view and the front view.
3. Locate surface (8) in the front view.
4. How many surfaces are to be finished?
5. Which line in the left-side view represents surface (3) ?
6. What is the center distance between holes (B) and (O) in the front view?
7. Determine distances at (4) (5) (6) and (11) .
8. Locate surface (J) in the top view.
9. Which surface of the left-side view does line (14) represent?
10. What point in the front view is represented by line (15) ?
11. What is the thickness of boss (E) ?
12. Locate surface (G) in the left-side view.
13. Locate point (K) in the top view.
14. Locate surface (D) in the top view.
15. Determine distances at (M) (N) (S) (T) .
16. What point or line in the top view is represented by point (16) ?
17. What is the maximum horizontal distance between holes in the front view?
18. What would be the (A) width, (B) height, and (C) depth of the part before machining?

ANSWERS

1. _____
2. L.V. _____
   F.V. _____
3. _____
4. _____
5. _____
6. _____
7. (4) _____
   (5) _____
   (6) _____
   (11) _____
8. _____
9. _____

10. _____
11. _____
12. _____
13. _____
14. _____
15. (M) _____
    (N) _____
    (S) _____
    (T) _____
17. _____
18. A. _____
    B. _____
    C. _____

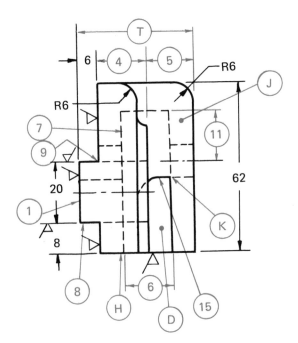

REVISIONS	1	MAR 4/78	C.J.
		28 DIMENSION WAS 30	

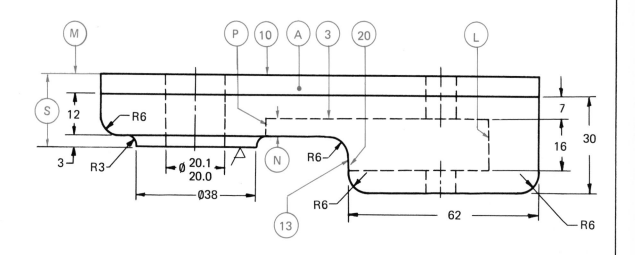

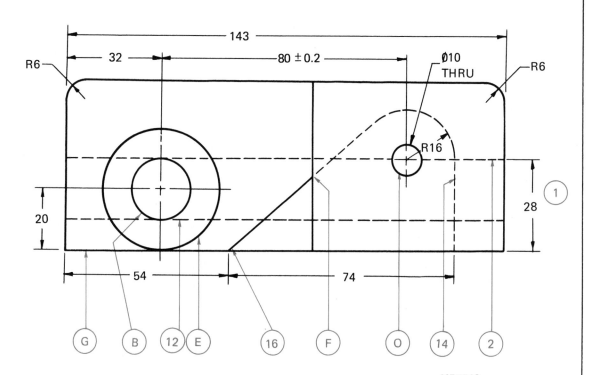

NOTE: UNLESS OTHERWISE SPECIFIED
TOLERANCES ON DIMENSIONS ± 0.5
$\frac{3.2}{2}\sqrt{}$ FINISH WHERE SHOWN AS $\sqrt{}$.

METRIC	
QUANTITY	500
MATERIAL	CI
SCALE	NOT TO SCALE
DRAWN	DATE

TRIP BOX

A-47

157

UNIT 22

CASTING DESIGN

Simplicity of Molding from Flat Back Patterns

Simple shapes such as the one shown in figure 22-1 are very easy to mold. In this case the flat face of the pattern is at the parting line and lies perfectly flat on the molding board. In this position no molding sand sifts under the flat surface to interfere with the drawing of the pattern. The simplicity with which flat back patterns of this type may be drawn from the mold is illustrated in figure 22-1.

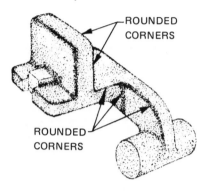

Fig. 22-2 Casting of offset bracket shown in drawing A-48

Irregular or Odd-Shaped Castings

When a casting is to be made for an odd-shaped piece such as the offset bracket, figure 22-2, it is necessary to make the pattern for the bracket in one or more parts to facilitate the making of the mold. The difficulties with this are mostly due to the removal of the pattern from the sand mold.

The pattern for the bracket is made in two parts as shown in figure 22-3. The two

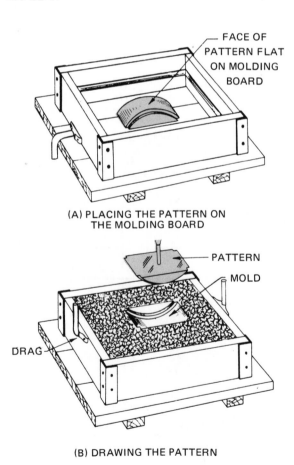

(A) PLACING THE PATTERN ON THE MOLDING BOARD

(B) DRAWING THE PATTERN

Fig. 22-1 Making a mold of a flatback pattern

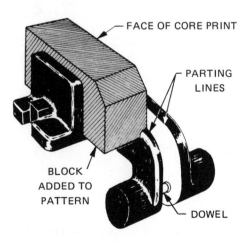

Fig. 22-3 Two-piece pattern of offset bracket

adjacent flat surfaces of the divided pattern come together at the parting line.

Set Cores

In examining the illustration of the off-set bracket, it will be noted that there are rounded corners. In order to make it possible to mold these rounded corners, a block must be added to the pattern for ease in its removal from the mold.

This block, which becomes an integral part of the pattern, also acts as a core print for a *set core* (a core that has been baked hard), figure 22-4. It is made to conform to the shape of the faces of the casting, including the rounded corners.

The face of the core print also forms the parting line for one side of the two-part pattern. When making the mold for the bracket casting, this face, which corresponds to the flat face of a flat back pattern, is laid on the molding board with the drag of the flask in position. The sand is then rammed around the pattern.

When the drag is reversed, the cope, or upper part of the flask, is placed in position. The other half of the pattern is then joined with the first part, and the sand is rammed into the cope to flow around that part of the pattern which projects into it.

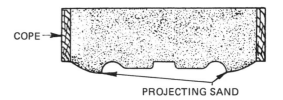

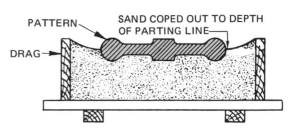

Fig. 22-5 Coping down

After the pattern is removed, the set core, which is formed in a core box and baked hard, is set in the impression in the mold made by the core print of the pattern. When poured into the mold, the molten metal fills the cavity made by the pattern and the faces of the set core as shown in figure 22-4 at **A** and **B** to form the casting.

Coping Down

The core print is made as part of the pattern to avoid removing molding sand in the drag which would correspond to the shape of the core print.

If this sand were dug out or *coped out,* as shown in figure 22-5, the remaining cavity would be again filled with molding sand when the cope was rammed. The sand would then hang below the parting line of the cope down into the drag.

When the mold is made by *coping down,* the hanging portion of the cope is supported by *soldiers* or *gaggers* embedded within the sand to hold the projecting part in position for subsequent operations, figure 22-6.

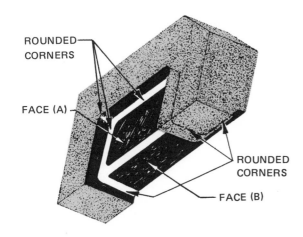

Fig. 22-4 Set core for offset bracket

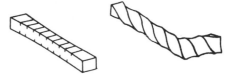

Fig. 22-6 Soldiers and gaggers

Coping down requires skill and takes time. The set core principle is at times preferred to coping down to avoid delay and assure a more even parting line on the casting.

Split Patterns

Irregularly shaped patterns which cannot be drawn from the sand are sometimes split so that one half of the pattern may be rammed in the drag as a simple flat back pattern while the cope half, when placed in position on the drag half, forms the mold in the cope. Patterns of this type are called *split patterns* and do not require coping down to the parting line which would be necessary if the pattern were made solid.

The pattern for the casting shown in figure 22-7, when made without a print for a core, can be drawn from the sand only by splitting the pattern on the parting line as illustrated.

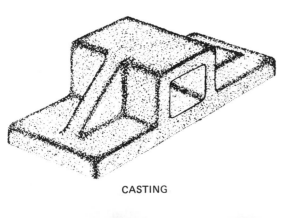

CASTING

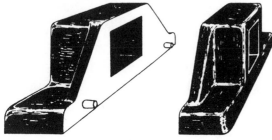

SPLIT PATTERN

Fig. 22-7 Application of split pattern

The drawing must be examined to determine how the pattern should be constructed. This is important because the parting line must be located in a position which permits the halves of the pattern to be drawn from the sand without interference.

CORED CASTINGS

Cored castings have certain advantages over solid castings. Where practical, castings are designed with cored holes or openings for economy, appearance, and accessibility to interior surfaces.

Cored openings often improve the appearance of a casting. In most instances, cored castings are more economical than solid castings because of the savings in metal. While cored castings are lighter, they are designed without sacrificing strength. The openings cast in the part eliminate unnecessary machining.

Hand holes may also be formed by coring in order to provide an opening through which the interior of the casting may be reached. These openings also permit machining an otherwise inaccessible surface of the part as shown in figure 22-8.

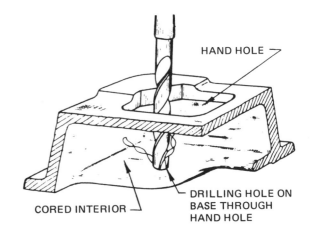

HAND HOLE

DRILLING HOLE ON BASE THROUGH HAND HOLE

CORED INTERIOR

Fig. 22-8 Section of cored casting

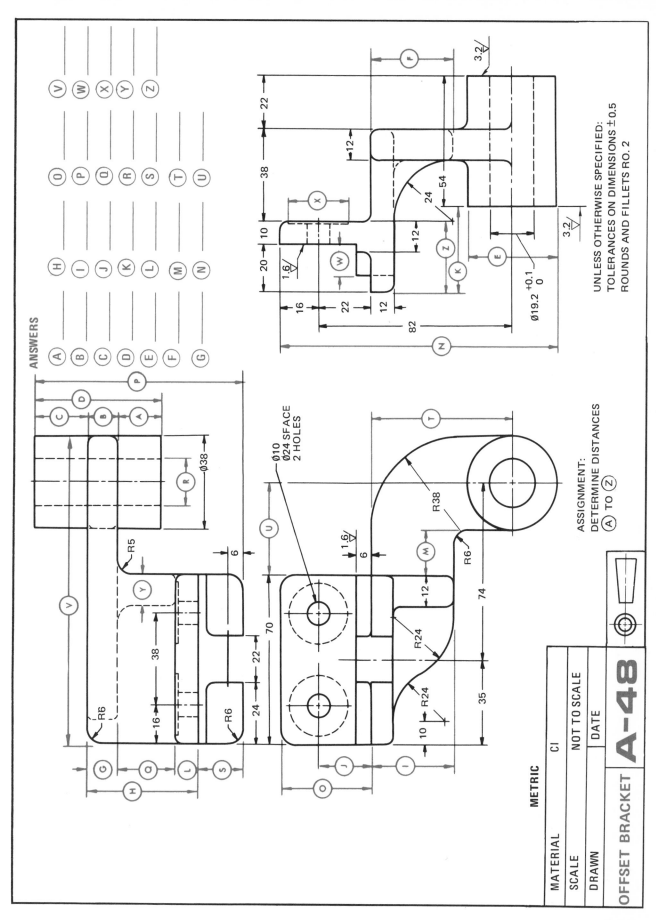

ANSWERS

Ⓐ Ⓑ Ⓒ Ⓓ Ⓔ Ⓕ Ⓖ

Ⓗ Ⓘ Ⓙ Ⓚ Ⓛ Ⓜ Ⓝ

Ⓞ Ⓟ Ⓠ Ⓡ Ⓢ Ⓣ Ⓤ

Ⓥ Ⓦ Ⓧ Ⓨ Ⓩ

UNLESS OTHERWISE SPECIFIED:
TOLERANCES ON DIMENSIONS ± 0.5
ROUNDS AND FILLETS R0. 2

ASSIGNMENT:
DETERMINE DISTANCES
Ⓐ TO Ⓩ

Ø10
Ø24 SFACE
2 HOLES

MATERIAL	CI	
SCALE	NOT TO SCALE	
DRAWN	DATE	

METRIC

OFFSET BRACKET **A-48**

161

QUESTIONS

1. Of what material is the base made?
2. How many finished surfaces are indicated?
3. Which circled letters on the drawing indicate the spaces that were cored when the casting was made?
4. Locate the surface in the top view that is represented by line ⑥ .
5. What is the height of the cored area Ⓧ ?
6. What surface finish is used on the top pad?
7. What is the horizontal length of the pad ⑤ ?
8. What reason might be given for openings Ⓠ Ⓢ Ⓨ Ⓩ ?
9. Determine distance Ⓦ .
10. What radius fillet would be used at Ⓡ ?
11. What is the total allowance added to the height for machining?
12. Determine distances Ⓒ through Ⓝ .

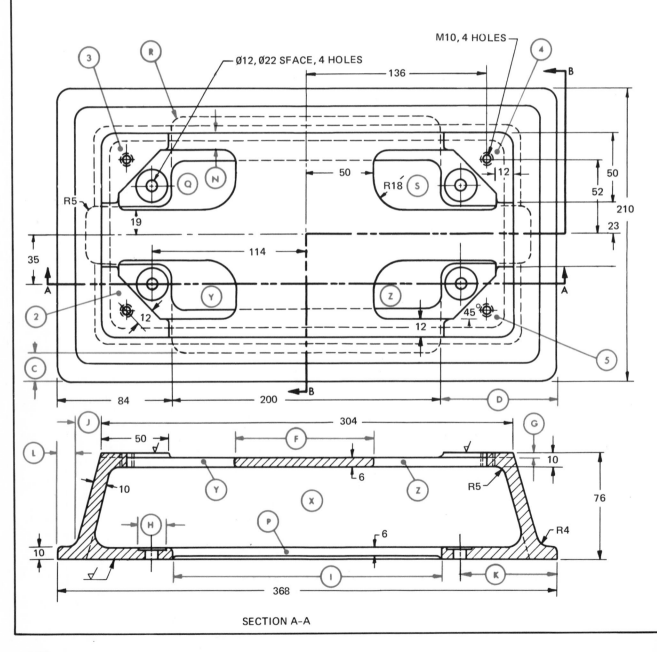

SECTION A-A

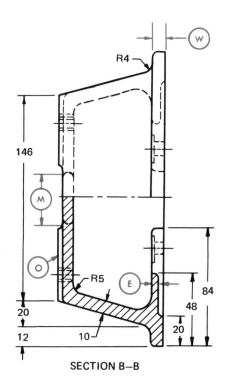

SECTION B—B

NOTE: 1. ALL DIMENSIONS SYMMETRICAL
AROUND THE CENTER LINES.
2. UNLESS OTHERWISE SPECIFIED:
 – TOLERANCES ON DIMENSIONS ±0.5
 – ROUNDS AND FILLETS R6
 – FINISHES 3 $\overset{32}{\triangledown}$

METRIC

MATERIAL	CI
SCALE	NOT TO SCALE
DRAWN	DATE

AUXILIARY PUMP
BASE

A-49

163

UNIT
23

MACHINING LUGS

It is difficult to hold and machine certain parts without using lugs. This is because of the design and nature of certain parts. The lugs are an integral part of the casting. They are sometimes removed to avoid interference with the functioning of the part. Lugs are usually represented in phantom outline.

In each corner of the top and front views of the drive housing (drawing A-50) rectangles are outlined with phantom lines to represent the lugs which are molded on the casting. The lugs keep the part in position during machining operations and are later machined off.

Another example of a machining lug is shown in figure 23-1. This lug is used to provide a flat surface on the end of the casting for centering and also to give a uniform center bearing for other machining operations. Both the function of the lug and its appearance determine whether or not it is removed after machining.

SURFACE COATINGS

Machined parts are frequently finished either to protect the surfaces from oxidation or for appearance. The type of finish depends on the use of the part. The finish commonly applied may be a protective coat of paint, lacquer, or a metallic plating. In some cases only a surface finish such as polishing or buffing may be specified.

The finish may be applied before any machining is done, between the various stages of machining, or after the piece has been completed. The type of finish is usually specified on the drawing in a notation similar to the one indicated on the drive housing (drawing A-50) which reads CASTING TO BE PAINTED WITH ALUMINUM BEFORE MACHINING.

The decorative finish on the drive housing should be added before machining so that there will be no accidental deposit of paint on the machined surfaces. This will insure the desired accuracy when assembled with other parts.

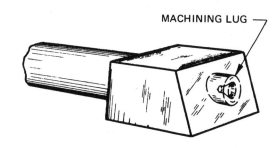

MACHINING LUG

Fig. 23-1 Application of machining lug

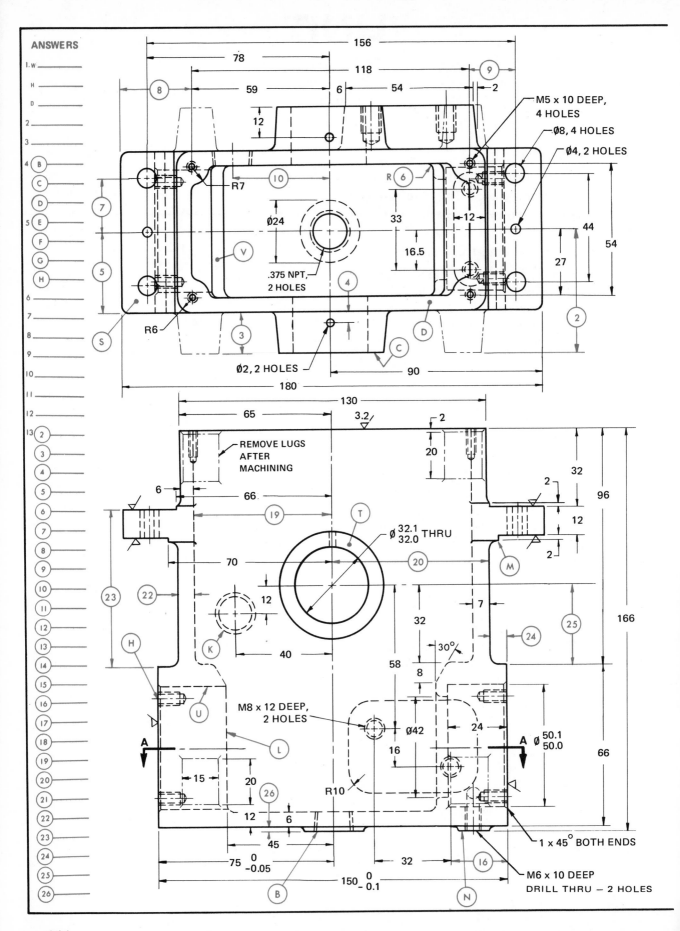

ANSWERS

1. W _____
H _____
D _____
2. _____
3. _____
4. _____
5. _____
6. _____
7. _____
8. _____
9. _____
10. _____
11. _____
12. _____
13. _____

166

Draw section A-A on the graph section provided.

1. What are the overall dimensions of the casting?
2. What size are the radii not dimensioned on the views?
3. From what material is the part made?
4. What finish is required on surfaces (B) (C) (D) ?
5. What is the thread or maximum diameter of holes (E) (F) (G) (H) ?
6. What is the thickness of material at (K) hole?
7. What type of line is used to depict the temporary lugs?
8. What is the tolerance on the 100 mm dimension on the end view?
9. What is the tolerance on hole (G) ?
10. How many countersunk holes are there?
11. At what angle is surface (V) to the horizontal?
12. What is the distance between the Ø2 holes?
13. Determine distances (2) through (26) . When limits are shown, use larger dimension for calculations.

NOTE:

1. DIMENSIONS SYMMETRICAL AROUND CENTER LINES.
2. UNLESS OTHERWISE STATED:
   (A) ±TOLERANCE ON DIMENSIONS
   (B) FINISHED SURFACES $\overset{6.3}{\triangledown}$
   (C) ROUNDS AND FILLETS R3
3. CASTING TO BE PAINTED ALUMINUM BEFORE MACHINING.

DRAW SECTION A-A HERE   (10 mm SQUARES)

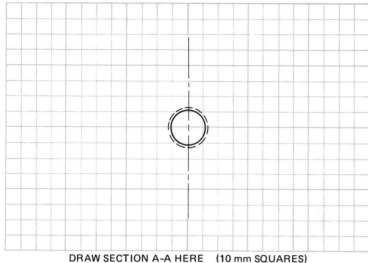

M4 x 8 DEEP, 8 HOLES

66 ±0.05

R6

R6

METRIC	
MATERIAL	MALLEABLE IRON
SCALE	NOT TO SCALE
DRAWN	DATE

DRIVE HOUSING   **A-50**

BEARINGS

All rotating machinery parts are supported by *bearings*. Each bearing type and style has its particular advantages and disadvantages. Bearings are classified into two groups: *plain bearings* and *antifriction bearings*.

Plain Bearings

Plain bearings have many uses. They are available in a variety of shapes and sizes, figure 24-1. Because of their simplicity, plain bearings are versatile. There are several plain bearing categories. The most common are *journal* (sleeve) bearings and *thrust* bearings. These are available in a variety of standard sizes and shapes.

Journal or Sleeve Bearings. Journal bearings are the simplest and most economical means of supporting moving parts. Journal bearings are

usually made of one or two pieces of metal enclosing a shaft. They have no moving parts. The journal is the supporting portion of the shaft.

Speed, mating materials, clearances, temperature, lubrication and type of loading affect the performance of bearings. The maintenance of an oil film between the bearing surfaces is important. The oil film reduces friction, dissipates heat, and retards wear by minimizing metal-to-metal contact, figure 24-2. Starting and stopping are the most critical periods of operation, because the load may cause the bearing surfaces to touch each other.

The shaft should have a smooth finish and be harder than the bearing material. The bearing will perform best with a hard, smooth shaft. For practical reasons, the length of the bearing should be between one and two times the shaft diameter. The outside diameter should be approximately 25 percent larger than the shaft diameter.

Cast bronze and porous bronze are usually used for journal bearings.

SLEEVE

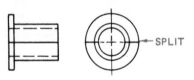

FLANGED

(A) JOURNAL TYPE

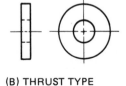

(B) THRUST TYPE

Fig. 24-1 Plain bearings

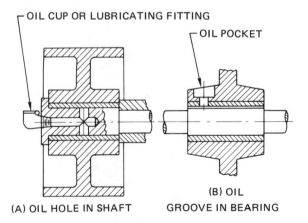

(A) OIL HOLE IN SHAFT   **(B) OIL GROOVE IN BEARING**

Fig. 24-2 Common methods of
lubricating journal bearings

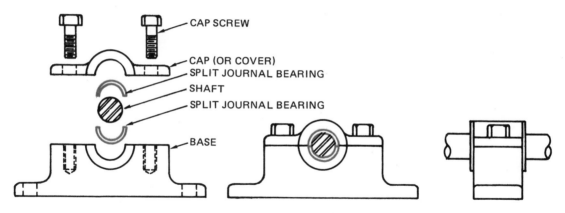

Fig. 24-3 Pillow block with split journal bearing

Bearings are sometimes split. This design feature facilitates assembly and permits ajustment and replacement of worn parts. Split bearings allow the shaft to be set in one half of the bearing while the other half, or cover, is later secured in position, figure 24-3.

If the bearings shown in figure 24-3 are to be made from two parts, they must be fastened together before the hole is bored or reamed. This will facilitate the machining operation and make a perfectly round bearing. For an incorrectly assembled bearing, see figure 24-4.

One method that gives longer life to the bearing is the insertion of very thin strips of metal between the base and cover halves before boring. These thin strips of varying thickness are called *shims.*

When a bearing is shimmed, the same number of pieces of corresponding thickness are used on both sides of the bearing.

As the hole wears, one or more pairs of these shims may be removed for wear compensation.

Thrust Bearings. Plain thrust bearings or thrust washers are available in various materials, including: sintered metal, plastic, woven TFE fabric on steel backing, sintered teflon-bronze-lead on metal backing, aluminum alloy on steel, aluminum alloy, and carbon-graphite.

REFERENCES AND SOURCE MATERIAL
1. A.O. De Hart, "Basic Bearing Types" *Machine Design,* 40, No. 14 (1968).
2. W.A. Glaeser, "Plain Bearings" *Machine Design,* 40, No. 14 (1968).

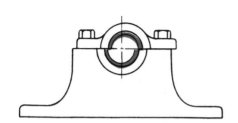

Fig. 24-4 Bearing halves incorrectly matched

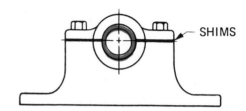

Fig. 24-5 "Shimmed" bearing

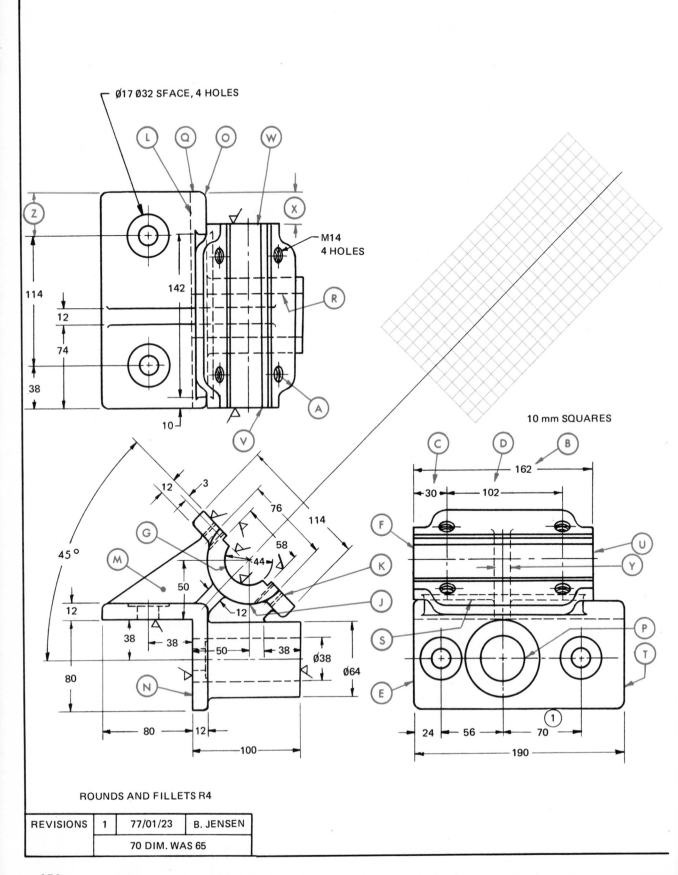

Ø17 Ø32 SFACE, 4 HOLES

M14
4 HOLES

10 mm SQUARES

ROUNDS AND FILLETS R4

REVISIONS	1	77/01/23	B. JENSEN
		70 DIM. WAS 65	

170

1. How many definite finished surfaces are on the casting?
2. What is the size of (P) hole?
3. What are the number and size of bored holes?
4. What are the number and size of the mounting holes?
5. Note that surface (E) is not finished, but surface (F) is to be finished. Allowing 3 mm for finishing, what would be the depth of the rough casting?
6. Would (G) hole be bored before or after bearing cap is assembled?
7. Which line in the side view shows surface (M) ?
8. In which view, and by what line, is the surface represented by (N) shown?
9. Which line or surface in the side view shows the projection of point (J) ?
10. Which point or surface in the side view does line (R) represent?
11. Locate surface (T) in the top view.
12. Locate surface (U) in the top view.
13. Locate (V) in the side view.
14. What are dimensions (X) , (Y) , and (Z) ?
15. What size is the round (O) ?
16. Determine dimension (A) and place it correctly on the sketch of the auxiliary view.
17. Place dimensions (B) , (C) , and (D) correctly on the sketch of the auxiliary view.

MAKE A FREEHAND SKETCH IN THE GRAPH AREA SHOWN AT LEFT, SHOWING THE AUXILIARY VIEW OF THE SPLIT BEARING SURFACES (K) AND (G). SEE QUESTIONS 16 AND 17.

ANSWERS

1 _____
2 _____
3 _____
4 _____
5 _____
6 _____
7 _____
8 _____
9 _____
10 _____
11 _____
12 _____
13 _____
14 (X) _____
   (Y) _____
   (Z) _____
15 _____
16 _____

NOTE: UNLESS OTHERWISE SPECIFIED:
 − TOLERANCE ON DIMENSIONS ±0.5
 − TOLERANCE ON ANGLES ±0.5°
 − ∇ TO BE 1·6∇.

METRIC		
MATERIAL	CI	
SCALE	NOT TO SCALE	
DRAWN		DATE

CORNER BRACKET **A-51**

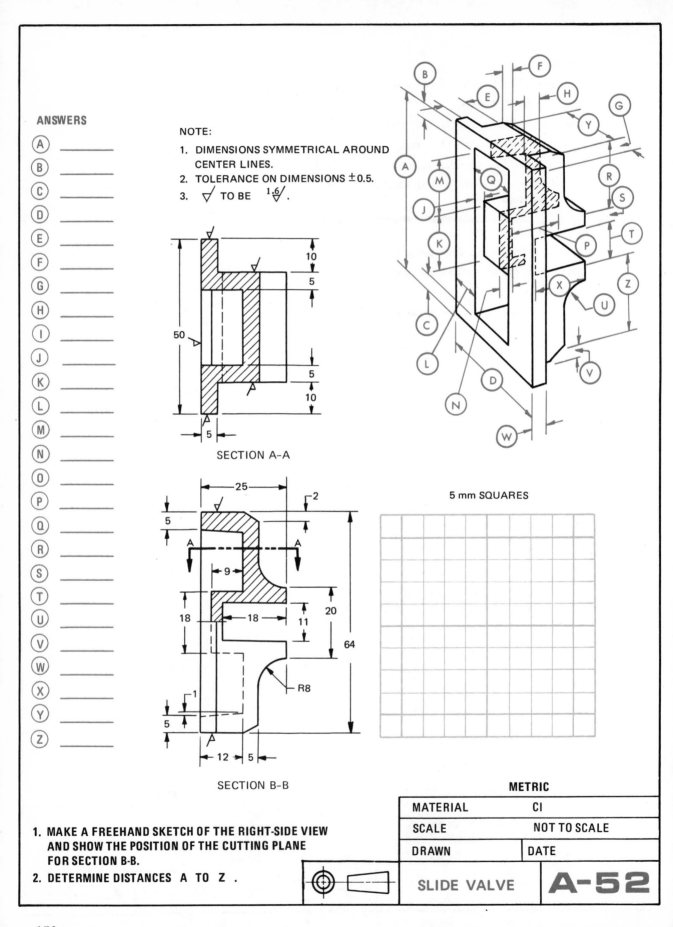

ANSWERS

A _____
B _____
C _____
D _____
E _____
F _____
G _____
H _____
I _____
J _____
K _____
L _____
M _____
N _____
O _____
P _____
Q _____
R _____
S _____
T _____
U _____
V _____
W _____
X _____
Y _____
Z _____

NOTE:

1. DIMENSIONS SYMMETRICAL AROUND CENTER LINES.
2. TOLERANCE ON DIMENSIONS ±0.5.
3. ▽ TO BE 1.6▽.

SECTION A-A

SECTION B-B

5 mm SQUARES

1. MAKE A FREEHAND SKETCH OF THE RIGHT-SIDE VIEW AND SHOW THE POSITION OF THE CUTTING PLANE FOR SECTION B-B.
2. DETERMINE DISTANCES A TO Z.

METRIC		
MATERIAL	CI	
SCALE	NOT TO SCALE	
DRAWN	DATE	
SLIDE VALVE		A-52

UNIT
25

SIMPLIFIED DRAFTING

The challenge of modern industry is to produce more and better goods at competitive prices. Drafting, like all branches of industry, must share in the responsibility for increasing productivity. The traditional concept of drafting is that of producing an elaborate drawing, complete with all the lines, projected views, and sections. This concept is giving way to a simplified method of drafting in some companies. This new simplified method of drafting embraces many modern economical drafting practices while surrendering little in either *clarity* of presentation or *accuracy* of dimensions, figures 25-1 and 25-2.

When many similarly sized holes must be made in a part, the person producing the part may misinterpret a conventional drawing as shown in figure 25-3(A), page 174. To simplify the drawing and reduce the chance of errors, simplified drafting practices such as those shown in figure 25-3(B) may be used.

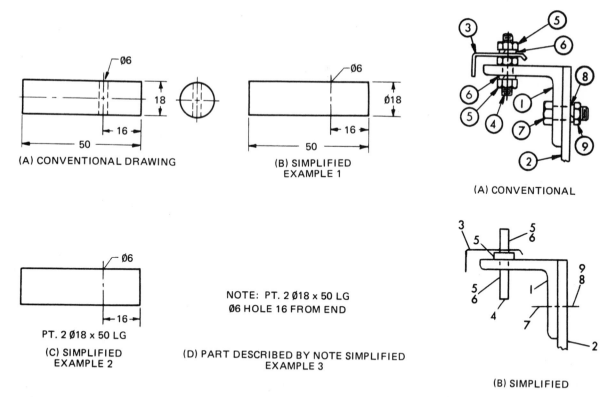

(A) CONVENTIONAL DRAWING

(B) SIMPLIFIED EXAMPLE 1

(A) CONVENTIONAL

NOTE: PT. 2 Ø18 x 50 LG
Ø6 HOLE 16 FROM END

PT. 2 Ø18 x 50 LG

(C) SIMPLIFIED EXAMPLE 2

(D) PART DESCRIBED BY NOTE SIMPLIFIED EXAMPLE 3

(B) SIMPLIFIED

Fig. 25-1 Simplified drafting practices for detailed parts

Fig. 25-2 Simplified drafting practices for assembly drawings

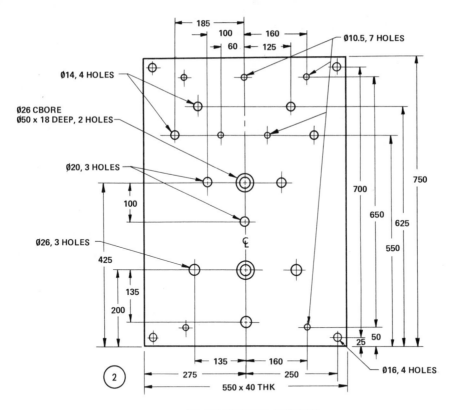

(A) CONVENTIONAL DRAFTING

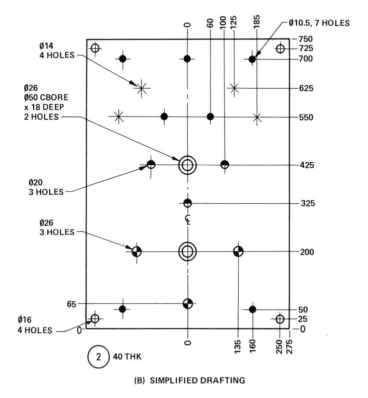

(B) SIMPLIFIED DRAFTING

Fig. 25-3 Comparison between conventional and simplified drawings

Arrowless dimensioning removes many of the lines cluttering the drawing, and hole symbols simplify the location of similarly sized holes.

The use of the abbreviation $\mathcal{C_L}$ indicates that all dimensions are symmetrical around the line indicated.

Hole Symbols

The symbols shown in figure 25-4 are recommended to designate hole sizes on the drawing. The method of producing holes should not be specified on the drawing because the holes should be made to suit factory planning.

On parts such as tubes the drawing should state NEAR SIDE ONLY, FAR SIDE ONLY or THRU HOLES to make sure the factory operator understands the orders.

Fig. 25-4 Recommended hole symbols

Other Timesaving Practices

- Show only partial views of symmetrical objects.

- Avoid the use of elaborate pictorial or repetitive detail.

- Omit detail of nuts, bolt heads, and other hardware.

- Avoid the use of unnecessary hidden lines which do not aid clarification.

- Use description wherever it is practical to eliminate drawings.

- Use symbols instead of words.

- Within limits, a small drawing is made more easily and quickly than a large drawing.

- Eliminate repetitive data by use of general notes.

- Omit part number circles and arrows on leader lines when it will not cause confusion with other data on the drawing.

- Omit center lines except when necessary for processing.

- Eliminate views where the shape can be given by description; for example, HEX, SQ, Ø, ON $\mathcal{C_L}$, THK, etc.

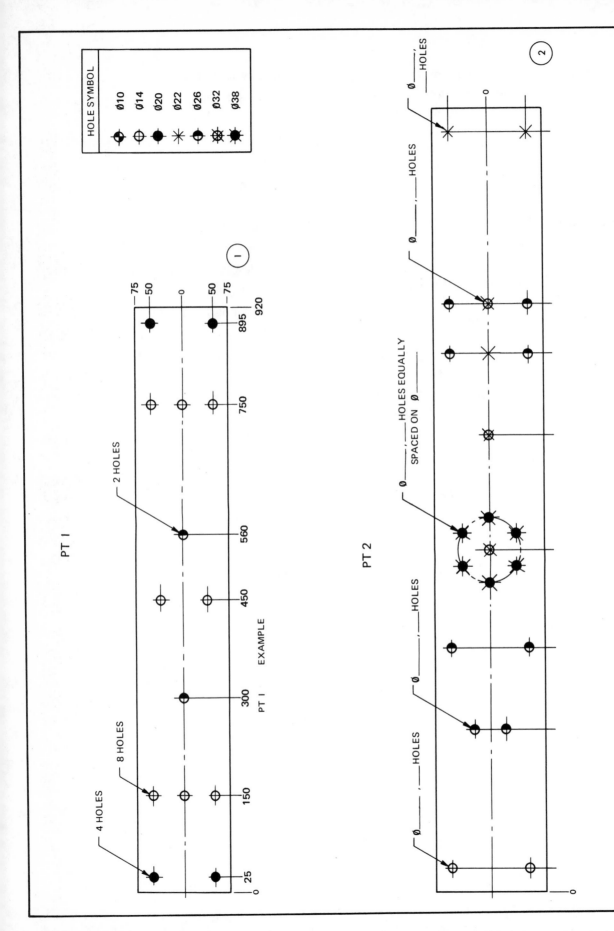

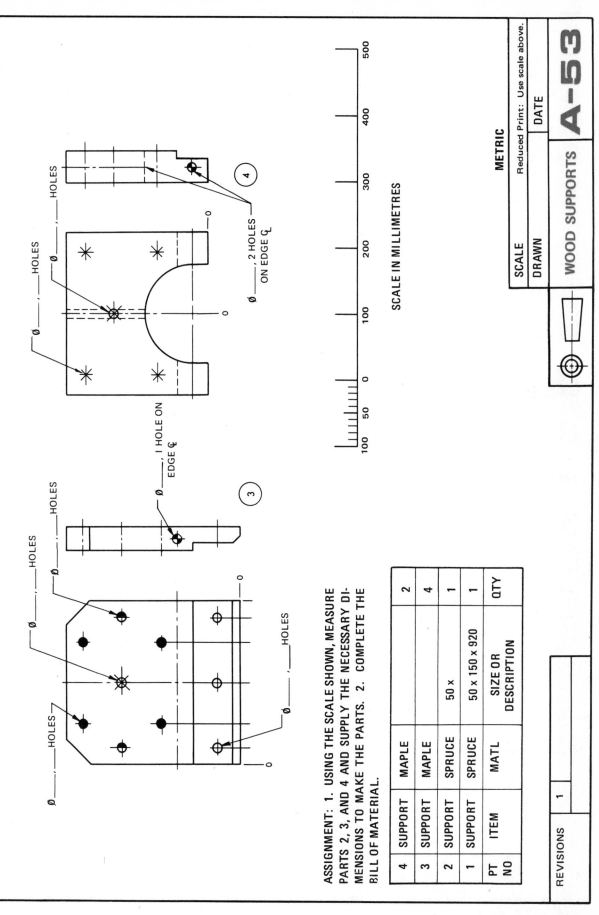

ASSIGNMENT: 1. USING THE SCALE SHOWN, MEASURE PARTS 2, 3, AND 4 AND SUPPLY THE NECESSARY DIMENSIONS TO MAKE THE PARTS. 2. COMPLETE THE BILL OF MATERIAL.

SCALE IN MILLIMETRES

HOLES

Ø____, 2 HOLES
ON EDGE ₵

Ø____, I HOLE ON
EDGE ₵

METRIC

SCALE	Reduced Print: Use scale above.	
DRAWN		DATE
WOOD SUPPORTS		**A-53**

4	SUPPORT	MAPLE		2
3	SUPPORT	MAPLE		4
2	SUPPORT	SPRUCE	50 x	1
1	SUPPORT	SPRUCE	50 x 150 x 920	1
PT NO	ITEM	MATL	SIZE OR DESCRIPTION	QTY

| REVISIONS | 1 | |

177

ASSIGNMENT: IN THE GRAPH AREA PRO-
VIDED, SHOW THE SIMPLIFIED METHOD OF
DETAILING THE PARTS.

GRAPH SIZE – 5 mm SQUARES

CONVENTIONAL

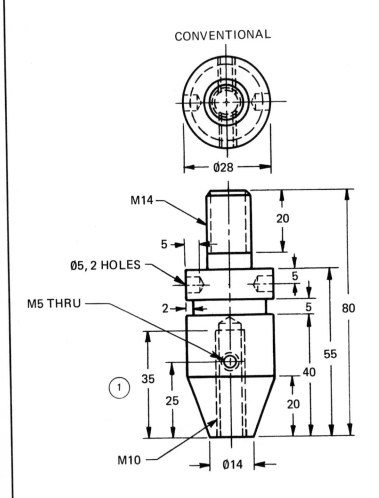

PT I COUPLING
MAT – SAE 1020 – 2 REQD

REVISION	1	JAN. 23/78	P. JENSEN
		35 DIM. WAS 30	

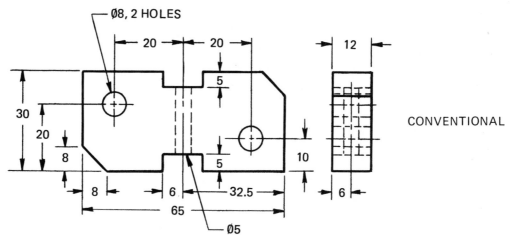

Ø8, 2 HOLES

20 20

5

30

20

8

8 6 32.5

65

Ø5

10

CONVENTIONAL

12

6

PT 2   SUPPORT MATL — SAE 1120 — 4 REQD

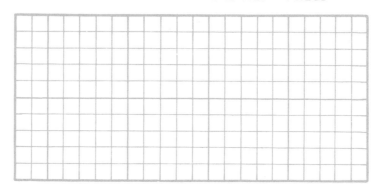

SIMPLIFIED

M10 BOTH ENDS

CONVENTIONAL

Ø10

25 15

75

PT 3   STUD MATL — SAE 1020 — 8 REQD

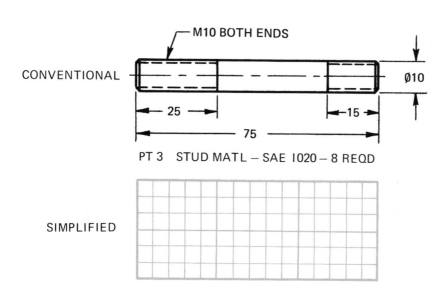

SIMPLIFIED

METRIC	
SCALE	1:1
DRAWN	DATE
PUNCH DETAILS	A-54

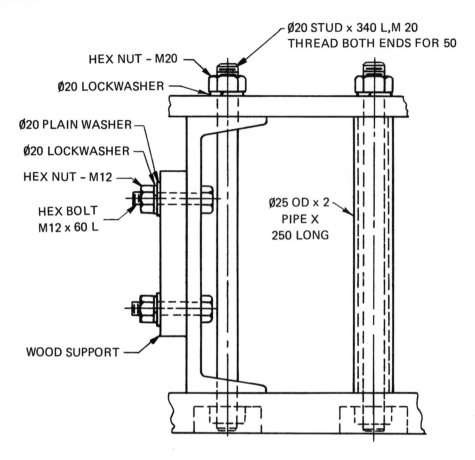

Ø20 STUD x 340 L,M 20
THREAD BOTH ENDS FOR 50

HEX NUT – M20

Ø20 LOCKWASHER

Ø20 PLAIN WASHER

Ø20 LOCKWASHER

HEX NUT – M12

HEX BOLT
M12 x 60 L

Ø25 OD x 2
PIPE X
250 LONG

WOOD SUPPORT

CONVENTIONAL DRAWING
CLAMP ASSEMBLY

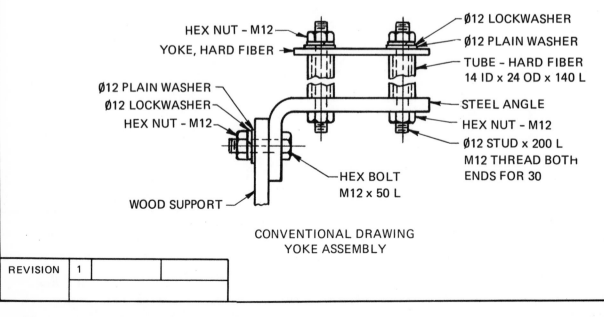

HEX NUT – M12
YOKE, HARD FIBER

Ø12 LOCKWASHER
Ø12 PLAIN WASHER
TUBE – HARD FIBER
14 ID x 24 OD x 140 L
STEEL ANGLE
HEX NUT – M12
Ø12 STUD x 200 L
M12 THREAD BOTH
ENDS FOR 30

Ø12 PLAIN WASHER
Ø12 LOCKWASHER
HEX NUT – M12

HEX BOLT
M12 x 50 L

WOOD SUPPORT

CONVENTIONAL DRAWING
YOKE ASSEMBLY

REVISION	1		

180

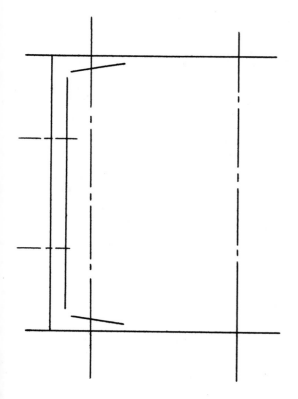

SIMPLIFIED DRAWING
CLAMP ASSEMBLY

ASSIGNMENT: IN THE SPACES PROVIDED,
MAKE SIMPLIFIED ASSEMBLY DRAWINGS OF
THE CONVENTIONAL DRAWINGS SHOWN.

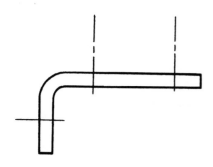

SIMPLIFIED DRAWING
YOKE ASSEMBLY

METRIC		
SCALE	1:1	
DRAWN		DATE
CLAMP AND YOKE ASSEMBLIES	A-55	

181

ARRANGEMENT OF VIEWS

The shape of an object and its complexity influence the possible choices and arrangement of views for that particular object. Since one of the main purposes of making drawings is to furnish the worker with enough information to be able to make the object, only the views which will aid in the interpretation of the drawing should be drawn.

The drafter chooses the view of the object which gives the viewer the clearest idea of the purpose and general contour of the object. He calls this the front view. This choice of the front view may have no relationship to the actual front of the piece when it is used. The *front view* does not have to be the *actual front* of the object itself.

The index pedestal which follows is an object of a shape that is best drawn as the views are shown on drawing A-56. These views could be called front, right-side, and bottom views, as illustrated in figure 26-1. Or, these views could be designated as front, top, and right-side, as shown in figure 26-2. The designation is not of major importance. What is important is that these views, in the opinion of the drafter or designer, give the necessary information in the most understandable way. The views may also be arranged as shown in figure 26-3 or 26-4. While these arrangements would be acceptable, they are not as easily read.

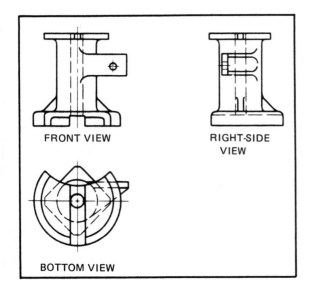

FRONT VIEW　　　RIGHT-SIDE
　　　　　　　　　　VIEW

BOTTOM VIEW

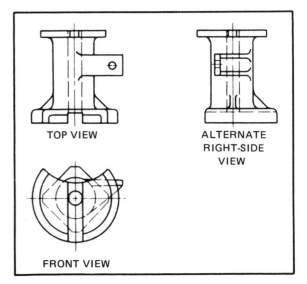

TOP VIEW　　　ALTERNATE
　　　　　　　RIGHT-SIDE
　　　　　　　VIEW

FRONT VIEW

Fig. 26-1 Arrangement A

Fig. 26-2 Arrangement B

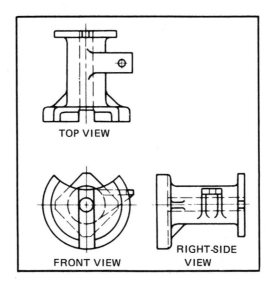

Fig. 26-3 Arrangement C

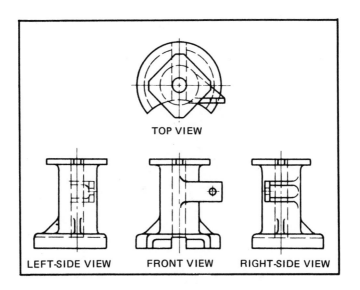

Fig. 26-4 Arrangement D

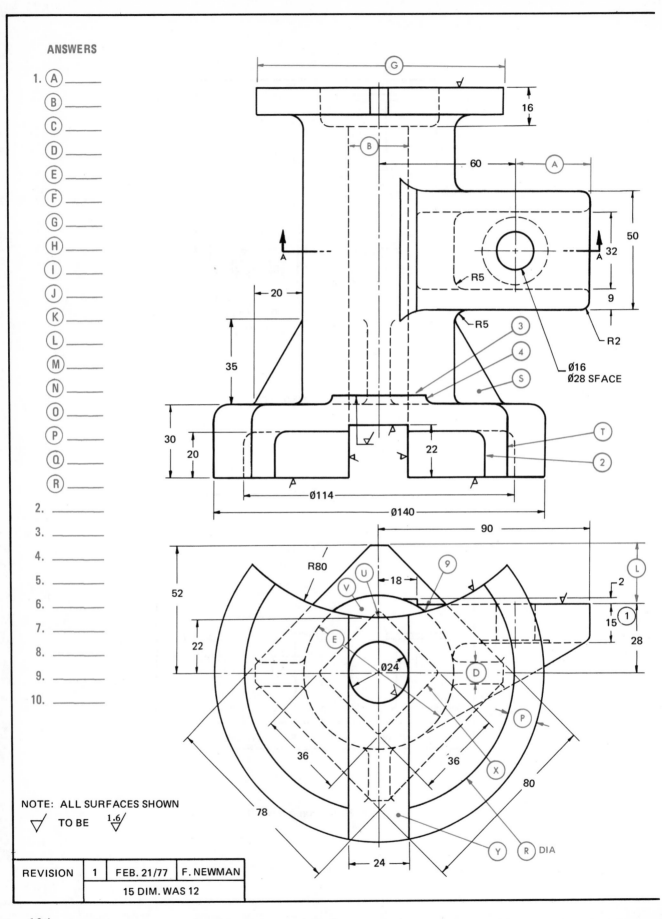

ANSWERS

1. (A) _____
   (B) _____
   (C) _____
   (D) _____
   (E) _____
   (F) _____
   (G) _____
   (H) _____
   (I) _____
   (J) _____
   (K) _____
   (L) _____
   (M) _____
   (N) _____
   (O) _____
   (P) _____
   (Q) _____
   (R) _____
2. _____
3. _____
4. _____
5. _____
6. _____
7. _____
8. _____
9. _____
10. _____

NOTE: ALL SURFACES SHOWN
      ▽ TO BE ¹·⁶▽

REVISION	1	FEB. 21/77	F. NEWMAN
15 DIM. WAS 12			

184

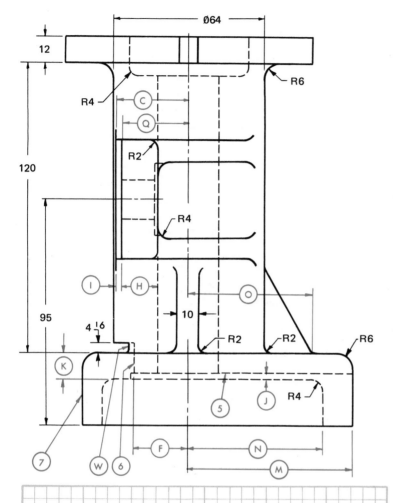

Ø64

12

R6

R4

C

Q

R2

120

R4

I H

O

95

10

4 !6

R2

R2

R6

K

R4

5 J

7 W 6

F

N

M

SKETCH PATTERN FOR RIB(S) BELOW

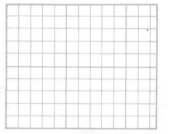

METRIC

MATERIAL		CI
SCALE		
DRAWN BY		DATE

INDEX PEDESTAL

A-56

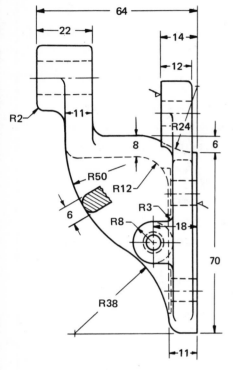

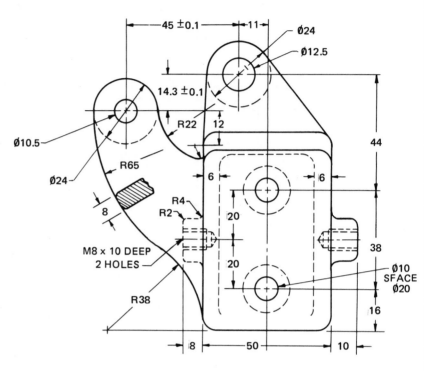

NOTE:

1. ROUNDS AND FILLETS R5 EXCEPT WHERE OTHER-
   WISE NOTED.

2. ∇ TO BE ³·²∇ MACHINING.

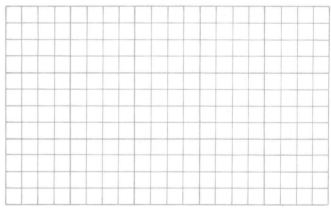

ASSIGNMENT: MAKE A FREEHAND SKETCH ON THE
GRAPH SECTION PROVIDED, SHOWING THE BOTTOM
VIEW OF THE SHAFT INTERMEDIATE SUPPORT. IF
CERTAIN DIMENSIONS CAN BE SHOWN BETTER ON
THE BOTTOM VIEW, DUPLICATE THEM.

METRIC

MATERIAL	CI
SCALE	NOT TO SCALE
DRAWN	DATE

SHAFT INTERMEDIATE SUPPORT	A-57

UNIT

27

ALIGNMENT OF PARTS AND HOLES

Two important factors which must be considered when drawing an object are the number of views to be drawn and the time required to draw them. If possible use time saving devices and less space for making the drawing, being sure to maintain consistency and clarity for easy reading.

To simplify the representation of common features, a number of conventional drawing practices are used, figure 27-1. Many conventions deviate from true projection for the purpose of clarity; others are used to save drafting time. These conventions must be executed carefully; clarity is even more important than speed.

Foreshortened Projection

When the true projection of ribs or arms results in confusing foreshortening, these parts should be rotated until parallel to the line of the section or projection.

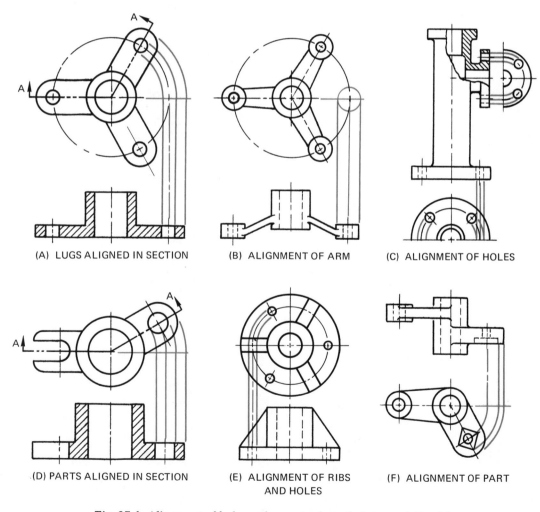

(A) LUGS ALIGNED IN SECTION

(B) ALIGNMENT OF ARM

(C) ALIGNMENT OF HOLES

(D) PARTS ALIGNED IN SECTION

(E) ALIGNMENT OF RIBS AND HOLES

(F) ALIGNMENT OF PART

Fig. 27-1 Alignment of holes and parts to show their true relationship

Holes Revolved to Show True Center Distance

Drilled flanges in elevation or section should show the holes at their true distance from center, rather than the true projection.

NAMING OF VIEWS FOR SPARK ADJUSTER

The drawing of the spark adjuster, drawing A-58, illustrates several violations of true projection. The names of the views of the spark adjuster could be questionable. The importance lies not in the names, however, but in the relationship of the views to each other. This means that the right view must be on the right side of the front, the left view must be on the left side of the front view, etc. Any combination in figure 27-2 may be used for naming the views of the spark adjuster.

DRILL SIZES

Twist drills are the most common tools used in drilling. They are made in many sizes. Twist drills are grouped according to millimetre sizes; by number sizes, from 1 to 80 which

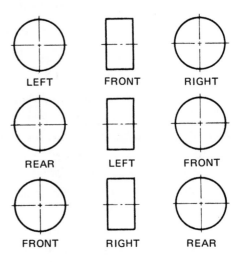

**Fig. 27-2 Naming of views
for spark adjuster, drawing A-58**

correspond to the Stubbs steel wire gauge; by letter sizes A to Z; and by fractional sizes from 1/64th up or their decimal equivalents in inches. Twist drill sizes are listed in tables 4 and 5 of the Appendix.

QUESTIONS

1. What is radius (A) ?

2. What surface is line (2) in the rear view?

3. Locate surface (B) in section A-A.

4. Locate surface (C) in section A-A.

5. Locate line (D) in the rear view.

6. Locate a surface or line in the front view that represents line (F) .

7. Which point or line in the front view does line (G) represent?

8. What size cap screw would be used in (N) hole?

9. What is the diameter of (K) hole?

10. Locate point or line in the rear view from which line (I) is projected.

11. Locate surface (H) in the rear view.

12. Locate (Z) in section A-A.

13. How thick is lug (7) ?

14. What are the diameters of holes (L) , (M) , and (N) ?

15. Determine angles (O) , and (P) .

16. Determine distances (Q) through (U) .

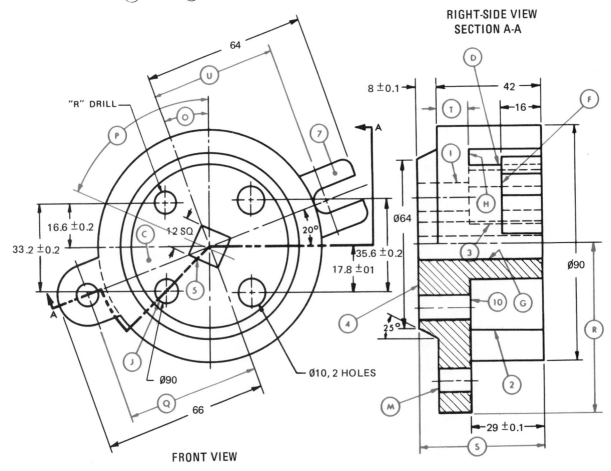

FRONT VIEW

RIGHT-SIDE VIEW
SECTION A-A

ANSWERS

1 _____
2 _____
3 _____
4 _____
5 _____
6 _____
7 _____
8 _____
9 _____
10 _____
11 _____
12 _____
13 _____
14 L _____
   M _____
   N _____
15 O _____
   P _____
16 Q _____
   R _____
   S _____
   T _____
   U _____

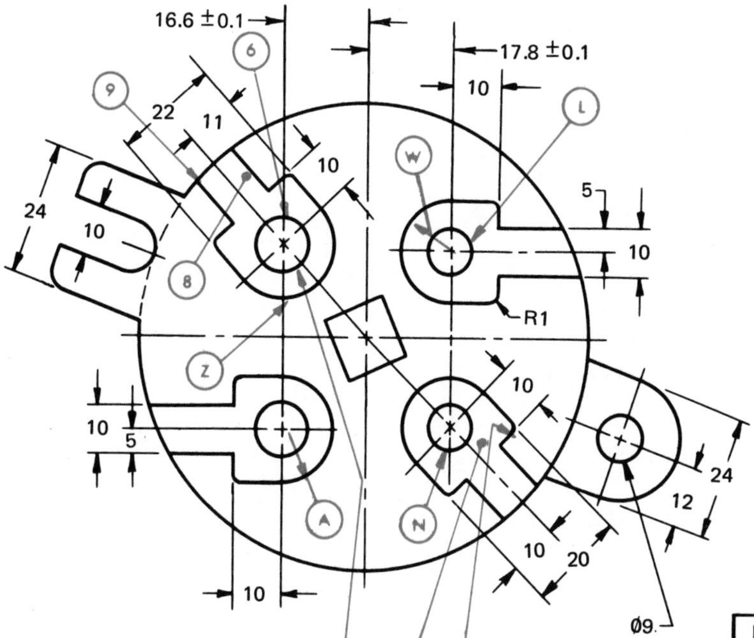

16.6 ±0.1
17.8 ±0.1
10
22
11
10
24
10
5
10
R1
10
5
10
Z
10
10
24
12
10
20
Ø9
10

BACK VIEW

METRIC	
MATERIAL	BAKELITE
SCALE	1:1
DRAWN	DATE

SPARK ADJUSTER A-58

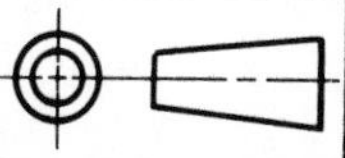

BROKEN-OUT AND PARTIAL SECTIONS

When certain internal and external features of an object should be shown without drawing another view, broken-out and partial sections are used, figure 28-1. A cutting-plane line or a broken line is used to indicate where the section is taken. In the front view of the following assignment, the raise block, drawing A-59, two partial sections are used. Although this method of showing a partial section is unusual, it is an accepted practice.

SECTIONING OF WEBS

The conventional method of representing a section of a part having webs or parti-tions is shown in figure 28-2. This method is preferred to drawing the section in true projection. While the conventional method is a violation of true projection, it is preferred over true projection because of clarity and ease in drawing.

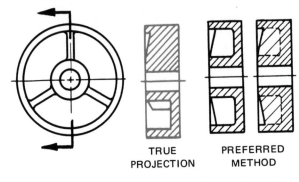

TRUE PROJECTION   PREFERRED METHOD

Fig. 28-2 Conventional methods of sectioning webs

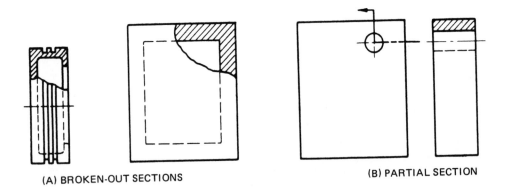

(A) BROKEN-OUT SECTIONS   (B) PARTIAL SECTION

Fig. 28-1 Broken-out and partial sections

1. What is the diameter of the largest unthreaded hole?

2. What size is the smallest threaded hole?

3. What size are the smallest unthreaded holes?

4. Which surface does ⑦ represent in the top view?

5. Which surface does ① represent in the top view?

6. Which line or surface does ⑥ represent in the left view?

7. Which line or surface does Ⓥ represent in the front view?

8. What was the original width of the part?

9. By which line or surface is Ⓗ represented in the left view?

10. Locate in the left view the line or surface that is represented by line Ⓖ .

11. Which line or surface represents Ⓕ in the top view?

12. Which line represents surface Ⓙ in the left view?

13. Determine the overall depth of the raise block.

14. Determine distances Ⓐ through Ⓔ .

15. Determine distances ⑧ through ⑲ .

NOTE: UNLESS OTHERWISE STATED
- TOLERANCE ON DIMENSIONS ±0.5
- TOLERANCE ON ANGLES ±0.5°
- ROUNDS AND FILLETS R3

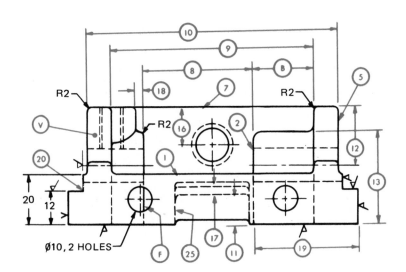

Ø10, 2 HOLES

REVISIONS	1	MAR. 4/77	R. KERR
		166 DIM. WAS 170	

Dowel Pins. Dowel pins or small straight pins have many uses. They are used to hold parts in alignment and to guide parts into desired positions. Dowel pins are most commonly used for the alignment of parts which are fastened with screws or bolts and must be accurately fitted together.

When two pieces are to be fitted together, as in the case of the part in figure 29-2, one method of alignment is to clamp the two pieces in the desired location; drill and ream the dowel holes; insert the dowel pins; and then drill and tap for the screw holes.

Drill jigs are frequently used when production in the interchangeability of parts is required or when the nature of the piece does not permit the transfer of the doweled holes from one piece to the other. A drill jig, figure 29-3, was used in drilling the dowel holes for the spider (drawing A-67).

Taper Pins. Holes for taper pins are usually sized by reaming. A through hole is formed by step drills and straight fluted reamers. The present trend is toward the use of helically fluted taper reamers which provide more accurate sizing and require only a pilot hole the

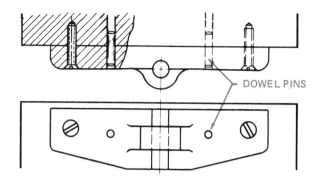

Fig. 29-2 Aligning parts wth dowel pins

size of the small end of the taper pin. The pin is usually driven into the hole until it is fully seated. The taper of the pin aids hole alignment in assembly.

A tapered hole in a hub and a shaft is shown in figure 29-4 (page 198). If the hub and shaft are drilled and reamed separately, a misalignment might occur as shown in figure 29-5 (page 198). To prevent misalignment, the hub and shaft should be drilled at the same time as the parts are assembled. Each of the detailed parts should carry a note similar to the following: DRILL AND REAM FOR NO. 1 TAPER PIN AT ASSEMBLY.

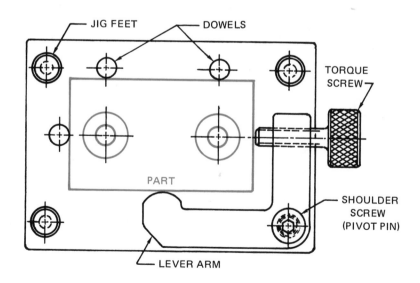

Fig. 29-3 Dowel pins used to align part during drillings

Cotter Pins.   The cotter pin is a standard machine pin commonly used as a fastener in the assembly of machine parts where great accuracy is not required, figure 29-6. There is no standard way to represent cotter pins in assembly drawings. The method of representation shown in figure 29-7 is, however, commonly used to indicate cotter pins.

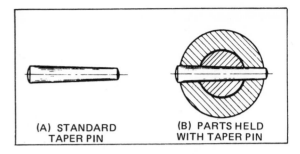

Fig. 29-4 Taper pin application

Fig. 29-5 Possibility of hole misalignment if holes are not drilled at assembly

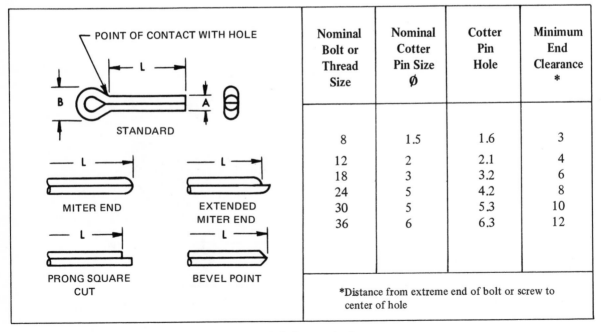

Nominal Bolt or Thread Size	Nominal Cotter Pin Size Ø	Cotter Pin Hole	Minimum End Clearance *
8	1.5	1.6	3
12	2	2.1	4
18	3	3.2	6
24	5	4.2	8
30	5	5.3	10
36	6	6.3	12
*Distance from extreme end of bolt or screw to center of hole			

Fig. 29-6 Cotter pin data

Radial-Locking Pins

Low cost, ease of assembly, and high resistance to vibration and impact loads are common attributes of this group of commercial pin devices designed primarily for semi-permanent fastening service. Two basic pin forms are used: solid with grooved surfaces, and hollow spring pins which may be either slotted or spiral wrapped, figure 29-8 (page 200). In assembly, radial forces produced by elastic action at the pin surface develop a secure, frictional-locking grip against the hole wall. These pins are reusable and can be removed and reassembled many times without appreciable loss of fastening effectiveness. Live spring action at the pin surface also prevents loosening under shock and vibration loads. The need for accurate sizing of holes is reduced since the pins accommodate variations.

Solid Pins with Grooved Surfaces. The locking action of groove pins is provided by parallel, longitudinal grooves uniformly spaced around the pin surface. Rolled or pressed into solid pin stock, the grooves expand the effective diameter of the pin. When the pin is driven into a drilled hole corresponding in size to the nominal pin diameter, elastic deformation of the raised groove edges produces a secure interference fit with the hole

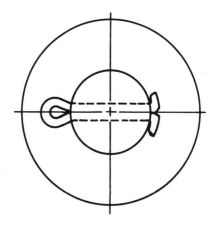

Fig. 29-7 Typical representation of a cotter pin in an assembly drawing

wall. Figure 29-8 shows the six standardized constructions of grooved pins.

Hollow Spring Pins. Spiral-wrapped and slotted-tubular pin forms are made to controlled diameters greater than the holes into which they are pressed. Compressed when driven into the hole, the pins exert spring pressure against the hole wall along their entire engaged length to develop a strong locking action.

REFERENCES AND SOURCE MATERIALS

1. F.W. Braendel, "Pin Fasteners," *Machine Design,* 39, (1967).

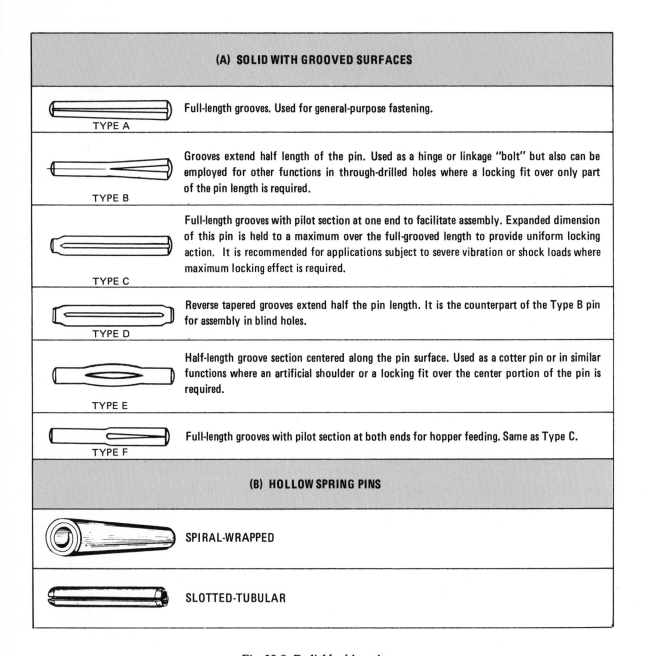

(A) SOLID WITH GROOVED SURFACES

TYPE A

Full-length grooves. Used for general-purpose fastening.

TYPE B

Grooves extend half length of the pin. Used as a hinge or linkage "bolt" but also can be employed for other functions in through-drilled holes where a locking fit over only part of the pin length is required.

TYPE C

Full-length grooves with pilot section at one end to facilitate assembly. Expanded dimension of this pin is held to a maximum over the full-grooved length to provide uniform locking action. It is recommended for applications subject to severe vibration or shock loads where maximum locking effect is required.

TYPE D

Reverse tapered grooves extend half the pin length. It is the counterpart of the Type B pin for assembly in blind holes.

TYPE E

Half-length groove section centered along the pin surface. Used as a cotter pin or in similar functions where an artificial shoulder or a locking fit over the center portion of the pin is required.

TYPE F

Full-length grooves with pilot section at both ends for hopper feeding. Same as Type C.

(B) HOLLOW SPRING PINS

SPIRAL-WRAPPED

SLOTTED-TUBULAR

Fig. 29-8 Radial locking pins

1. Which views are shown?

2. How many surfaces are to be finished?

3. How many scraped surfaces are indicated?

4. How many holes are to be tapped?

5. What is the purpose of tapped hole ⓇR ?

6. Which surface is ③3 in the left-side view?

7. Which surface is ②2 in the left-side view?

8. Which surface is ④4 in the front view?

9. Which surface is ⑭14 in the left-side view?

10. Which surface in the top view and front view is ⑨9 ?

11. Which surface in the top view and front view is ⑧8 ?

12. Which surface is ⑫12 in the top view?

13. What is the name of part ⓋV ?

14. What is the purpose of part ⓋV ?

15. What do dotted lines at ⓌW represent?

16. Which surface is line ⑥6 in the front view?

17. Which top view line indicates point ⓏZ ?

18. What is the depth of the tapped hole at ⓍX ?

19. Which edges or surfaces in the left and front views does line ⓉT represent?

20. What is the diameter of tap drill ⓎY ? (See Appendix.)

21. Determine dimensions or operations at ⓐA to ⓆQ ; 20 to 56

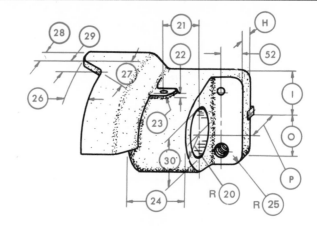

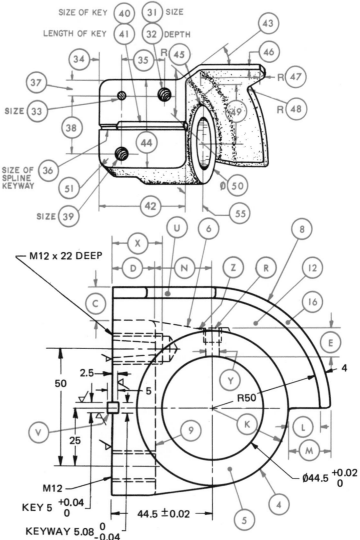

NOTE: UNLESS OTHERWISE SPECIFIED
—TOLERANCE ON DIMENSIONS ±0.5.
—TOLERANCE ON ANGLES ±0.5°.

REVISIONS | 1

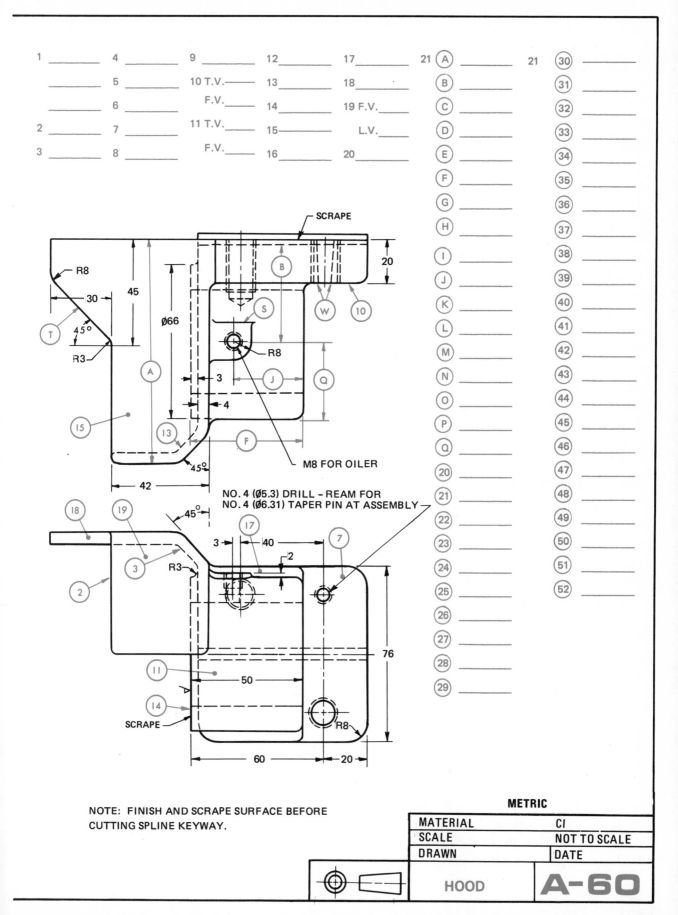

1 _____   4 _____   9 _____   12 _____   17 _____   21 Ⓐ _____   21 ㉚ _____

_____   5 _____   10 T.V. _____   13 _____   18 _____   Ⓑ _____   ㉛ _____

_____   6 _____   F.V. _____   14 _____   19 F.V. _____   Ⓒ _____   ㉜ _____

2 _____   7 _____   11 T.V. _____   15 _____   L.V. _____   Ⓓ _____   ㉝ _____

3 _____   8 _____   F.V. _____   16 _____   20 _____   Ⓔ _____   ㉞ _____

Ⓕ _____   ㉟ _____

Ⓖ _____   ㊱ _____

Ⓗ _____   ㊲ _____

Ⓘ _____   ㊳ _____

Ⓙ _____   ㊴ _____

Ⓚ _____   ㊵ _____

Ⓛ _____   ㊶ _____

Ⓜ _____   ㊷ _____

Ⓝ _____   ㊸ _____

Ⓞ _____   ㊹ _____

Ⓟ _____   ㊺ _____

Ⓠ _____   ㊻ _____

⑳ _____   ㊼ _____

㉑ _____   ㊽ _____

㉒ _____   ㊾ _____

㉓ _____   ㊿ _____

㉔ _____   51 _____

㉕ _____   52 _____

㉖ _____

㉗ _____

㉘ _____

㉙ _____

SCRAPE

R8

30

45

Ø66

Ⓣ

45°

R3

Ⓐ

⑮

⑬

42

45°

Ⓑ

Ⓢ

Ⓦ ⑩

20

R8

3

Ⓙ

4

Ⓠ

Ⓕ

M8 FOR OILER

NO. 4 (Ø5.3) DRILL – REAM FOR
NO. 4 (Ø6.31) TAPER PIN AT ASSEMBLY

⑱ ⑲

45°

⑰

3 → ← 40

⑦

② ③ R3

2

⑪ 50

⑭

SCRAPE

60 20

76

R8

NOTE: FINISH AND SCRAPE SURFACE BEFORE
CUTTING SPLINE KEYWAY.

METRIC	
MATERIAL	CI
SCALE	NOT TO SCALE
DRAWN	DATE

HOOD

A-60

203

UNIT

30

POINT-TO-POINT DIMENSIONING

Most linear dimensions are intended to apply on a point-to-point basis. *Point-to-point* dimensions are applied directly from one feature to another, as shown in figure 30-1(A). Such dimensions locate surfaces and features directly between the points indicated, or between corresponding points on the indicated surfaces.

For example, a diameter applies to all diameters of a cylindrical surface (not merely to the diameter at the end where the dimension is shown), a thickness applies to all opposing points on the surfaces, and a hole-locating dimension applies from the hole axis perpendicular to the edge of the part on the same center line.

DATUM DIMENSIONING

When several dimensions extend from a common data point or points on a line or surface, figure 30-1 (B), this is called *datum,* or base line, dimensioning. This form of dimensioning is preferred for parts to be manufactured by numerical control machines.

A *datum* is a feature (point, line, or surface) from which other features are located. A datum may be the center of a circle, the axis of a cylinder, or the axis of symmetry.

The location of a series of holes or step features as shown in figure 30-1(B) is an example of the use of datum line dimensioning. Figures 30-2 and 30-3 illustrate more complex uses of datum dimensioning. Without this type of dimensioning the distances between the first and last steps could vary considerably because of the buildup of tolerances permitted between the adjacent holes. Dimensioning from a datum line controls the

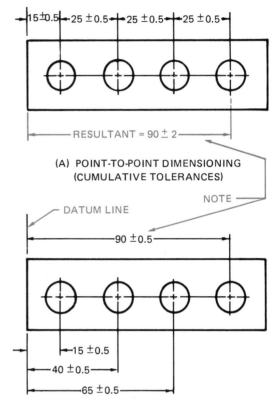

(A) POINT-TO-POINT DIMENSIONING
(CUMULATIVE TOLERANCES)

(B) DATUM LINE DIMENSIONING
(NON-CUMULATIVE TOLERANCES)

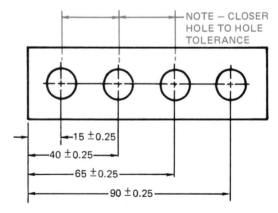

(C) MAINTAINING SAME DIMENSION BETWEEN HOLES AS (A) OR (B) BUT WITH CLOSER TOLERANCE

Fig. 30-1 Comparison between point-to-point and datum dimensions

tolerances between the holes to the basic general tolerance of ±0.5 mm.

In this system, the tolerance from the common point to each of the features must be held to half the tolerance acceptable between individual features. For example, in figure 30-1 (C), if a tolerance between two individual holes of ± 0.5 mm were desired, each of the dimensions shown would have to be held to ± 0.25 mm.

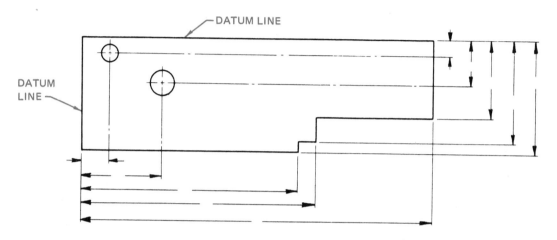

Fig. 30-2 Application of datum dimensioning

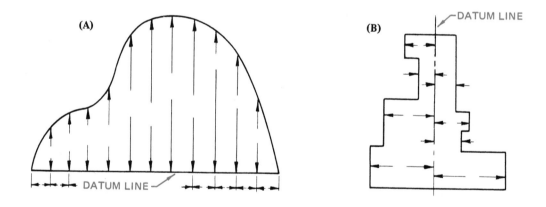

Fig. 30-3 Applications of datum dimensioning

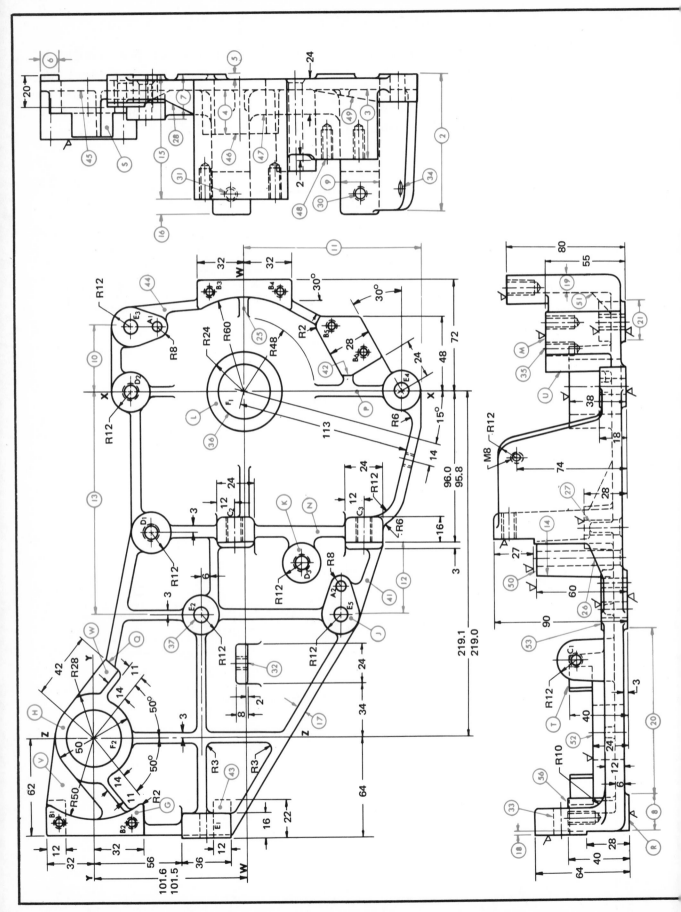

QUESTIONS

NOTE: Where limit dimensions are given use larger limit.

1. What are the overall dimensions of the casting?
2. Show a metric roughness symbol other than the one given on the drawing.
3. Identify surfaces (G) to (U) on one of the other views.
4. Locate rib (25) on the front view.
5. Locate rib (26) on the top view.
6. Locate rib (27) on the side view.
7. Locate rib (28) on the top view.
8. Give the sizes of the following holes: (30), (31), (32), (33), and (34).

9. How deep is hole (35)?
10. What is the tolerance on hole (36)?
11. How deep is hole (37)?
12. Determine distances (2) to (21). (Where limit dimensions are shown, use maximum size.)

ANSWERS

1. W _____
   D _____
   H _____
2. _____
3. (G) _____
   (H) _____
   (J) _____
   (K) _____
   (L) _____
   (M) _____
   (N) _____
   (P) _____
   (Q) _____
   (R) _____
   (S) _____
   (T) _____
   (U) _____
4. _____
5. _____
6. _____
7. _____

8. (30) _____
   (31) _____
   (32) _____
   (33) _____
   (34) _____
9. _____
10. _____
11. _____
12. (2) _____
    (4) _____
    (6) _____
    (8) _____
    (10) _____
    (12) _____
    (14) _____
    (16) _____
    (18) _____
    (20) _____

    (3) _____
    (5) _____
    (7) _____
    (9) _____
    (11) _____
    (13) _____
    (15) _____
    (17) _____
    (19) _____
    (21) _____

HOLE SIZE AND LOCATION

HOLE	DISTANCE FROM					SIZE
	V-V	W-W	X-X	Y-Y	Z-Z	
A_1		59	41			⌀ 6.2 / 6.0
A_2		64	124			
B_1				24	54	
B_2				24	54	
B_3		24	64			M8 x
B_4		24	64			20 DEEP
B_5		58	41			
B_6		78	24			
C_1	35				46	
C_2	76	70				M10
C_3	76	80				
D_1		64	90			⌀ 9.53 / 9.52
D_2		76	0			
D_3		38	108			CSK 2 x 45°
E_1	46			75		
E_2		30	141			
E_3		76	41			⌀ 10
E_4		104	0			
E_5		64	141			
F_1		0	0			⌀ 34.96 / 34.90
F_2				0	0	

NOTE:
- ALL FILLETS R2 UNLESS OTHERWISE SHOWN.
- ALL RIBS AND WEBS 6 THICK.
- TOLERANCE ON DIMENSIONS ±0.5.
- TOLERANCE ON ANGLES ±0.5°.

SURFACES MARKED TO BE 6.3

METRIC

MATERIAL	CI
SCALE	NOT TO SCALE
DRAWN	DATE

 INTERLOCK BASE **A-61**

ASSIGNMENT: DETERMINE DISTANCES OR DIMENSIONS OF THE FOLLOWING:

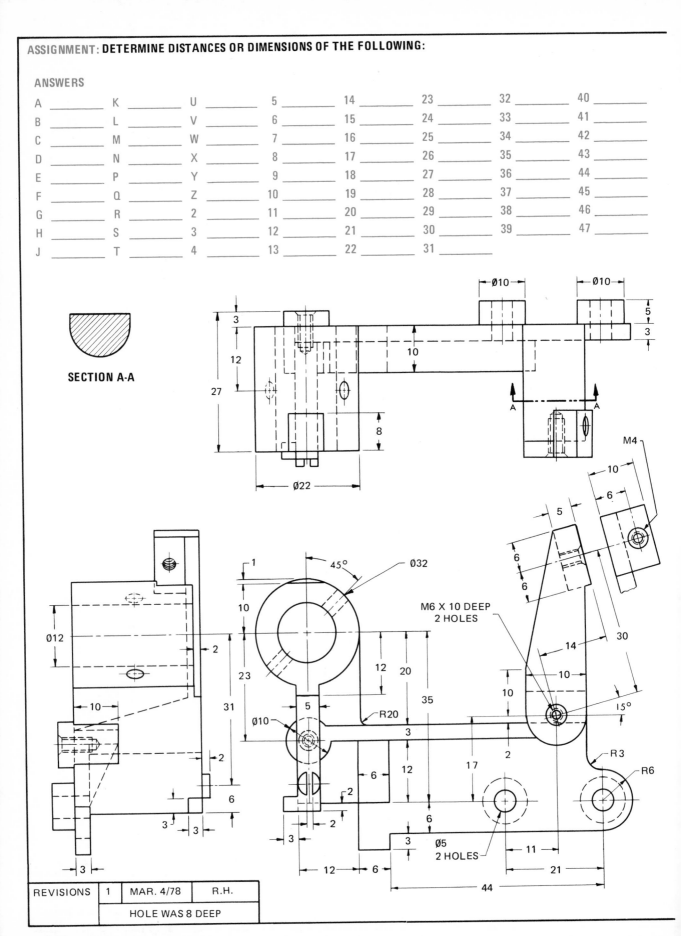

ANSWERS

A ___ K ___ U ___ 5 ___ 14 ___ 23 ___ 32 ___ 40 ___
B ___ L ___ V ___ 6 ___ 15 ___ 24 ___ 33 ___ 41 ___
C ___ M ___ W ___ 7 ___ 16 ___ 25 ___ 34 ___ 42 ___
D ___ N ___ X ___ 8 ___ 17 ___ 26 ___ 35 ___ 43 ___
E ___ P ___ Y ___ 9 ___ 18 ___ 27 ___ 36 ___ 44 ___
F ___ Q ___ Z ___ 10 ___ 19 ___ 28 ___ 37 ___ 45 ___
G ___ R ___ 2 ___ 11 ___ 20 ___ 29 ___ 38 ___ 46 ___
H ___ S ___ 3 ___ 12 ___ 21 ___ 30 ___ 39 ___ 47 ___
J ___ T ___ 4 ___ 13 ___ 22 ___ 31 ___

SECTION A-A

REVISIONS 1 MAR. 4/78 R.H.

HOLE WAS 8 DEEP

208

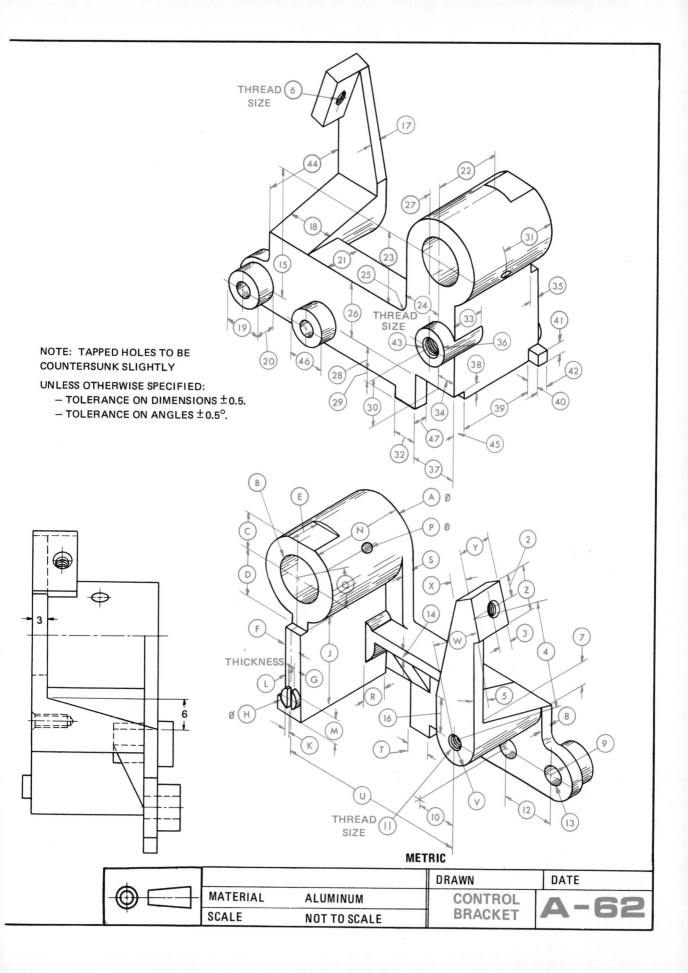

THREAD SIZE ⑥

⑰ ⑪ ㉒ ㉗ ㉛ ⑭ ㉔ ㉓ ㉕ ㉖ ㉑ ⑮ ⑲ ⑳ ㊻ ㉘ ㉙ ㉚ ㉜ ㊼ ㊲ ㉞ ㉟ ㊺ ㊵ ㊷ ㊶ ㉟ ㊳ ㊴ ㉝ ㊱ ㊸ ㉟ ㉟

THREAD SIZE

NOTE: TAPPED HOLES TO BE
COUNTERSUNK SLIGHTLY

UNLESS OTHERWISE SPECIFIED:
   – TOLERANCE ON DIMENSIONS ±0.5.
   – TOLERANCE ON ANGLES ±0.5°.

B E A Ø
C N P Ø
D S Y ②
F X ⑭ Z
J W ③ ⑦
L G R ④ ⑤ ⑧ ⑨
Ø H M ⑯ T
K U V ⑩ ⑫ ⑬ ⑪

THICKNESS

3

6

THREAD SIZE ⑪

METRIC

	DRAWN	DATE
MATERIAL	ALUMINUM	CONTROL BRACKET
SCALE	NOT TO SCALE	A-62

UNIT

31

ASSEMBLY DRAWINGS

The term *assembly drawing* refers to the type of drawing in which the various parts of a machine or structure are drawn in their relative positions in the completed unit.

In addition to showing how the parts fit together, the assembly drawing is used mainly:

- To represent the proper working relationships of the mating parts of a machine or structure and the function of each.

- To give a general idea of how the finished product should look.

- To aid in securing overall dimensions and center distances in assembly.

- To give the detailer data needed to design the smaller units of a larger assembly.

- To supply illustrations which may be used for catalogs, maintenance manuals, or other illustrative purposes.

In order to show the working relationship of interior parts, the principles of projection may be violated and details omitted for clarity. Assembly drawings should not be overly detailed because precise information describing part shapes is provided on detail drawings.

Detail dimensions which would confuse the assembly drawing should be omitted. Only such dimensions as center distances, overall dimensions, and dimensions showing the relationship of the parts as they apply to the mechanism as a whole should be included. There are times when a simple assembly drawing may be dimensioned so that no other detail drawings are needed. In such a case the assembly drawing becomes a working assembly drawing.

Sectioning is used more extensively on assembly drawings than on detail drawings. The conventional method of section lining is used on assembly drawings to show the relationship of the various parts, figure 31-1.

Subassembly Drawings

Subassembly drawings are often made of smaller mechanical units. When combined in final assembly, they make a single machine. For a lathe, subassembly drawings would be furnished for the headstock, the apron, and other units of the carriage. These units might be machined and assembled in different departments by following the subassembly drawings. The individual units would later be combined in final assembly according to the assembly drawing.

Identifying Parts of an Assembly Drawing

When a machine is designed, an assembly drawing or design layout is first drawn to visualize clearly the method of operation, shape, and clearances of the various parts. From this assembly drawing, the detail drawings are made and each part is given a part number.

To assist in the assembly of the machine, item numbers corresponding with the part numbers of various details are placed on the assembly drawing. Small circles, 8 mm to 12 mm in diameter, that contain the part number are then attached to the corresponding part with a leader.

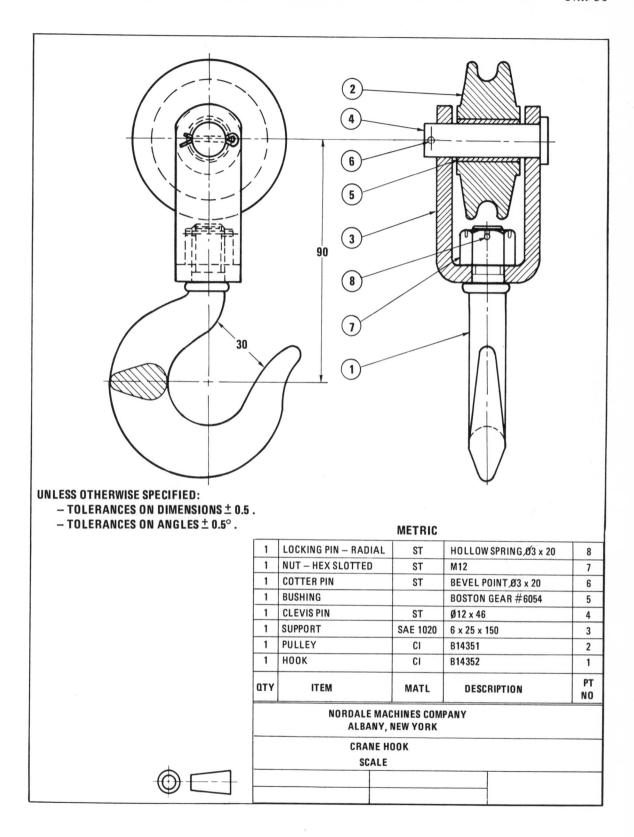

UNLESS OTHERWISE SPECIFIED:
  – TOLERANCES ON DIMENSIONS ± 0.5 .
  – TOLERANCES ON ANGLES ± 0.5° .

METRIC

QTY	ITEM	MATL	DESCRIPTION	PT NO
1	LOCKING PIN – RADIAL	ST	HOLLOW SPRING Ø3 x 20	8
1	NUT – HEX SLOTTED	ST	M12	7
1	COTTER PIN	ST	BEVEL POINT Ø3 x 20	6
1	BUSHING		BOSTON GEAR #6054	5
1	CLEVIS PIN	ST	Ø12 x 46	4
1	SUPPORT	SAE 1020	6 x 25 x 150	3
1	PULLEY	CI	B14351	2
1	HOOK	CI	B14352	1

NORDALE MACHINES COMPANY
ALBANY, NEW YORK

CRANE HOOK
SCALE

Fig. 31-1 A typical assembly drawing

BILL OF MATERIAL

A *bill of material* is an itemized list of all the components shown on an assembly drawing or a detail drawing, figure 31-2. Often, a bill of material is placed on a separate sheet of paper for handling and duplicating. For castings, a pattern number would appear in the size column instead of the physical size of the part.

Standard components, which are purchased rather than fabricated, including bolts, nuts, and bearings, should have a part number and appear on the bill of material. There should be sufficient information in the descriptive column to enable the purchasing agent to order these parts.

Standard components are incorporated in the design of machine parts for economical production. These parts are specified on the drawing according to the manufacturer's specification. The use of manufacturers' catalogs is essential for determining detailing standards, characteristics of a special part, methods of representation, etc. However, it should be pointed out that manufacturer's catalogs are very unreliable for specifying parts; they should be used as a guide only. To protect the integrity of a design, a purchase part drawing must be made. This overcomes the frequent problem whereby the component supplier makes changes, unknown to the user, which frequently affect the design.

The four-wheel trolly (drawing A-69) includes many standard parts. Grease cups, lock washers, Hyatt roller bearings, rivets, and nuts, all of which are standard purchased items, are used. These parts are not detailed but are described in the bill of material. However, the special countersunk head bolts and the taper washers, commonly called *Dutchmen,* are not standard parts and must therefore be made especially for this particular assembly.

SECTION THROUGH SHAFTS, PINS, AND KEYS

Shafts, bolts, nuts, rods, rivets, keys, pins, and similar solid parts, the axes of which lie on the cutting plane, are sectioned only when a broken-out section of the shaft is used to clearly indicate the key, keyseat, or pin, figure 31-3.

SWIVELS AND UNIVERSAL JOINTS

A *swivel* is composed of two or more pieces constructed so that each part rotates in relation to the other around a common axis, figure 31-4(A).

A *universal joint* is composed of three or more pieces designed to permit the free rotation of two shafts whose axes deviate from a straight line, figure 31-4(B).

In practice, universal joints are constructed in many different forms. The design of the universal joints is influenced by the requirements of the working mechanism and the cost of the parts.

Universal joints are used by the designer where a rotating or swinging motion is wanted or where power must be delivered along shafts which are not in a straight line.

4	NUT-HEX-REG	ST	MIO	7
4	BOLT-HEX-REG	ST	MIO x 1.5 x 40	6
1	KEY	MS	WOODRUFF 608	5
2	BEARINGS	SKF	RADIAL BALL 620	4
1	SHAFT	CRS	Ø 25 x 160	3
1	SUPPORT	MS	10 x 50 x 110	2
1	BASE	CI	PATTERN – A3154	1
QTY	ITEM	MATL	DESCRIPTION	PT NO

Fig. 31-2 Bill of material for assembly drawing

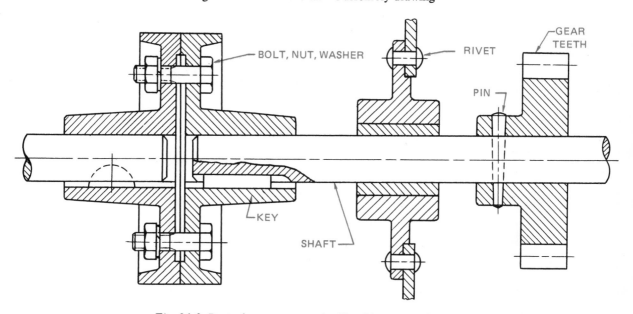

Fig. 31-3 Parts that are not section lined in section drawings

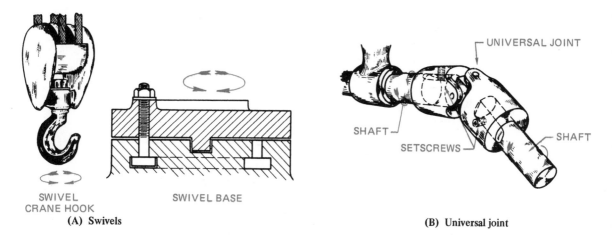

(A) Swivels

(B) Universal joint

Fig. 31-4 Swivels and universal joints

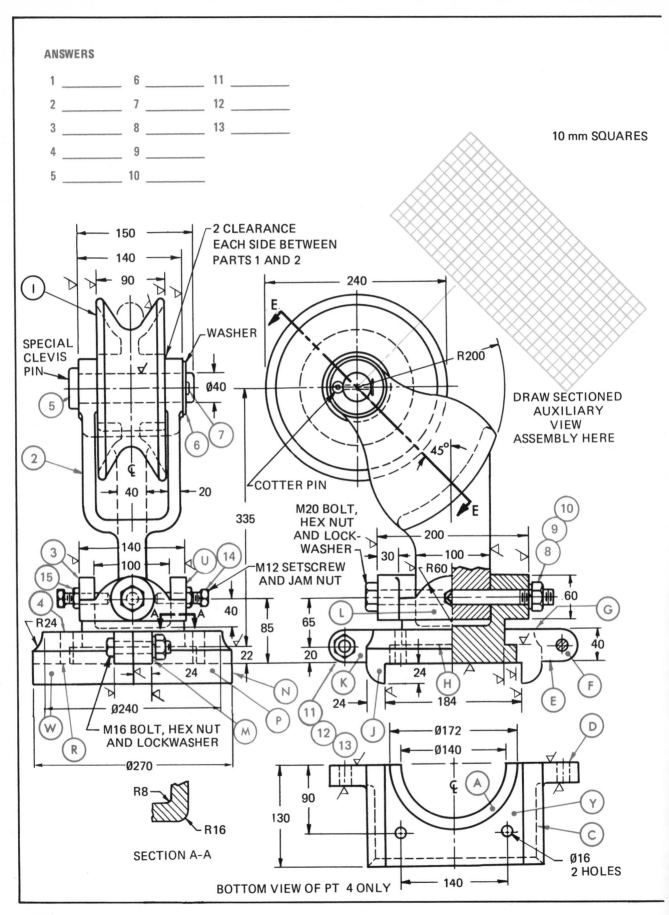

ANSWERS

1 _____   6 _____   11 _____
2 _____   7 _____   12 _____
3 _____   8 _____   13 _____
4 _____   9 _____
5 _____   10 _____

10 mm SQUARES

150
140
90

2 CLEARANCE
EACH SIDE BETWEEN
PARTS 1 AND 2

SPECIAL
CLEVIS
PIN

WASHER

Ø40

240

E

R200

DRAW SECTIONED
AUXILIARY
VIEW
ASSEMBLY HERE

45°

E

COTTER PIN

40   20

335

M20 BOLT,
HEX NUT
AND LOCK-
WASHER

200

140
100

M12 SETSCREW
AND JAM NUT

30   100

R60

60

40   85

65
20

22

R24

24

Ø240

M16 BOLT, HEX NUT
AND LOCKWASHER

Ø270

24   184

R8

R16

SECTION A-A

Ø172
Ø140

90

130

Ø16
2 HOLES

140

BOTTOM VIEW OF PT 4 ONLY

214

SKETCHING ASSIGNMENT

1. Make a sectioned auxiliary view assembly taken on cutting-plane line **E-E**.
2. Make a three-view working drawing of part 3 in the space provided.

BILL OF MATERIAL

Complete the bill of material for the universal trolley assembly and place the part numbers on the assembly drawing. Refer to manufacturer's catalogs and drafting manuals for standard components. Two lines in the bill of material may be used for one part if required. Sizes for cast iron parts need not be shown.

QUESTIONS

1. Locate surface $\widehat{F}$ on the bottom view.
2. Is surface $\widehat{J}$ shown in the bottom view?
3. Which line or surface in the left-side view represents surface $\widehat{J}$?
4. How many different parts are used to make the trolley?
5. What is the total number of parts used to make the trolley?
6. How many tapped holes (excluding the nuts) are there?
7. What is the total number of holes in parts $\widehat{1}$ to $\widehat{4}$? Note: Count through holes as 2 holes.
8. Which line or surface in the bottom view represents point $\widehat{G}$?
9. Locate surface $\widehat{A}$ in the front view.
10. Locate surface $\widehat{Y}$ in the left-side view.
11. What is the overall height of the assembly?
12. Locate surface $\widehat{L}$ in the left-side view.
13. What is the total number of surfaces that are to be finished for parts $\widehat{1}$ to $\widehat{4}$? Note: One or more surfaces can be on the same plane. Where one finish mark is shown on matched surfaces of castings, this implies that both surfaces are to be finished.

NOTE: UNLESS OTHERWISE SPECIFIED
— TOLERANCE ON DIMENSIONS ± 0.5
— TOLERANCE ON ANGLES ± 0.5°
— SURFACES $\sqrt{}$ TO BE $\overset{6.3}{\sqrt{}}$

METRIC

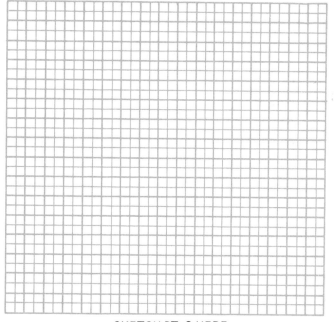

SKETCH PT 3 HERE

10 mm SQUARES

QTY	ITEM	MATL	DESCRIPTION	PT NO
	STAND	CI		4
	SWIVEL	CI		3
	SUPPORT	CI		2
	TROLLEY	CI		1
QTY	ITEM	MATL	DESCRIPTION	PT NO

SCALE		1:5	
DRAWN		DATE	

UNIVERSAL TROLLEY

A-63

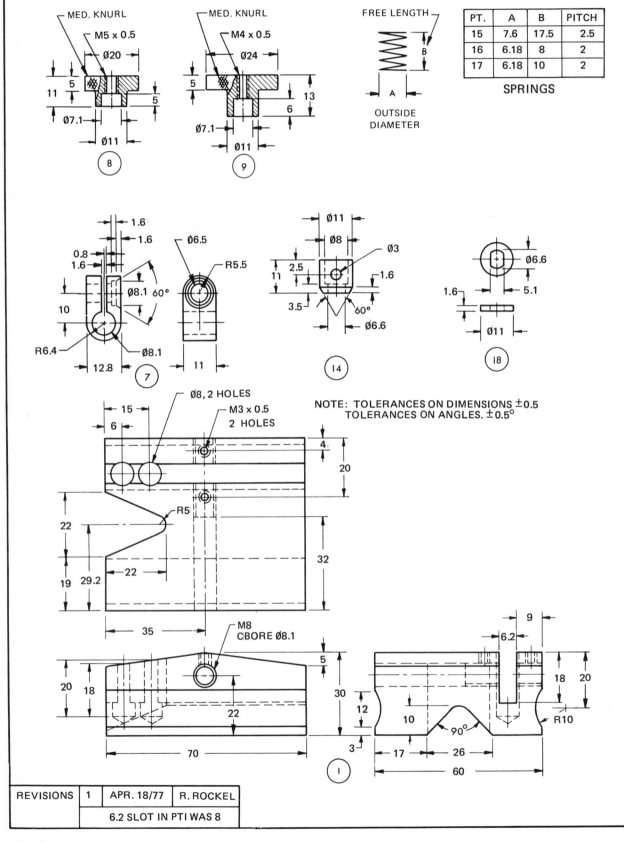

MED. KNURL
M5 x 0.5
Ø20
5
11
5
Ø7.1
Ø11
8

MED. KNURL
M4 x 0.5
Ø24
5
13
6
Ø7.1
Ø11
9

FREE LENGTH
B
A
OUTSIDE
DIAMETER

PT.	A	B	PITCH
15	7.6	17.5	2.5
16	6.18	8	2
17	6.18	10	2

SPRINGS

1.6
1.6
0.8
1.6
Ø8.1 60°
Ø6.5
R5.5
10
R6.4
Ø8.1
12.8
11
7

Ø11
Ø8
Ø3
2.5
1.6
11
3.5
60°
Ø6.6
14

Ø6.6
1.6
5.1
Ø11
18

Ø8, 2 HOLES
M3 x 0.5
2 HOLES
15
6
4
20
R5
22
32
19
29.2
22
35

NOTE: TOLERANCES ON DIMENSIONS ±0.5
TOLERANCES ON ANGLES. ±0.5°

M8
CBORE Ø8.1
5
20
18
30
22
12
70
3

9
6.2
18
20
10
90°
R10
17
26
60
1

REVISIONS | 1 | APR. 18/77 | R. ROCKEL
6.2 SLOT IN PTI WAS 8

216

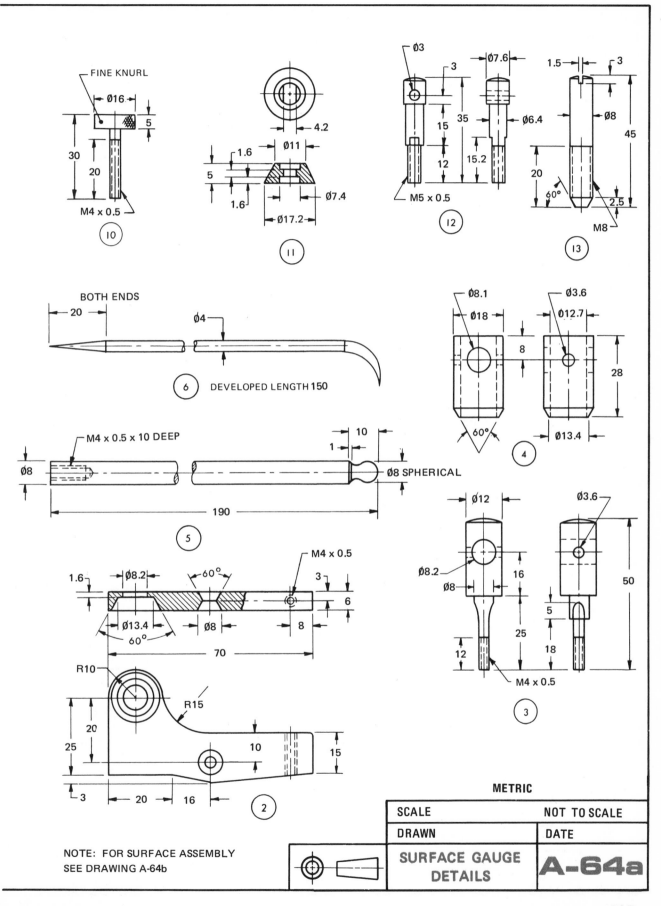

FINE KNURL

Ø16

5

30

20

M4 x 0.5

⑩

4.2

Ø11

1.6

5

1.6

Ø7.4

Ø17.2

⑪

Ø3

3

15

12

35

M5 x 0.5

⑫

Ø7.6

Ø6.4

15.2

1.5

3

Ø8

45

20

60°

2.5

M8

⑬

BOTH ENDS

20

Ø4

⑥ DEVELOPED LENGTH 150

Ø8.1

Ø18

Ø3.6

Ø12.7

8

28

60°

Ø13.4

④

M4 x 0.5 x 10 DEEP

Ø8

10

1

Ø8 SPHERICAL

190

⑤

Ø12

Ø3.6

Ø8.2

Ø8

16

5

25

18

50

12

M4 x 0.5

③

1.6

Ø8.2

60°

M4 x 0.5

3

6

Ø13.4

Ø8

8

60°

70

R10

R15

20

25

10

15

3

20

16

②

NOTE: FOR SURFACE ASSEMBLY
SEE DRAWING A-64b

METRIC

SCALE	NOT TO SCALE
DRAWN	DATE

SURFACE GAUGE
DETAILS

A-64a

217

1. Each part of the surface gauge is designated by a part number on the working drawing and in the bill of materials. Identify each of these parts on the assembly drawing.

2. Complete the bill of materials.

QUESTIONS

1. Which part maintains the frictional contact between part ④ and part ⑤ when thumb nut ⑨ is loosened?

2. Which part is actuated so that scriber ⑥ may be adjusted on spindle ⑤ ?

3. Which part keeps spindle ⑤ from being pulled through parts ③ and ④ ?

4. Which part actuates micrometer adjustment of rocker ② to raise or lower scriber ⑥ ?

5. Which part or parts keep rocker ② in tension when adjustment is being made with thumbscrew ⑩ ?

6. Which part or parts serve as pivots for rocker ② ?

7. What is the length of the setscrews at Ⓐ ?

8. What is the outside diameter of the spring at Ⓑ ?

9. What is the diameter of Ⓒ hole?

10. Determine radius Ⓛ .

11. What is the size of thread on the thumbscrew Ⓗ ?

12. What is the developed length of part ⑥ ?

13. What is the diameter of the spring wire for part Ⓙ ?

14. What is the length of part Ⓚ ?

15. Calculate distances Ⓓ , Ⓔ , Ⓕ , and Ⓖ .

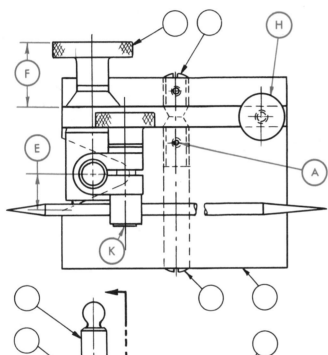

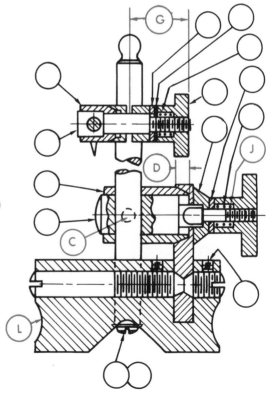

REVISION	1	APR. 8/77	S. HEINES
	THREAD AND SECTION LINING CHANGED		

ANSWERS

ANSWERS

1. _____

2. _____

3. _____

4. _____

5. _____

6. _____

7. _____

8. _____

9. _____

10. _____

11. _____

12. _____

13. _____

14. _____

15. (D) _____

(E) _____

(F) _____

(G) _____

QTY.	ITEM	MATL.	DESCRIPTION	PT. NO.
				30
				29
				28
				27
				26
				25
				24
1	MACHINE SCREW	ST	PANHEAD M4 x 0.5 x 10L	23
2	SETSCREW CUP POINT	ST	SLOTTED HEADLESS M3 x 5 L	22
1	SETSCREW	ST	SLOTTED HEADLESS CUP POINT M8 x 12L	21
1	WASHER, FLAT	ST	M5	20
2	WASHER, FLAT	ST	M4	19
1	WASHER, FLAT	ST		18
1	SPRING	ST	NO. 17 AM. STD. MUSIC WIRE	17
1	SPRING	ST	NO. 17 AM. STD. MUSIC WIRE	16
2	SPRING	ST	NO. 23 AM. STD. MUSIC WIRE	15
1	SPACER	ST		14
1	STUD	ST		13
1	LOCK SCREW	ST		12
1	SPACER	ST		11
1	THUMBSCREW	ST		10
1	THUMB NUT	ST		9
1	THUMB NUT	ST		8
1	SWIVEL	ST		7
1	SCRIBER	ST		6
1	SPINDLE	ST		5
1	SPACER	ST		4
1	SCREW ARM	ST		3
1	ROCKER	ST		2
1	BASE	ST		1
QTY.	ITEM	MATL.	DESCRIPTION	PT. NO.

SCALE	NOT TO SCALE
DRAWN	DATE

METRIC

SURFACE GAUGE ASSEMBLY

A-64 b

HELICAL SPRINGS

The *coil* or *helical spring* is commonly used in machine design and construction. It may be cylindrical or conical in shape or a combination of the two, figure 32-1.

Because of the labor and time involved, the true projection of a helical spring is usually not drawn. Instead, a schematic, or simplified drawing is preferred because of its simplicity. All the required information can be given on such a drawing.

On assembly drawings, springs are usually shown in section; either crosshatched lines or simplified, symbolic representation is recommended, depending on the size of the wire's diameter, figure 32-2.

The following information must be given on a drawing of a spring, figure 32-3:

- Size, shape, and type of material used in the spring

- Diameter (outside or inside)

- Pitch or number of coils

- Shape of ends

- Length

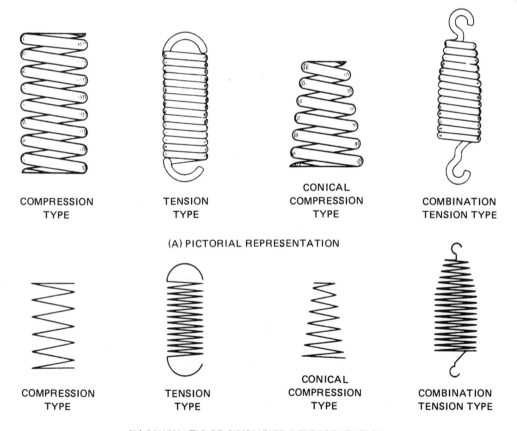

COMPRESSION TYPE   TENSION TYPE   CONICAL COMPRESSION TYPE   COMBINATION TENSION TYPE

(A) PICTORIAL REPRESENTATION

COMPRESSION TYPE   TENSION TYPE   CONICAL COMPRESSION TYPE   COMBINATION TENSION TYPE

(B) SCHEMATIC OR SIMPLIFIED REPRESENTATION

Fig. 32-1 Helical springs

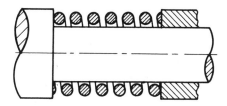

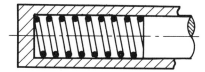

(A) LARGE SPRINGS

(B) SMALL SPRINGS

Fig. 32-2 Showing helical springs on assembly drawings

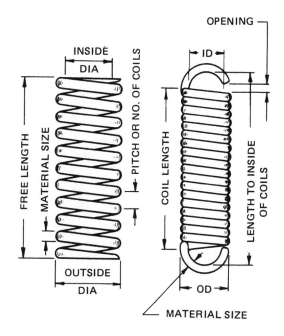

Fig. 32-3 Information given on spring drawings

For example, ONE HELICAL TENSION SPRING 75 mm L (OR NUMBER OF COILS), 4.12 ID, PITCH 4, 18 B & S GA. SPRING BRASS WIRE clearly states the required information.

The *pitch* of a coil spring is the distance from the center of one coil to the center of the next. The sizes of spring wires are desig- nated by millimetre sizes and also in gauge numbers. The tables for these are found in handbooks.

Springs are made to a dimension of either outside diameter (if the spring works in a hole) or inside diameter (if the spring works on a rod). In some cases the mean diameter is specified for computation purposes.

Sketch two views of each of parts 1 and 5 in the space provided on the drawing. Completely dimension the parts.

QUESTIONS

1. How many separate parts are shown on the valve assembly?

2. What is the length of the spring when the valve is closed?

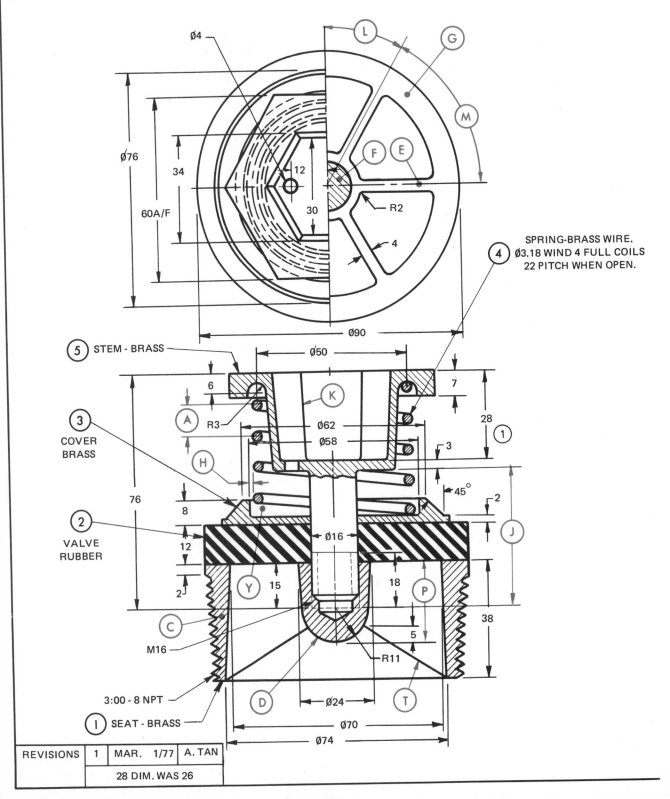

Ø4

Ø76

34

12

60A/F

30

R2

4

Ø90

L

G

M

F

E

SPRING-BRASS WIRE.
Ø3.18 WIND 4 FULL COILS
22 PITCH WHEN OPEN.

4

5 STEM - BRASS

3

COVER
BRASS

2

VALVE
RUBBER

76

8

12

2

M16

3:00 - 8 NPT

1 SEAT - BRASS

Ø50

6

A

R3

H

Y

15

C

D

Ø24

K

Ø62

Ø58

Ø16

18

P

5

R11

Ø70

Ø74

7

28

1

3

45°

2

J

T

38

REVISIONS	1	MAR. 1/77	A. TAN
		28 DIM. WAS 26	

3. Determine distances Ⓐ, Ⓙ, and Ⓟ.
4. What is the overall free length of spring?
5. Locate Ⓔ in the front view.
6. How many supporting ribs are there connecting
   Ⓓ to Ⓒ?
7. How thick are these ribs?
8. Identify the part numbers to which features
   Ⓕ and Ⓖ belong.
9. How many full threads are there on the stem
   (part 5)?
10. What is the nominal size of the pipe thread?
11. Give the length of the pipe thread.
12. Determine clearance distance Ⓗ.
13. Determine angles Ⓛ and Ⓜ.

NOTE: UNLESS OTHERWISE SPECIFIED
   — TOLERANCES ON DIMENSIONS ±0.2
   — TOLERANCES ON ANGLES ±0.5°

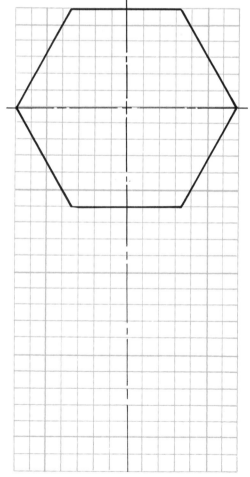

5 mm SQUARES

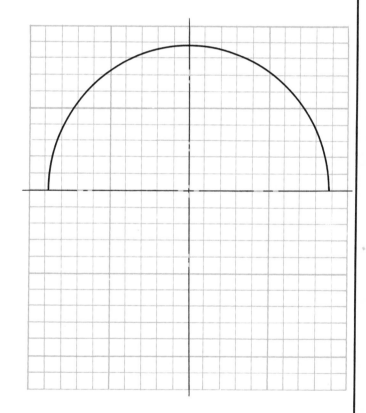

PARTIAL TOP VIEW AND A FRONT
SECTION VIEW OF SEAT, PT. 1

METRIC

SCALE	NOT TO SCALE
DRAWN	DATE

DRAW TOP VIEW AND FRONT VIEW IN
SECTION OF STEM, PT. 5. USE A CON-
VENTIONAL BREAK TO SHORTEN
HEIGHT OF FRONT VIEW.

FLUID PRESSURE VALVE

A-65

UNIT

33

DUAL DIMENSIONING

With much exchange of drawings occurring between North America and the rest of the world, it was for a time advantageous to show dimensions in both millimetres and inches. As a result, many companies developed systems of dual dimensioning. Today, however, this type of dimensioning is avoided. When both dimensioning systems are used, dimensions are shown in one system on the drawing and the corresponding values in the other system are shown in a chart, figure 33-1.

ARRANGEMENT OF VIEWS

Parts which are to be fitted over shafts as a single unit are sometimes made in two or more pieces. This is done for ease in assembly and replacement on the main structure of a machine rather than for ease in manufacture. Drawing **A-66** shows two parts which are bolted and doweled together to form one unit.

The arrangement of views of the spider is illustrated by the diagrams in figure 33-2. By comparing this figure with the drawing of the spider, note that the two halves together represent the top view. The right view is a full section of each half (**A** and **B**). The front view is a drawing of the front of part **A** only.

Although the front and side views are incomplete, the manner in which they are drawn and the arrangement of the views satisfies the demand for clearness and economy in time and space.

Inches	Millimetres	Inches	Millimetres
1/8	3.2	3 1/2	88.9
3/16	4.8	5 3/4	146.1
1/4	6.4	6 1/4	158.8
3/8	9.5	9	228.6
1/2	12.7	10	254.0
3/4	19.1	12	304.8
1	25.4	16	406.4
1 1/4	31.8	16 1/4	412.8
1 1/2	38.1	18	457.2
1 3/4	45.5	33	838.2
2	50.8	59	1498.6
2 1/8	54.0	63 1/2	1612.9
3	76.2	66	1676.4

Fig. 33-1 Chart of dimension
equivalents for dual dimensioning

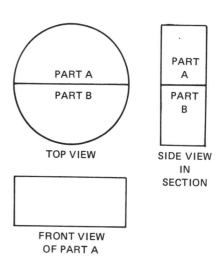

Fig. 33-2 Arrangement of views
for spider, Drawing A-66

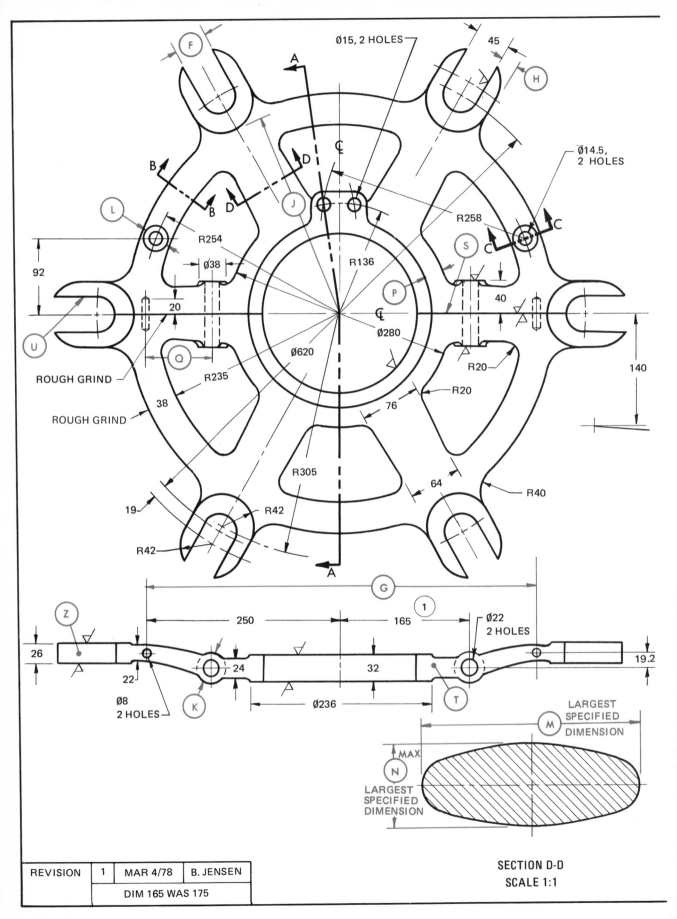

Ø15, 2 HOLES

45

(F)

(H)

Ø14.5,
2 HOLES

A

B

D

(L)

R254

ℂ

(J)

R258

(S)

C

C

92

Ø38

R136

(P)

Ø280

40

20

ROUGH GRIND

(U)

(Q)

R235

R20

140

ROUGH GRIND

38

R20

76

Ø620

R305

64

R40

19

R42

R42

(A)

(Z)

(G)

250

165

(1)

Ø22
2 HOLES

26

24

32

19.2

22

(K)

Ø236

(T)

Ø8
2 HOLES

LARGEST
SPECIFIED
DIMENSION

(M)

MAX

(N)

LARGEST
SPECIFIED
DIMENSION

REVISION	1	MAR 4/78	B. JENSEN
	DIM 165 WAS 175		

SECTION D-D
SCALE 1:1

226

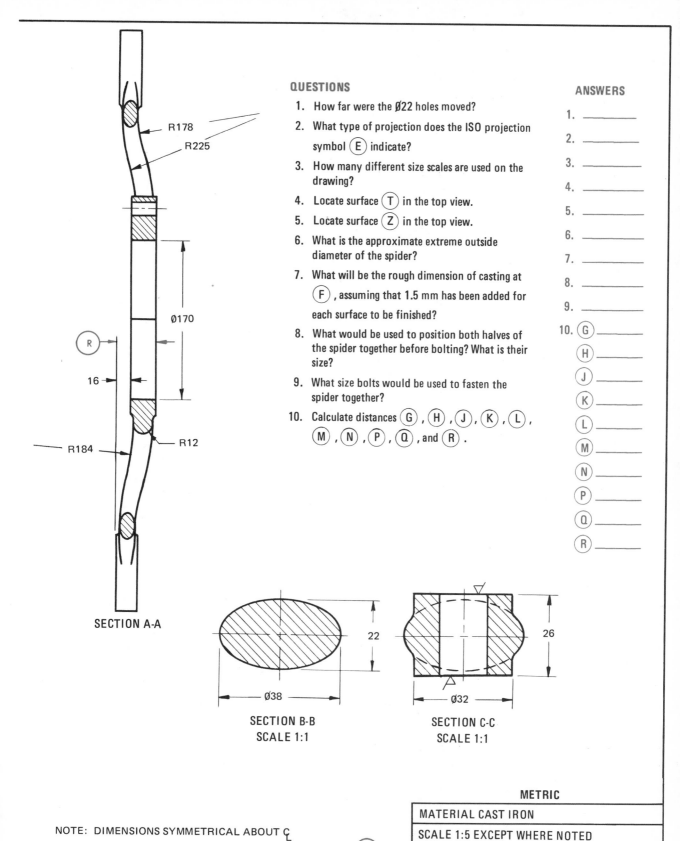

SECTION A-A

R178
R225
Ø170
R
16
R184
R12

QUESTIONS

1. How far were the Ø22 holes moved?
2. What type of projection does the ISO projection symbol (E) indicate?
3. How many different size scales are used on the drawing?
4. Locate surface (T) in the top view.
5. Locate surface (Z) in the top view.
6. What is the approximate extreme outside diameter of the spider?
7. What will be the rough dimension of casting at (F), assuming that 1.5 mm has been added for each surface to be finished?
8. What would be used to position both halves of the spider together before bolting? What is their size?
9. What size bolts would be used to fasten the spider together?
10. Calculate distances (G), (H), (J), (K), (L), (M), (N), (P), (Q), and (R).

ANSWERS

1. _____
2. _____
3. _____
4. _____
5. _____
6. _____
7. _____
8. _____
9. _____
10. (G) _____
    (H) _____
    (J) _____
    (K) _____
    (L) _____
    (M) _____
    (N) _____
    (P) _____
    (Q) _____
    (R) _____

SECTION B-B
SCALE 1:1

22
Ø38

SECTION C-C
SCALE 1:1

26
Ø32

NOTE: DIMENSIONS SYMMETRICAL ABOUT ℄
 — TOLERANCE ON DIMENSIONS ±0.5
 — TOLERANCE ON ANGLES ±0.5°
 — √ TO BE ⌄1.6⌄

METRIC	
MATERIAL CAST IRON	
SCALE 1:5 EXCEPT WHERE NOTED	
DRAWN	DATE
SPIDER	**A-66**

227

UNIT
34

PARTIAL VIEWS

A *partial view* is one in which only part of an object is shown, figure 34-1. However, sufficient information is given in the partial view to complete the description of the object. The partial view is limited by a break line. Partial views are used because:

- They save time in drawing.

- They conserve space which might otherwise be needed for drawing the object.

- They sometimes permit the drawing to be made to a scale large enough to bring out all details clearly, whereas if the whole view were drawn, lack of space might make it necessary to draw to a smaller scale, resulting in the loss of detail clarity.

- If the part is symmetrical, a partial view (referred to as a half or quarter view) may be drawn on one side of the center line as shown in figure 34-2. In the case of the coil frame (drawing **A-67**), a partial view was used so that the object could be drawn to a larger scale for clarity saving time and space.

RIBS IN SECTION

A true projection section view, figure 34-3 (A), would be misleading when the cutting plane passes longitudinally through the center of a rib. To avoid this impression of solidity, a preferred section not showing the ribs section-lined or crosshatched is used. When there is an odd number of ribs, figure 34-3 (B), the top rib is aligned with the bottom rib to show its true relationship with the hub and flange. If the rib is not aligned or revolved, it appears distorted on the section view and is misleading.

An alternative method of identifying ribs in a section view is shown in figure 34-3 (C), a base in section. If rib A of the base were not sectioned as previously mentioned, it would appear exactly like **B** in the section view and would be misleading. To distinguish between the ribs on the base, alternate section lining on the ribs is used. The line between the rib and solid portions is shown as a broken line.

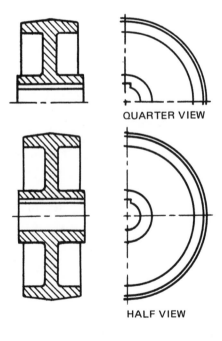

QUARTER VIEW

HALF VIEW

Fig. 34-2 Half and quarter views

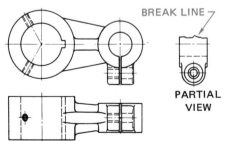

BREAK LINE

PARTIAL VIEW

Fig. 34-1 Partial views

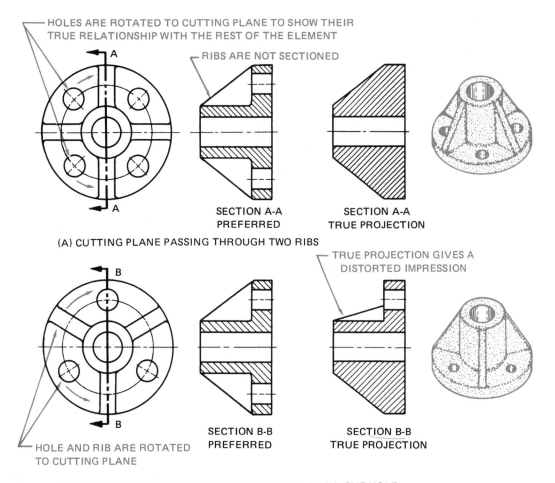

HOLES ARE ROTATED TO CUTTING PLANE TO SHOW THEIR
TRUE RELATIONSHIP WITH THE REST OF THE ELEMENT

RIBS ARE NOT SECTIONED

A

A

SECTION A-A
PREFERRED

SECTION A-A
TRUE PROJECTION

(A) CUTTING PLANE PASSING THROUGH TWO RIBS

B

B

TRUE PROJECTION GIVES A
DISTORTED IMPRESSION

SECTION B-B
PREFERRED

SECTION B-B
TRUE PROJECTION

HOLE AND RIB ARE ROTATED
TO CUTTING PLANE

(B) CUTTING PLANE PASSING THROUGH ONE RIB AND ONE HOLE

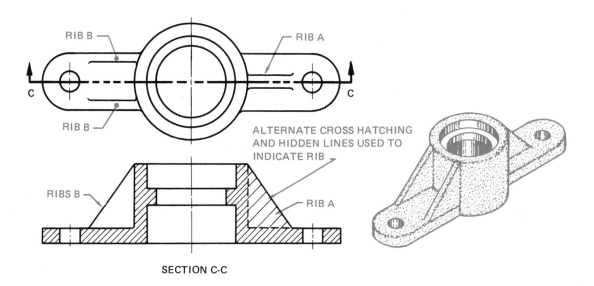

RIB B

RIB A

C

C

ALTERNATE CROSS HATCHING
AND HIDDEN LINES USED TO
INDICATE RIB

RIB B

RIBS B

RIB A

SECTION C-C

(C) ALTERNATE METHOD OF SHOWING RIBS IN SECTION

Fig. 34-3 Ribs in section

SPOKES IN SECTION

Spokes in section are represented in the same manner as ribs. Figure 34-4 shows the preferred method of identifying spokes in section for aligned and unaligned designs. Note that the spokes are not sectioned in either case.

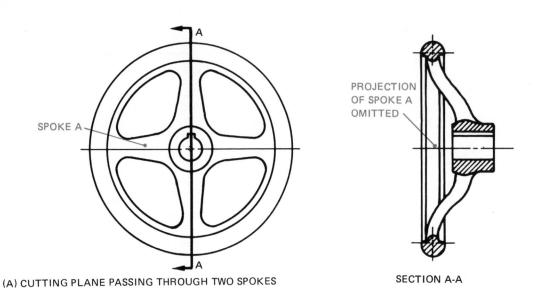

(A) CUTTING PLANE PASSING THROUGH TWO SPOKES

SECTION A-A

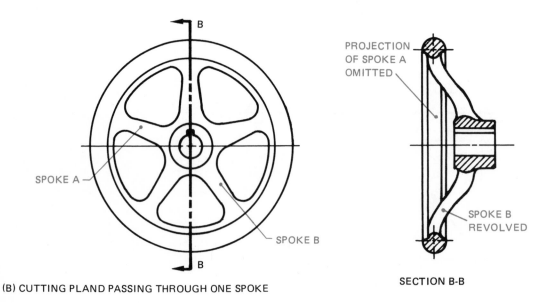

(B) CUTTING PLAND PASSING THROUGH ONE SPOKE

SECTION B-B

Fig. 34-4 Spokes in section

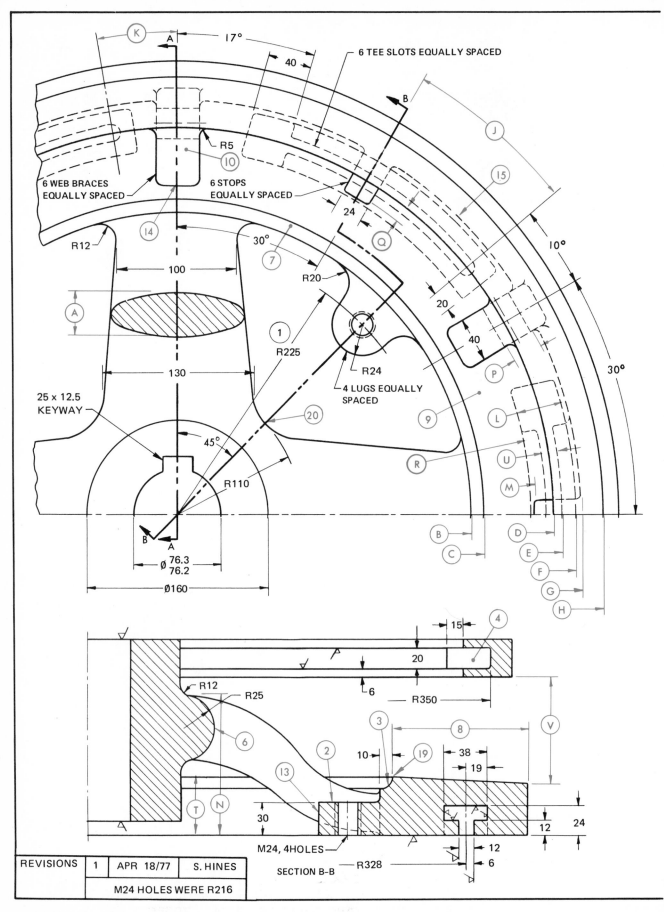

6 TEE SLOTS EQUALLY SPACED

17°

40

6 WEB BRACES EQUALLY SPACED

6 STOPS EQUALLY SPACED

R5

R12

100

130

25 x 12.5 KEYWAY

30°

R20

R225

R24

4 LUGS EQUALLY SPACED

45°

R110

Ø 76.3 / 76.2

Ø160

10°

30°

24

20

40

15 20 R350

R12 R25

6

30

M24, 4HOLES

SECTION B-B R328

3 8 V

10 19 38 19

12 24

12 6

REVISIONS | 1 | APR 18/77 | S. HINES

M24 HOLES WERE R216

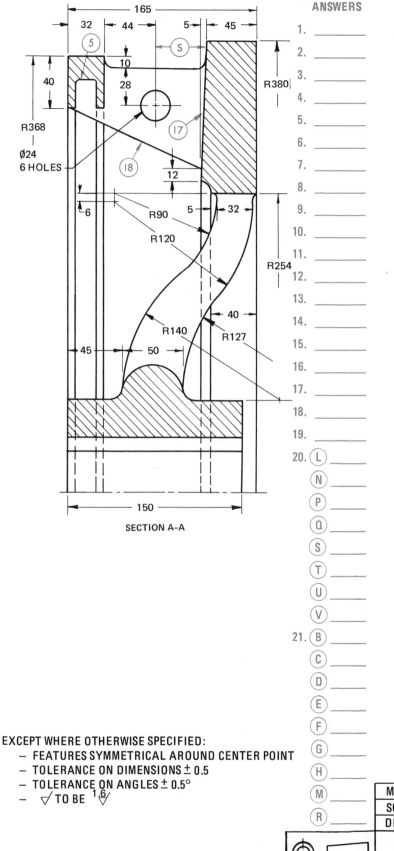

SECTION A-A

ANSWERS

1. _____
2. _____
3. _____
4. _____
5. _____
6. _____
7. _____
8. _____
9. _____
10. _____
11. _____
12. _____
13. _____
14. _____
15. _____
16. _____
17. _____
18. _____
19. _____
20. (L) _____
    (N) _____
    (P) _____
    (Q) _____
    (S) _____
    (T) _____
    (U) _____
    (V) _____
21. (B) _____
    (C) _____
    (D) _____
    (E) _____
    (F) _____
    (G) _____
    (H) _____
    (M) _____
    (R) _____

QUESTIONS

1. What surface texture is required on the machined surface?

2. What is the distance (A) , assuming the revolved section was taken at the middle of the arm?

3. What is the total quantity of feature (4) ?

4. Locate surface (5) in the front view.

5. Locate point (6) in the front view.

6. How far is line (14) from the center point?

7. How far is surface (13) from the center point?

8. Which point on section B-B represents radius (C) ?

9. Which line on section B-B represents surface (7) ?

10. What is the radius of surface (3) ?

11. Determine distance (8) .

12. Which line on section A-A represents surface (10) ?

13. Which surface in the front view represents line (17) ?

14. What is the angle at (J) ?

15. What is the angle at (K) ?

16. What is the thickness of the web brace?

17. What would be the length of the key used between the shaft and coil frame?

18. What is the thickness of the lugs?

19. What is the distance from the Ø24 hole to the center of the coil frame?

20. Determine distances (L) , (N) , (P) , (Q) , (S) , (T) , (U) , and (V) .

21. Determine radii (B) , (C) , (D) , (E) , (F) , (G) , (H) , (M) , and (R) .

EXCEPT WHERE OTHERWISE SPECIFIED:
– FEATURES SYMMETRICAL AROUND CENTER POINT
– TOLERANCE ON DIMENSIONS ± 0.5
– TOLERANCE ON ANGLES ± 0.5°
– ▽ TO BE ¹·⁶▽

METRIC

MATERIAL	CAST IRON
SCALE	NOT TO SCALE
DRAWN	DATE

COIL FRAME

A-67

UNIT
35

STRUCTURAL STEEL SHAPES

Structural steel is widely used in the metal trades for the fabrication of machine parts because the many standard shapes lend themselves to many different types of construction.

Remember that the steel produced at the rolling mills and shipped to the fabricating shop comes in a wide variety of shapes and forms (approximately 600). At this stage it is called plain material.

The great bulk of this material can be designated as one of the following and shown in figure 35-1.

- Welded wide flanged beams and columns designated as WWF (350 mm to 1200 mm), sometimes referred to as H shapes.

- Rolled wide flanged beams designated as W (100 mm to 900 mm) and M (100 mm to 350 mm).

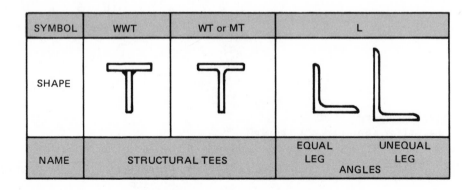

Fig. 35-1 Common structural steel shapes

- Rolled S beams (75 mm to 600 mm) formerly referred to as I beams because of their resemblance to the capital I.

- Standard channels designated as C (75 mm to 380 mm) and miscellaneous channels designated as MC (75 mm to 450 mm).

- Structural tees made by splitting the webs of beams and designated as WWT (split from WWF), WT (split from W beam), MT (split from M beam), thus forming two T-shapes from each beam.

- Angles, consisting of two equal or unequal legs, set at right angles (25 mm to 225 mm) and designated as L.

- Plates and round and rectangular bars, which are available in a variety of sizes.

Abbreviations

When structural steel shapes are designated on drawings, a standard method of abbreviating should be followed that will identify the group of shapes without reference to the manufacturer and without the use of millimetre or kilograms per metre (metric) or inches and pounds per foot (English).

Therefore, it is recommended that structural steel be abbreviated as illustrated in figures 35-2 and 35-3 (page 236).

The abbreviations shown are intended only for use on design drawings. When lists of materials are being prepared for ordering from the mills, the requirements of the respective mills from which the material is to be ordered should be observed.

S-shaped beams and all standard and miscellaneous channel have a slope on the inside flange of 16.67 percent. (16.67 percent slope is equivalent to 9 degrees 18′ or a bevel of 1:6.) All other beams have parallel face flanges.

PHANTOM OUTLINES

When parts or mechanisms not included in the actual detail or assembly drawing, but which, if shown, will clarify how the mechanism will connect with or operate from an adjacent part this must be shown by drawing light dash lines (one long line and two short dashes) in the operating position. Such a drawing of the extra part is known as a phantom drawing or view drawn *in phantom,* figure 35-4.

On the drawing of the four-wheel trolley (drawing **A-68**), the track the wheels run on is an S beam. The wheels are set at an angle to the vertical plane in order to ride upon the sloping bottom flange of the S beam. The outline of the beam is shown by dash lines and, while not an integral part of the trolley, the outline or phantom view of the S beam shows clearly how the trolley operates.

Shape	Metric Size Examples (See Note 1)	Inch Size Examples (See Note 2)	
		New Designation	Old Designation
Welded Wide Flange Shapes (WWF Shapes) —Beam —Columns	WWF1000 x 244 WWF350 x 315	WWF48 x 320	48WWF320
Wide Flange Shapes (W Shapes)	W600 x 114 W160 x 18	W24 x 76 W14 x 26	24WF76 14B26
Miscellaneous Shapes (M Shapes)	M200 x 56 M160 x 30	M8 x 18.5 M10 x 9	8M18.5 10JR9.0
Standard Beams (S Shapes)	S380 x 64	S24 x 100	24I100
Standard Channels (C Shapes)	C250 x 23	C12 x 20.7	12 ⊏20.7
Structural Tees —Cut from WWF shapes —Cut from W shapes —Cut from M shapes	WWT600 x 244 WT130 x 16 MT100 x 14	WWT24 x 160 WT12 x 38 MT4 x 5.25	ST24WWF160 ST12WF38 ST4M9.25
Bearing Piles (HP Shapes)	HP350 x 109	HP14 x 73	14BP73
Angles (L Shapes) (leg dimensions x thickness)	L75 x 75 x 6 L150 x 100 x 13	L6 x 6 x .75 L6 x 4 x .62	∠6 x 6 x 3/4 ∠6 x 4 x 5/8
Plates (Width x Thickness)	500 x 12	20 x .50	20 x 1/2
Square Bar (Side)	⊡ 25	⊡ 1.00	BAR 1 ⊡
Round Bar (Diameter)	⌀30	⌀1.25	BAR 1 1/4⌀
Flat Bar (Width x Thickness)	60 x 6	250 x .25	BAR 2 1/2 x 1/4
Round Pipe (Type of Pipe x OD x Wall Thickness)	XS 102 OD x 8	12.75 OD x .375	12 3/4 x 3/8
Square and Rectangular Hollow Structural Sections (Outside Dimensions x Wall Thickness)	HSS102 x 102 x 8	HSS4 x 4 x .375 HSS8 x 4 x .375	4 x 4 RT x 3/8 8 x 4 RT x 3/8
Steel Pipe Piles (OD x Wall Thickness)	320 OD x 6		

NOTE 1 – Values shown are nominal depth (millimetres) x mass per metre length (kilograms)
NOTE 2 – Values shown are nominal depth (inches) x weight per foot length (pounds)
NOTE 3 – Metric size examples shown are not necessarily the equivalents of the inch size examples shown.

Fig. 35-2 Abbreviations for shapes, plates, bars and tubes

(A) Metric designation

(B) Inch designation

Fig. 35-3 Structural steel callout

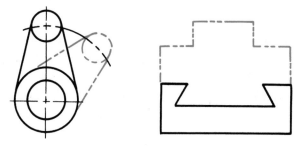

Fig. 35-4 Phantom Lines

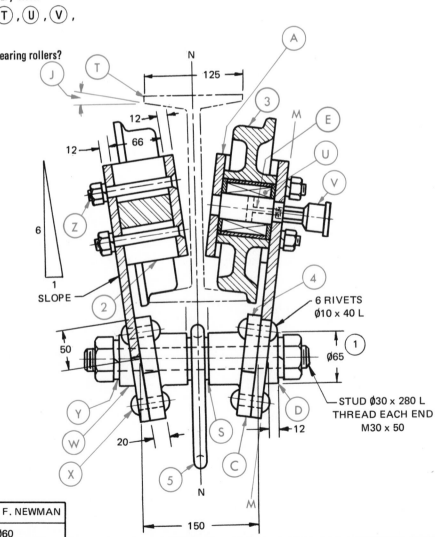

S BEAM — S250 x 52
WHEEL Ø200
SHAFT Ø35
BEARING 72 OD
— ROLLERS Ø14

6 RIVETS
Ø10 x 40 L

STUD Ø30 x 280 L
THREAD EACH END
M30 x 50

REVISIONS	1	APR. 7/77	F. NEWMAN
		Ø65 WAS Ø60	

238

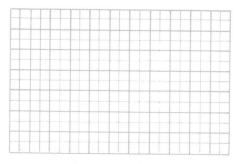

SKETCH PART D HERE
10 mm SQUARES

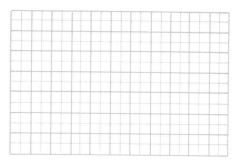

SKETCH PART C HERE
10 mm SQUARES

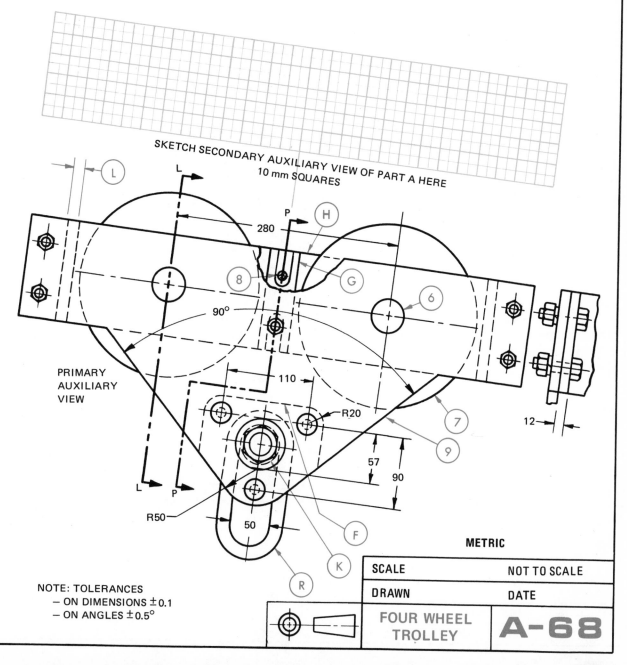

SKETCH SECONDARY AUXILIARY VIEW OF PART A HERE
10 mm SQUARES

L

P

280

H

8

G

6

90°

110

R20

7

9

PRIMARY
AUXILIARY
VIEW

L

P

57

90

R50

50

12

F

K

R

NOTE: TOLERANCES
— ON DIMENSIONS ±0.1
— ON ANGLES ±0.5°

METRIC

SCALE	NOT TO SCALE
DRAWN	DATE

FOUR WHEEL
TROLLEY

A-68

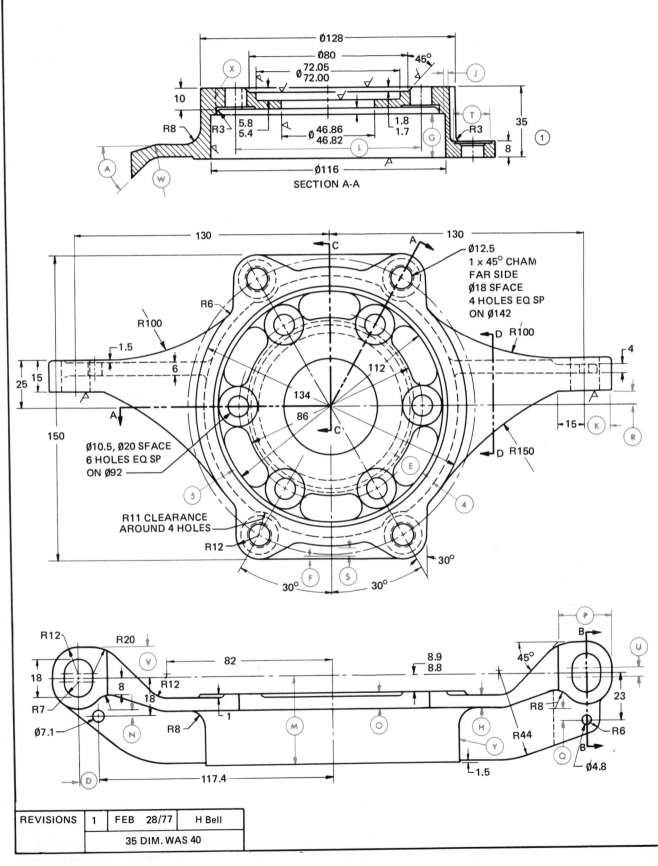

SECTION A-A

Ø128
Ø80
Ø 72.05 / 72.00
45°
10
5.8 / 5.4
Ø 46.86 / 46.82
1.8 / 1.7
R8
R3
R3
35
8
Ø116

130
C
A
Ø12.5
1 x 45° CHAM
FAR SIDE
Ø18 SFACE
4 HOLES EQ SP
ON Ø142
R6
R100
D R100
112
134
86
1.5
6
15
25
4
15
Ø10.5, Ø20 SFACE
6 HOLES EQ SP
ON Ø92
D R150
R11 CLEARANCE
AROUND 4 HOLES
R12
150
30°
30°
30°

R12
R20
R12
18
R7
Ø7.1
8
18
R8
1
82
8.9 / 8.8
45°
R8
23
R44
R6
Ø4.8
117.4
1.5

REVISIONS	1	FEB	28/77	H Bell
		35 DIM. WAS 40		

240

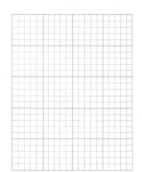

SECTION B-B
10 mm SQUARES

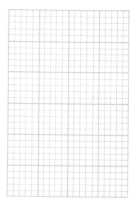

SECTION C-C
10 mm SQUARES

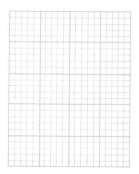

SECTION D-D
10 mm SQUARES

SKETCHING ASSIGNMENT

In the spaces provided, draw sections **B-B**, **C-C** and **D-D**.

QUESTIONS

1. How many through holes are shown?
2. What was the original height of the 35 dimension?
3. Determine angle (A) .
4. Determine radius (W) .
5. Locate (4) in the front view.
6. Which line in section A-A indicates the surface represented by line (5) ?
7. Determine distances (D) to (V) . (Where limits are given, use maximum sizes.)

ANSWERS

1. _____	7. (D) _____	7. (K) _____	7. (Q) _____
2. _____	(E) _____	(L) _____	(R) _____
3. _____	(F) _____	(M) _____	(S) _____
4. _____	(G) _____	(N) _____	(T) _____
5. _____	(H) _____	(O) _____	(U) _____
6. _____	(J) _____	(P) _____	(V) _____

NOTE: UNLESS OTHERWISE SPECIFIED:
FEATURES SYMMETRICAL AROUND CENTER POINT
 — TOLERANCE ON DIMENSIONS ±0.5
 — TOLERANCE ON ANGLES ±0.5°
 — ▽ TO BE

METRIC	
MATERIAL	
SCALE	1:2
DRAWN	DATE

CASE COVER

A-69

241

WELDING DRAWINGS

The primary importance of welding is the uniting of pieces of metal so they will operate properly as a unit to support the loads to be carried. In order to design and build such a structure, to be economical and efficient, a basic knowledge of welding is essential. Figure 36-1 illustrates many basic welding terms.

The introduction of welding symbols on a drawing enables the designer to indicate clearly the type and size of weld required to meet the design requirements. It is becoming increasingly important for the designer to indicate the required type of weld correctly. Basic welding joints are shown in figure 36-2. Points which must be made clear are the type of weld, the joint penetration, the weld size, the root gap (if any), and the degree of penetration required. These points can be clearly indicated on the drawing by the welding symbol.

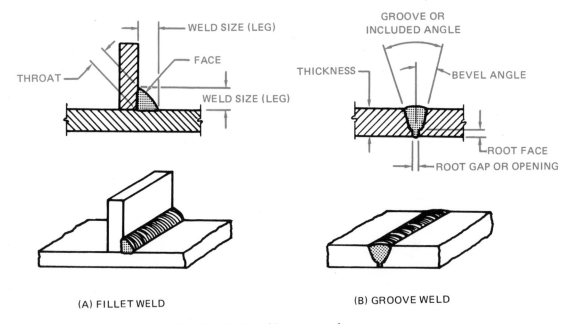

(A) FILLET WELD

(B) GROOVE WELD

Fig. 36-1 Basic welding nomenclature

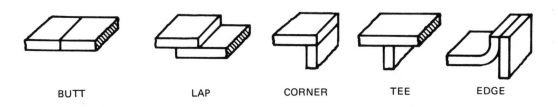

BUTT   LAP   CORNER   TEE   EDGE

Fig. 36-2 Basic welding joints

WELDING SYMBOLS

Welding symbols are a shorthand language. They save time and money and insure understanding and accuracy. Welding symbols should be a universal language; for this reason the symbols of the American Welding Society have been adopted.

The basic welding symbol consists of a *reference line,* and *arrow,* and occasionally, a *tail,* figure 36-3. Information is added to the basic welding symbol through the use of abbreviations, dimensions, and other symbols. In this manner the welder is provided with the information necessary for construction of the proper weld. Figure 36-4 illustrates the position of the weld symbols and other information in relation to the basic welding symbol. The various weld symbols which may be

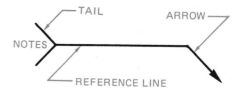

Fig. 36-3 Basic welding symbol

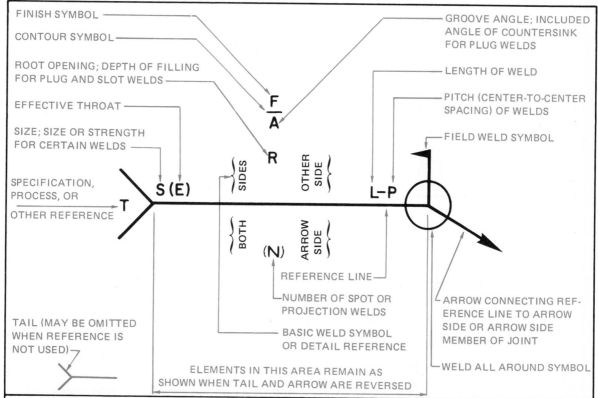

NOTE: SIZE, WELD SYMBOL, LENGTH OF WELD AND SPACING MUST READ IN THAT ORDER FROM LEFT TO RIGHT ALONG THE REFERENCE LINE. NEITHER ORIENTATION OR REFERENCE LINE NOR LOCATION ALTER THIS RULE.

THE PERPENDICULAR LEG OF $\triangle$, V , $\triangleright$, $\triangleright$ WELD SYMBOLS MUST BE AT LEFT.

ARROW AND OTHER SIDE WELDS ARE OF THE SAME SIZE UNLESS OTHERWISE SHOWN.

SYMBOLS APPLY BETWEEN ABRUPT CHANGES IN DIRECTION OF WELDING UNLESS GOVERNED BY THE "ALL AROUND" SYMBOL OR OTHERWISE DIMENSIONED.

THESE SYMBOLS DO NOT EXPLICITLY PROVIDE FOR THE CASE THAT FREQUENTLY OCCURS IN STRUCTURAL WORK, WHERE DUPLICATE MATERIAL (SUCH AS STIFFENERS) OCCURS ON THE FAR SIDE OF THE WEB OR GUSSET PLATE. THE FABRICATING INDUSTRY HAS ADOPTED THIS CONVENTION: THAT WHEN THE BILLING OF THE DETAIL MATERIAL DISCLOSES THE IDENTITY OF FAR SIDE WITH NEAR SIDE, THE WELDING SHOWN FOR THE NEAR SIDE SHALL SO BE DUPLICATED ON THE FAR SIDE.

Fig. 36-4 Standard location of elements of a welding symbol

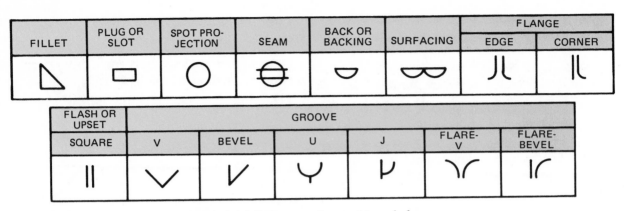

FILLET	PLUG OR SLOT	SPOT PRO-JECTION	SEAM	BACK OR BACKING	SURFACING	FLANGE	
						EDGE	CORNER

Fig. 36-5 Basic arc and gas weld symbols

FLASH OR UPSET	GROOVE					
SQUARE	V	BEVEL	U	J	FLARE-V	FLARE-BEVEL

applied to the basic welding symbol are shown in figure 36-5. Figure 36-6 shows the actual shape of many of the weld types symbolized in figure 36-5.

Supplementary symbols may also be added to the basic welding symbol. The supplementary symbols are illustrated in figure 36-7.

Any welding joint which is indicated by a symbol will always have an *arrow side* and an *other side*. The words *arrow side, other side,* and *both sides* are used accordingly to locate the weld with respect to the joint. The *tail* of the symbol is used for designating the welding specifications, procedures, or other supplementary information to be used when making the weld, figure 36-8. The notation to be placed in the *tail* of the symbol is to indicate the process, the type of filler metal to be used and whether or not peening or root chipping is required. If notations are not used, the *tail* of the symbol may be omitted.

The process indicated by the notation may be given as a letter designation. Figure 36-9 lists most of the welding processes and their appropriate letter designations.

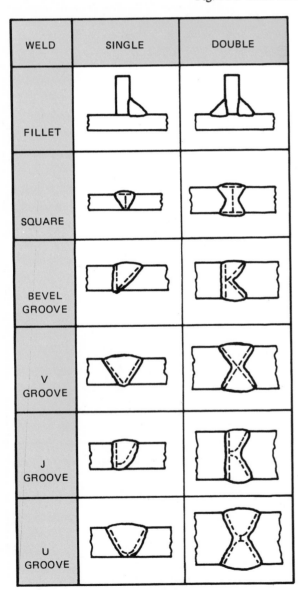

WELD	SINGLE	DOUBLE
FILLET		
SQUARE		
BEVEL GROOVE		
V GROOVE		
J GROOVE		
U GROOVE		

Fig. 36-6 Types of welds

WELD ALL AROUND	FIELD WELD	MELT-THRU	CONTOUR		
			FLUSH	CONVEX	CONCAVE

Fig. 36-7 Supplementary symbols

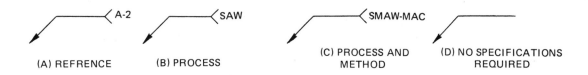

(A) REFRENCE   (B) PROCESS   (C) PROCESS AND METHOD   (D) NO SPECIFICATIONS REQUIRED

Fig. 36-8 Location of specifications, processes, and other references on welding symbols

Welding Process	Welding Process (Specific)	Letter Designation
Brazing	Infrared Brazing	IRB
	Torch Brazing	TB
	Furnace Brazing	FB
	Induction Brazing	IB
	Resistance Brazing	RB
	Dip Brazing	DB
Gas Welding	Oxyacetylene Welding	OAW
	Oxyhydrogen Welding	OHW
	Pressure Gas Welding	PGW
Resistance Welding	Resistance-Spot Welding	RSW
	Resistance-Seam Welding	RSEW
	Projection Welding	RPW
	Flash Welding	FW
	Upset Welding	UW
	Percussion Welding	PEW
Arc Welding	Stud Welding	SW
	Plasma-Arc Welding	PAW
	Submerged Arc Welding	SAW
	Gas Tungsten-Arc Welding	GTAW
	Gas Metal-Arc Welding	GMAW
	Flux Cored Arc Welding	FCAW
	Shielded Metal-Arc Welding	SMAW
	Carbon-Arc Welding	CAW
Other Processes	Thermit Welding	TW
	Laser Beam Welding	LBW
	Induction Welding	IW
	Electroslag Welding	EW
	Electron Beam Welding	EBW
Solid State Welding	Ultrasonic Welding	USW
	Friction Welding	FRW
	Forge Welding	FOW
	Explosion Welding	EXW
	Diffusion Welding	DFW
	Cold Welding	CW

Cutting Method	Letter Designation
Arc Cutting	AC
Air Carbon-Arc Cutting	AAC
Carbon-Arc Cutting	CAC
Metal-Arc Cutting	MAC
Plasma-Arc Cutting	PAC
Oxygen Cutting	OC
Chemical Flux Cutting	FOC
Metal Powder Cutting	POC
Oxygen-Arc Cutting	AOC

Fig. 36-9 Designation of welding process by letters

Welds on the *arrow side* of the joint are shown by placing the weld symbol on the bottom side of the reference line. Welds on the *other side* of the joint are shown by placing the weld symbol on the top side of the reference line. Welds on *both sides* of the joint are shown by placing the weld symbol on both sides of the reference line. A weld extending completely around a joint is indicated by means of a *weld-all-around* symbol placed at the intersection of the *reference line* and the *arrow*.

Field welds (welds not made in the shop or at the initial place of construction) are indicated by means of the *field weld* symbol placed at the intersection of the *reference line* and the *arrow*.

All weld dimensions on a drawing may be subject to a general note. Such a note might state: ALL FILLET WELDS 8mm UNLESS OTHERWISE NOTED.

Only the basic fillet and groove welds will be discussed here. Figure 36-10 illustrates several typical welding symbols and the resulting welds.

Fillet Welds

1. Dimensions of fillet welds are shown on the same side of the reference line and to the left of the weld symbol.

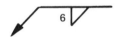

2. When both sides of a joint have the same size fillet welds, one is dimensioned.

3. When both sides of a joint have different size fillet welds, both are dimensioned.

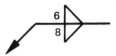

4. The dimension does not need to be shown when a general note is placed on the drawing to specify the dimension of fillet welds.

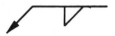

NOTE: SIZE OF FILLET WELDS 6 UNLESS OTHERWISE SPECIFIED.

5. The *length* of a fillet weld, when indicated on the welding symbol, is shown to the right of the weld symbol.

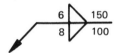

6. The *pitch* (center-to-center spacing) of an intermittent fillet weld is shown as the distance between centers of increments on one side of the joint. It is shown to the right of the length dimension.

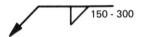

7. Staggered intermittent fillet welds are illustrated by staggering the weld symbols.

Groove Welds

1. Dimensions of groove welds are shown on the same side of the reference line as the weld symbol.

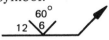

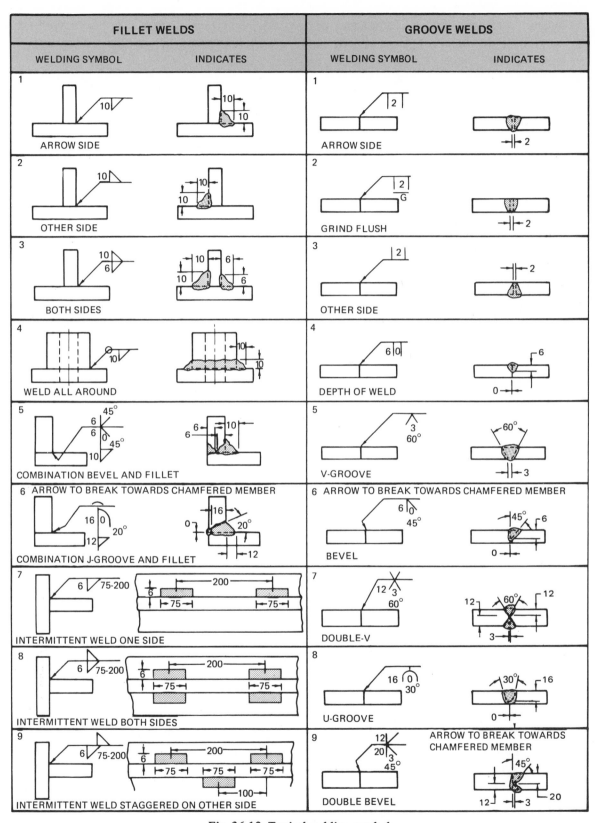

Fig. 36-10 Typical welding symbols

2. When both sides of a double-groove weld have the same dimensions, one is dimensioned.

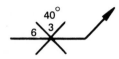

3. When both sides of a double-groove weld differ in dimensions, both are dimensioned.

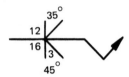

4. Groove welds do not need to be dimensioned when a general note appears on the drawing to govern the dimensions of groove welds, such as ALL V-GROOVE WELDS ARE TO HAVE A 60° ANGLE UNLESS OTHERWISE NOTED.

5. The size of groove welds is shown to the left of the weld symbol.

6. When the single-groove and symmetrical double-groove welds extend completely through the member or members being joined, the size of the weld need not be shown on the welding symbol.

7. When the groove welds extend only partly through the member being joined, the size of the weld is shown on the weld symbol.

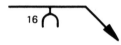

REFERENCES AND SOURCE MATERIAL

1. American Welding Society and Canadian Welding Bureau.

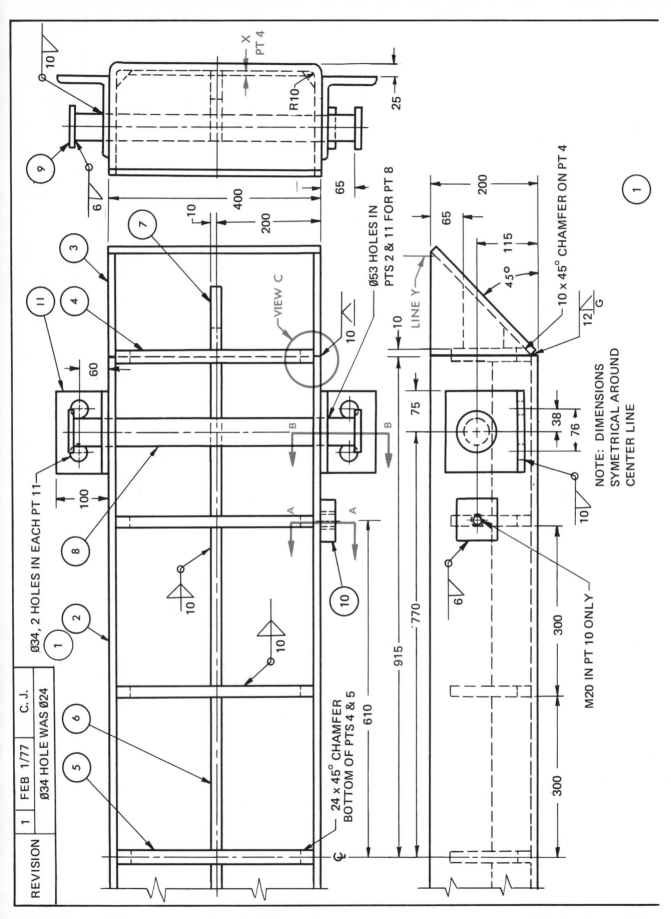

NOTE: DIMENSIONS
SYMETRICAL AROUND
CENTER LINE

M20 IN PT 10 ONLY

Ø53 HOLES IN
PTS 2 & 11 FOR PT 8

24 x 45° CHAMFER
BOTTOM OF PTS 4 & 5

Ø34, 2 HOLES IN EACH PT 11

10 x 45° CHAMFER ON PT 4

VIEW C

Ø34 HOLE WAS Ø24

| REVISION | 1 | FEB 1/77 | C.J. |

SKETCHING ASSIGNMENT

On the graph sections, complete the sketches and darken in welds. Square sizes are designated.

QUESTIONS

1. How many Ø34 holes are there in the complete assembly?

2. How deep is the M20 hole?

3. What was the original size of the Ø34?

4. What is the overall height of the assembly?

5. Determine the distance from line **Y** to the center line of the assembly.

6. Determine distance **X**.

7. What is the developed width of part 2? Use inside dimensions of channel.

8. What is the length of (A) pt. 2, (B) pt. 4, (C) pt. 7, (D) pt. 5?

9. What is the allowance for fitting on the length of pt. 5?

10. What is the difference between pt. 4 and pt. 5?

11. How many 24 chamfers are needed?

12. What does G mean on 12 bevel weld symbols?

13. What type of weld is used to fasten pt. 11 to pt. 2?

14. What type of weld is used to join pt. 3 to pt. 2 at the sides?

15. How many parts make up the assembly?

UNLESS OTHERWISE SPECIFIED:
- TOLERANCE ON DIMENSIONS ± 1 EXCEPT Ø OF HOLES ± 0.1
- TOLERANCE ON ANGLES ± 0.5°

5mm SQUARES

ENLARGED VIEW C

10mm SQUARES

SECTION B–B

10mm SQUARES

SECTION A–A

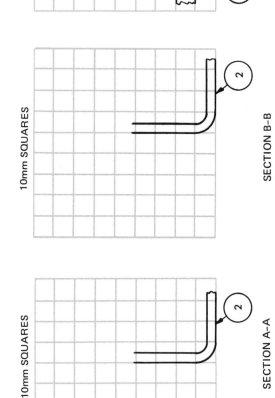

ANSWERS

1. _____
2. _____
3. _____
4. _____
5. _____
6. _____
7. _____
8. A. _____
   B. _____
8. C. _____   10. _____
D. _____   11. _____
9. _____   12. _____
13. _____
14. _____
15. _____

QTY	ITEM	MATL		PT NO
4	LOCATING ANGLE	ST	L150 × 100 × 12 × 150L	11
2	GROUND BAR	ST BAR	25 × 80 × 80	10
4	RETAINER	ST PL	12 × Ø 80	9
2	DRAW BAR	ST RD	Ø 50 ×	8
2	GUSSET	ST BAR	20 × 80 ×	7
6	GUSSET	ST BAR	20 × 80 × 280	6
5	GUSSET	ST BAR	20 × 160 × 370	5
2	GUSSET	ST BAR	20 × 160 ×	4
2	END PLATE	ST PL	12 × 270 × 650	3
1	BASE	ST PL	12 × ×	2
1	SKID ASS'Y			1

BASE SKID

METRIC

A-70

251

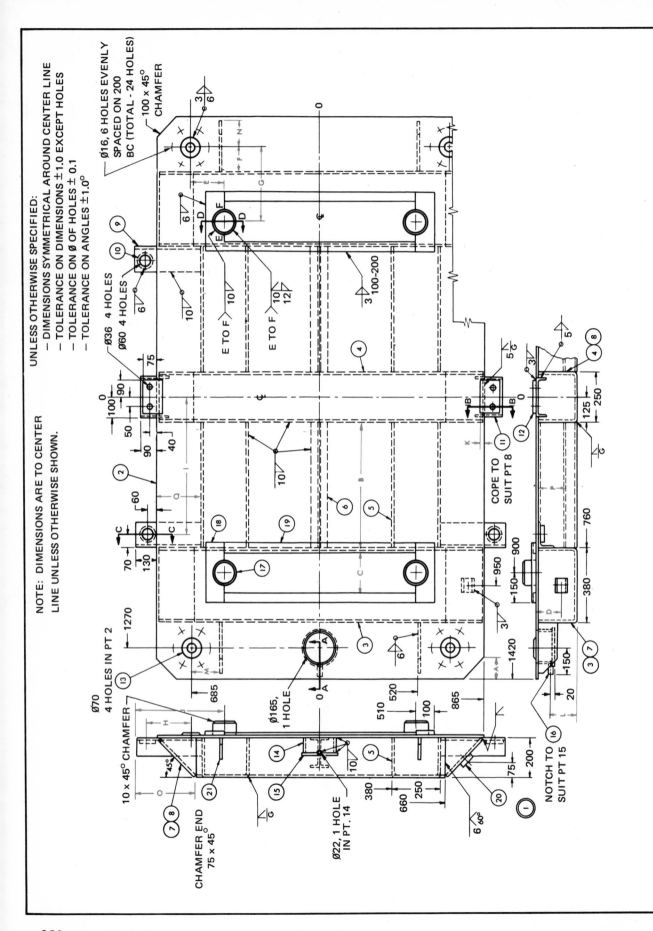

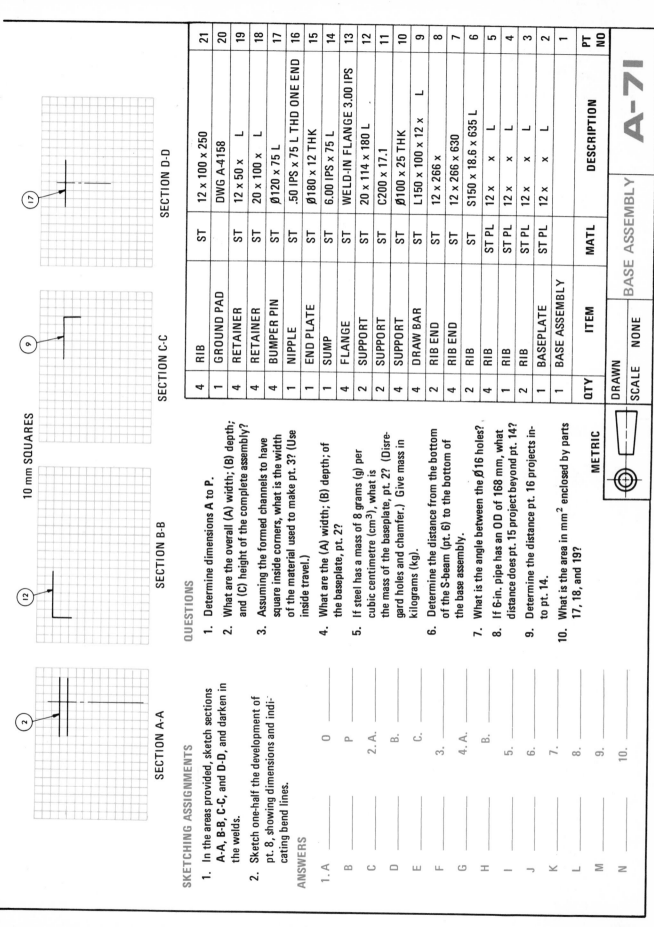

SECTION D-D

SECTION C-C

SECTION B-B

10 mm SQUARES

SECTION A-A

QTY	ITEM	MATL	DESCRIPTION	PT NO
4	RIB	ST	12 x 100 x 250	21
1	GROUND PAD		DWG A-4158	20
4	RETAINER	ST	12 x 50 x L	19
4	RETAINER	ST	20 x 100 x L	18
4	BUMPER PIN	ST	Ø120 x 75 L	17
1	NIPPLE	ST	.50 IPS x 75 L THD ONE END	16
1	END PLATE	ST	Ø180 x 12 THK	15
1	SUMP	ST	6.00 IPS x 75 L	14
4	FLANGE	ST	WELD-IN FLANGE 3.00 IPS	13
2	SUPPORT	ST	20 x 114 x 180 L	12
2	SUPPORT	ST	C200 x 17.1	11
4	SUPPORT	ST	Ø100 x 25 THK	10
4	DRAW BAR	ST	L150 x 100 x 12 x L	9
2	RIB END	ST	12 x 266 x	8
4	RIB END	ST	12 x 266 x 630	7
2	RIB	ST	S150 x 18.6 x 635 L	6
4	RIB	ST PL	12 x x L	5
1	RIB	ST PL	12 x x L	4
2	RIB	ST PL	12 x x L	3
1	BASEPLATE	ST PL	12 x x L	2
1	BASE ASSEMBLY			1

DRAWN

SCALE   NONE

METRIC

BASE ASSEMBLY

A-71

SKETCHING ASSIGNMENTS

1. In the areas provided, sketch sections **A-A**, **B-B**, **C-C**, and **D-D**, and darken in the welds.

2. Sketch one-half the development of pt. 8, showing dimensions and indicating bend lines.

ANSWERS

1. A _____   O _____
   B _____   P _____
   C _____   2. A. _____
   D _____   B. _____
   E _____   C. _____
   F _____   3. _____
   G _____   4. A. _____
   H _____   B. _____
   I _____   5. _____
   J _____   6. _____
   K _____   7. _____
   L _____   8. _____
   M _____   9. _____
   N _____   10. _____

QUESTIONS

1. Determine dimensions **A** to **P**.

2. What are the overall (A) width; (B) depth; and (C) height of the complete assembly?

3. Assuming the formed channels to have square inside corners, what is the width of the material used to make pt. 3? (Use inside travel.)

4. What are the (A) width; (B) depth; of the baseplate, pt. 2?

5. If steel has a mass of 8 grams (g) per cubic centimetre (cm^3), what is the mass of the baseplate, pt. 2? (Disregard holes and chamfer.) Give mass in kilograms (kg).

6. Determine the distance from the bottom of the S-beam (pt. 6) to the bottom of the base assembly.

7. What is the angle between the Ø16 holes?

8. If 6-in. pipe has an OD of 168 mm, what distance does pt. 15 project beyond pt. 14?

9. Determine the distance pt. 16 projects into pt. 14.

10. What is the area in mm^2 enclosed by parts 17, 18, and 19?

CAMS

The cam is invaluable in the design of automatic machinery. Cams make it possible to impart any desired motion to another mechanism.

A *cam* is a rotating, oscillating, or reciprocating machine element which has a surface or groove formed to impart special or irregular motion to a second part called a *follower*. The follower rides against the curved surface of the cam. The distance that the follower rises and falls in a definite period of time is determined by the shape of the cam profile.

Types of Cams

The type and shape of cam used is dictated by the required relationship of the parts and the motions of both, figure 37-1. The

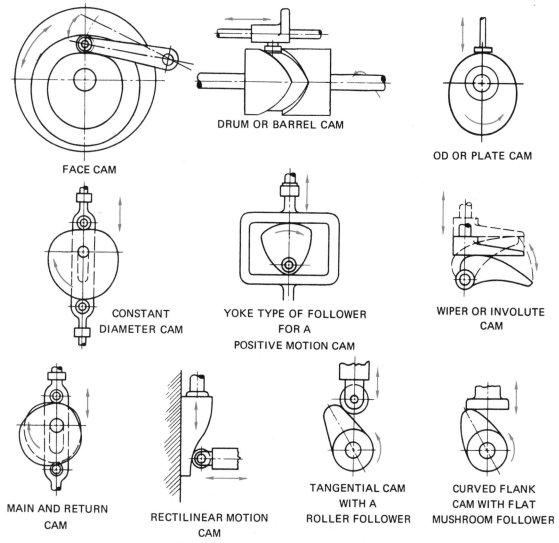

FACE CAM

DRUM OR BARREL CAM

OD OR PLATE CAM

CONSTANT DIAMETER CAM

YOKE TYPE OF FOLLOWER FOR A POSITIVE MOTION CAM

WIPER OR INVOLUTE CAM

MAIN AND RETURN CAM

RECTILINEAR MOTION CAM

TANGENTIAL CAM WITH A ROLLER FOLLOWER

CURVED FLANK CAM WITH FLAT MUSHROOM FOLLOWER

Fig. 37-1 Common types of cams

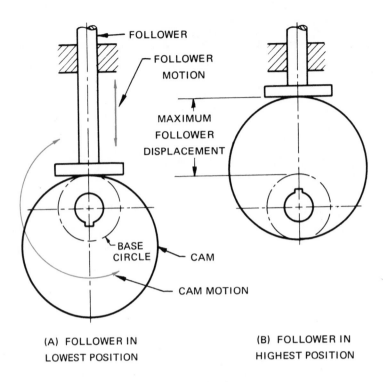

FOLLOWER

FOLLOWER
MOTION

MAXIMUM
FOLLOWER
DISPLACEMENT

BASE
CIRCLE

CAM

CAM MOTION

(A) FOLLOWER IN
LOWEST POSITION

(B) FOLLOWER IN
HIGHEST POSITION

Fig. 37-2 Eccentric plate cam

cams which are generally used are either radial or cylindrical. The follower of a radial or face cam moves in a plane perpendicular to the axis of the cam, while in the cylindrical type of cam the movement of the follower is parallel to the cam axis.

A simple OD (outside diameter) or plate cam is shown in figure 37-2. The hole in the plate is bored off-center, causing the follower to move up and down as it revolves. The follower can be any type that will roll or slide on the surface of the cam. The follower used with this cam is called a flat face follower.

The cam shown in figure 37-3 is a drum or barrel-type cam that transmits motion transversely to a lever connected to a conical follower which rides in the groove as the cam revolves.

Cam Displacement Diagrams

In preparing cam drawings, a cam displacement diagram is drawn first to plot the motion of the follower. The curve on the drawing represents the path of the follower, not the face of the cam. The diagram may be any convenient length, but often it is drawn equal to the circumference of the base circle of the cam and the height is drawn equal to the follower displacement. The lines drawn on the motion diagram are shown as radial lines on the cam drawing, the sizes

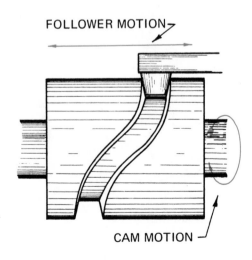

FOLLOWER MOTION

CAM MOTION

Fig. 37-3 Drum or barrel type cam

are transferred from the motion diagram to the cam drawing.

Figure 37-4 shows a cam displacement diagram having a modified uniform type of motion plus two dwell periods. Most cam displacement diagrams have 360 degree cam displacement angles. For drum or cylindrical grooved cams, the displacement diagram is often replaced by the developed surface of the cam.

The cylindrical feeder cam (drawing A-72) is a drum or barrel cam. In addition to the working views of the cam, a development of the contour of the grooves is shown. This development aids the machinist in scribing and laying out the contour of the cam action lobes on the surface of the cam blank preparatory to machining grooves.

Regardless of the type of cam or follower, the purpose of all cams is to impart motion to other mechanisms in various directions in order to actuate machines to do specific jobs.

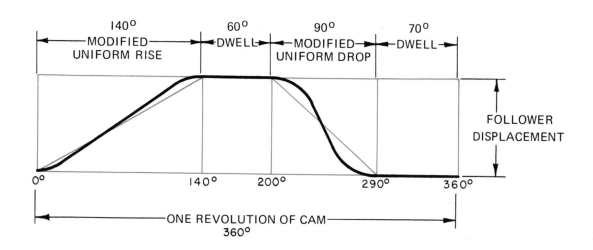

Fig. 37-4 Cam displacement diagram

QUESTIONS

1. Through what thickness of metal is hole ②
   drilled?
2. Locate line ④ in the right-side view.
3. Locate line ⑪ in the cam displacement
   diagram.
4. Locate line ⑥ in another view.
5. Locate line ⑨ in the right-side view.
6. What is the maximum permissible diameter of
   hole ② ?
7. Locate line ⑧ in the front view.
8. Locate line ⑮ in the left-side view.
9. What is the total follower displacement of
   the cam follower for (A) the finishing cut,
   (B) the roughing cut?
10. Assuming a 1.6 allowance for machining,
    what would be the outside diameter of the
    cam before finishing?
11. What is the total number of through holes?
12. Locate lines ㉚ to ㊶ on other view.
13. Determine distances Ⓐ to Ⓜ .

ANSWERS

1. _____	12. ㉚ _____	13. Ⓐ _____
2. _____	㉛ _____	Ⓑ _____
3. _____	㉜ _____	Ⓒ _____
4. _____	㉝ _____	Ⓓ _____
5. _____	㉞ _____	Ⓔ _____
6. _____	㉟ _____	Ⓕ _____
7. _____	㊱ _____	Ⓖ _____
8. _____	㊲ _____	Ⓗ _____
9. A. _____	㊳ _____	Ⓙ _____
B. _____	㊴ _____	Ⓚ _____
10. _____	㊵ _____	Ⓛ _____
11. _____	㊶ _____	Ⓜ _____

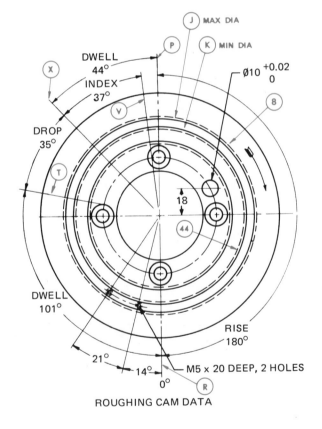

ROUGHING CAM DATA

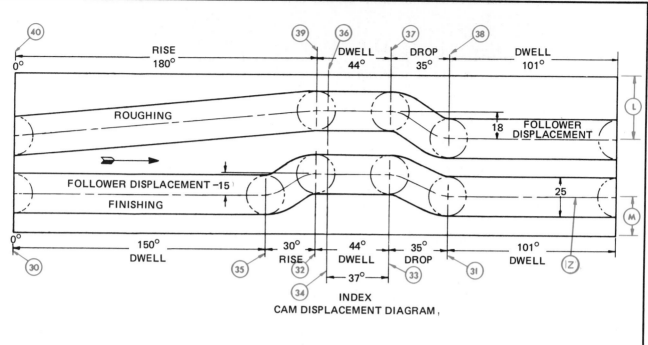

CAM DISPLACEMENT DIAGRAM

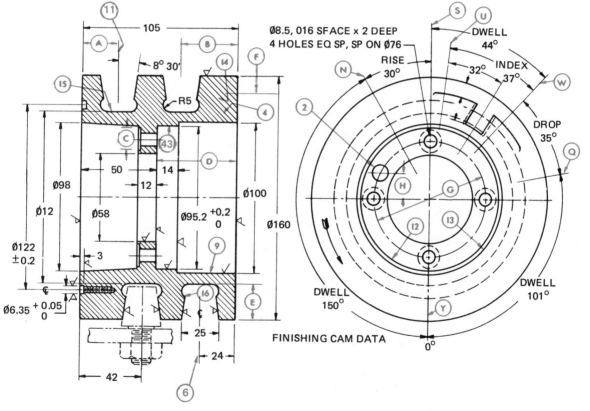

FINISHING CAM DATA

NOTE: UNLESS OTHERWISE SHOWN:
  — TOLERANCE ON DIMENSIONS ±0.5
  — TOLERANCE ON ANGLES ±0.5°
  — SURFACES √ TO BE √ 1.6

METRIC

MATERIAL	
SCALE	NOT TO SCALE
DRAWN	DATE

CYLINDRICAL
FEEDER CAM

A-72

GEAR DRIVES

The function of a *gear* is to transmit rotary or reciprocating motion from one machine part to another. Gears are often used to reduce or increase the rev/min of a shaft. Gears are rolling cylinders or cones. They have teeth on their contact surfaces to insure the transfer of motion, figure 38-1.

There are many kinds of gears; they may be grouped according to the position of the shafts they connect. *Spur gears* connect parallel shafts; *bevel gears* connect shafts having intersecting axes; and *worm gears* connect shafts having axes which do not intersect. A spur gear with a *rack* converts rotary motion to reciprocating or linear motion. The smaller of two gears is the *pinion.*

A simple gear drive consists of a toothed driving wheel meshing with a similar driving wheel. Tooth forms are designed to insure uniform angular rotation of the driven wheel during tooth engagement.

Spur gears, which are used for drives between parallel shafts, have teeth on the rim of the wheel, figure 38-2. A pair of spur gears operates as though it consists of two

Fig. 38-1 Gears

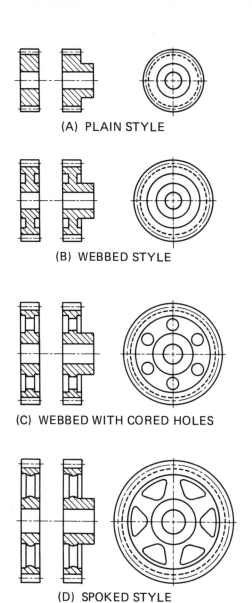

(A) PLAIN STYLE

(B) WEBBED STYLE

(C) WEBBED WITH CORED HOLES

(D) SPOKED STYLE

Fig. 38-2 Stock spur gear styles

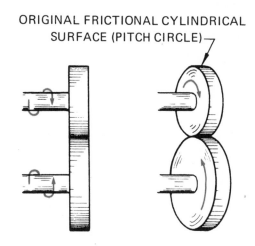

ORIGINAL FRICTIONAL CYLINDRICAL SURFACE (PITCH CIRCLE)

PRINCIPLE AS APPLIED TO SPUR GEARS

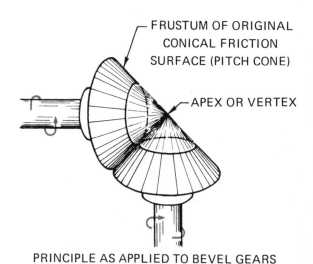

FRUSTUM OF ORIGINAL CONICAL FRICTION SURFACE (PITCH CONE)

APEX OR VERTEX

PRINCIPLE AS APPLIED TO BEVEL GEARS

Fig. 38-3 Comparing pitch circles of spur gears with pitch cones of bevel gears

cylindrical surfaces with formed teeth which maintain constant speed ratio between the driving and the driven wheel.

Gear design is complex, dealing with such problems as strength, wear, noise, and material selection. Usually, a designer selects a gear from a catalog. Most gears are made of cast iron or steel. However, brass, bronze, and fiber are used when factors such as wear or noise must be considered.

Theoretically, the teeth of a spur gear are built around the original frictional cylindrical surface called the *pitch circle,* while the teeth of a bevel gear are formed around the frustum of the original conical surface called *pitch cone,* figure 38-3.

The angle between the direction of pressure between contacting teeth and a line tangent to the pitch circle is the *pressure angle.*

The *14.5 degree pressure angle* has been used for many years and remains useful for duplicate or replacement gearing.

The *20-degree pressure angle* has become the standard for new gearing because of its smoother and quieter operation and greater load-carrying ability.

One type of commonly used bevel gear is the *miter gear*. The term miter gear refers to a pair of bevel gears of the same size that transmit motion at right angles.

Gear Train Terms and Calculations

The following terms are used in spur gear train calculations, figures 38-4 and 38-5.

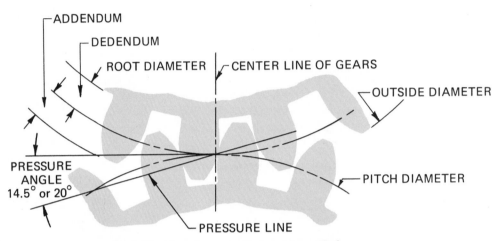

Fig. 38-4 Meshing of gear teeth showing symbols

Term	Symbol	Formula Based On	
		Millimetre Sizes	Inch Sizes
Pitch Diameter	D	$D = N \times Mo$	$D = \dfrac{N}{P}$
Number of Teeth	N	$N = \dfrac{D}{Mo}$	$N = D \times P$
Module	Mo	$Mo = \dfrac{D}{N}$	
Diametral Pitch	P		$P = \dfrac{N}{D}$
Addendum	A	$A = M$	$A = \dfrac{1}{P}$
Dedendum	B	$B = 1.157Mo$	$B = \dfrac{1.157}{P}$
Whole Depth	WD	$WD = 2.157Mo$	$WD = \dfrac{2.157}{P}$
Outside Diameter	OD	$OD = D + 2A = Mo(N + 2)$	$OD = D + 2A = \dfrac{N + 2}{P}$
Root Diameter	RD	$RD = D - 2B = D - 2.314Mo$	$RD = D - 2B = \dfrac{N - 2.314}{P}$

Fig. 38-5 Spur gear symbols and formulas

Pitch Diameter (D). The diameter of an imaginary circle on which the gear tooth is designed.

Number of Teeth (N). The number of teeth on the gear.

Module (Mo). The length in millimetres of the pitch diameter per tooth. Mo = D/N mm. In the English (inch) system, the diametral pitch is used in place of the module.

Diametral Pitch (P). The diametral pitch is a ratio of the number of teeth (N) to a unit length of pitch diameter, P = N/D.

Note: The module is equal to the reciprocal of the diametral pitch and thus is not its metric dimensional equivalent. In order for gears to mesh, they must have the same module or diametral pitch.

If the value of the module is calculated for gears designed with standard diametral pitches, the module values will be those shown in figure 38-6. Preferred module sizes (whole numbers) will eventually replace the modules sizes converted from diametral pitches. If the diametral pitch of a gear is known, the module can be obtained by dividing 25.4 by the diametral pitch.

Outside Diameter (OD). The overall gear diameter.

Root Diameter (RD). The diameter at the bottom of the tooth.

Addendum (A). The radial distance from the pitch circle to the top of the tooth.

Dedendum (B). The radial distance from the pitch circle to the bottom of the tooth.

Whole Depth (WD). The overall height of the tooth.

Center Distance. The center distance between the two shaft centers is determined by adding the pitch diameter of the two gears together and dividing the sum by 2.

MODULE	PITCH	PRESSURE ANGLE	
		14.5	20
6.35	4		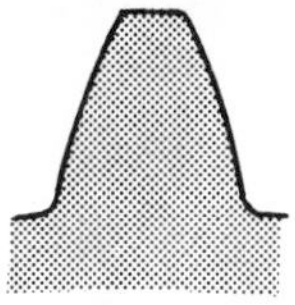
3.175	8		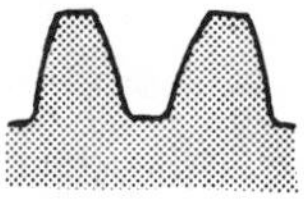
2.117	12		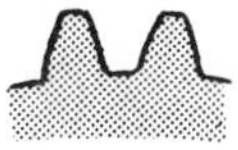
1.27	20		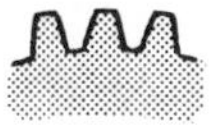

Fig. 38-6 Gear teeth sizes

Example: A 3.176 module, 24-tooth pinion mates with a 96-tooth gear. Find the center distance.

Pitch diameter of pinion = N x Mo = 24 x 3.175 = 76.2 mm

Pitch diameter of gear = N x Mo = 96 x 3.175 = 304.8 mm

Sum of the two pitch diameters = 76.2 + 304.8 = 381 mm.

Center distance = 1/2 sum of the two pitch diameters = $\frac{381}{2}$ = 190.5 mm

Ratio. The ratio of gears is a relationship between any of the following:

• The rev/min of the gears

• The number of teeth on the gears

• The pitch diameter of the gears

The ratio is obtained by dividing the larger value of any of the three by the corresponding smaller value.

Examples:

1. A gear rotates at 90 rev/min and the pinion at 360 rev/min.

    Ratio = $\frac{360}{90}$ = 4 or ratio = 4:1

2. A gear has 72 teeth; the pinion, 18 teeth.

    Ratio = $\frac{72}{18}$ = 4 or ratio = 4:1

3. A gear with a pitch diameter of 220 mm meshes with a pinion having a pitch diameter of 55 mm.

    Ratio = $\frac{\text{D of gear}}{\text{D of pinion}}$ = $\frac{220}{55}$ = 4 or ratio = 4:1

Determining Pitch Diameter (D) and Outside Diameter (OD).

The pitch diameter of a gear can easily be found if the number of teeth and diametral pitch are known. The outside diameter is equal to the pitch diameter plus two addendums. The addendum for a 14.5- or 20-degree spur gear tooth is equal to Mo.

Examples:

1. A 14.5-degree spur gear has an M of 6.35 and 34 teeth.
    Pitch diameter = N x Mo = 34 x 6.35 = 215.9 mm.
    OD = D + 2A = 215.9 + 2(6.35) = 238.6 mm.

2. The outside diameter of a 14.5-degree spur gear is 165.1 mm. The gear has 24 teeth.
    OD = Mo (N + 2) = 26Mo = 165.1

    Mo = $\frac{165.1}{26}$ = 6.35

    Addendum = Mo = 6.35
    Pitch diameter = OD – 2A
    $= 165.1 - 2 (6.35)$
    $= 165.1 - 12.7$
    $= 152.4$ mm.

Figure 38-7 illustrates how this type of information would be shown on an engineering sketch.

Motor Drive

Drawing **A-75** shows a motor drive similar to the type used to operate a load-ratio control switch on a power transformer.

The load-ratio control switch is operated by a small motor with a speed of 1080 rev/min. The shaft speed at the switch is reduced to 9 rev/min by a series of spur and miter gears.

When the operator pushes a button, the motor is activated until the circuit breaker pointer rotates 90 degrees and one of the arms depresses the roller and breaks the circuit. During this time the load-ratio control switch shaft will rotate 360 degrees, moving the contactor in the load-ratio control switch one position. This will be shown on the dial by the position indicator.

To simplify the assembly, only the pitch diameters of the gears are shown and much of the hardware has been omitted.

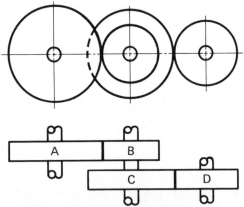

GEAR	D	N	Mo	REV/MIN	CENTER DISTANCE
A	152.4	24	6.35	300	
					114.3
B	76.2	12	6.35	600	
C	142.8	45	3.175	600	
					95.2
D	47.6	15	3.175	1800	

Fig. 38-7 Gear train data

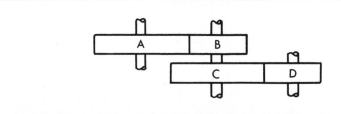

GEAR	D	N	Mo	DIRECTION	REV/MIN	CENTER DISTANCE
A	177.8		6.35	CLOCKWISE	300	
B		12				
C	152.4					
D		12	8.467			

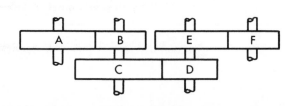

GEAR	D	N	Mo	DIRECTION	REV/MIN	CENTER DISTANCE
A	190.5		6.35	COUNTER-CLOCKWISE	240	
B		18				
C	254					
D	81.28	16				
E	203.2		4.233			
F		40				

ASSIGNMENT:
FILL IN THE MISSING INFORMATION.

A-73

GEAR DATA

GEAR	NO. OF TEETH	PITCH DIAMETER	MODULE	REV/MIN
G_1	24		1.27	
G_2		121.92		
G_3	20	25.4		
G_4	100			
G_5		25.4	1.27	
G_6		152.4		
G_7	18			
G_8		182.82		
G_9	72		1.27	
G_{10}		63.5	2.54	
G_{11}	25			

SHAFT DATA

SHAFT	GEARS ON SHAFT	REV/MIN	SHAFT ROTATION *
S_1			
S_2			
S_3			
S_4			
S_5			
S_6			
S_7			COUNTER-CLOCKWISE

* AS VIEWED FROM FRONT OR BOTTOM OF MOTOR DRIVE ASSEMBLY

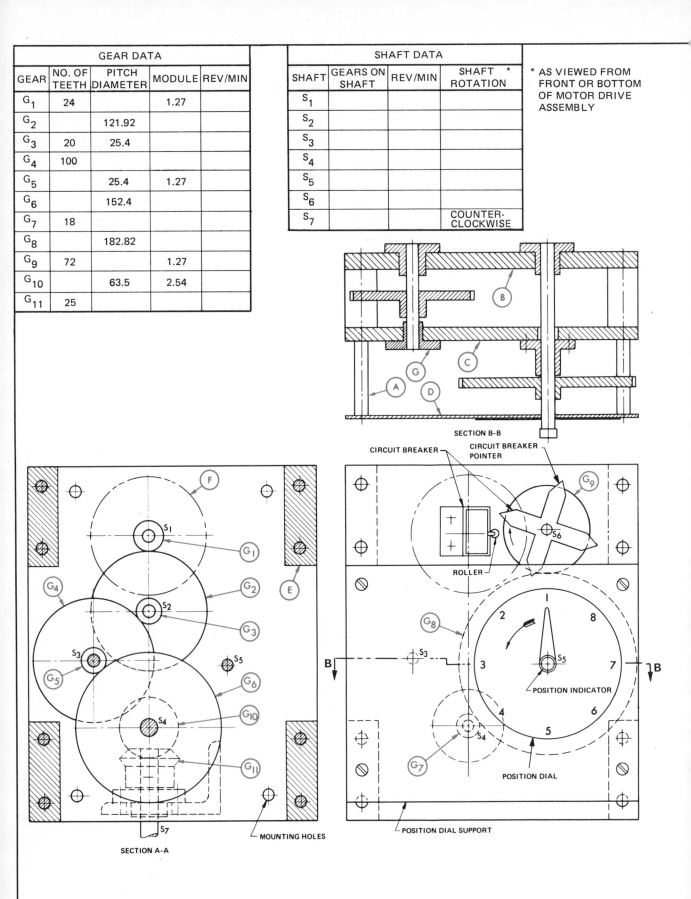

SECTION B-B

CIRCUIT BREAKER

CIRCUIT BREAKER POINTER

ROLLER

POSITION INDICATOR

POSITION DIAL

POSITION DIAL SUPPORT

SECTION A-A

MOUNTING HOLES

ASSIGNMENT

Complete the information shown in the gear and shaft tables.

QUESTIONS

1. What are the names of parts (A) to (G) ?
2. How many spur gears are shown?
3. How many miter gears are shown?
4. How many gear shafts are there?
5. What is the ratio between the following gears? (A) G_1 and G_2, (B) G_3 and G_4, (C) G_5 and G_6, (D) G_7 and G_8, (E) G_8 and G_9, (F) G_{10} and G_{11}.
6. What is the center-to-center distance between the following shafts? (A) S_1 and S_2, (B) S_2 and S_3, (C) S_3 and S_4, (D) S_4 and S_5, (E) S_5 and S_6.
7. How many seconds does it take to turn the load ratio control switch one position?
8. How many seconds does it take the position indicator to move continuously from position 4 to position 7?
9. What is the rev/min ratio between the motor and the switch?
10. If the switch shaft S_7 rotates 1800 degrees, how many degrees does the position indicator rotate?

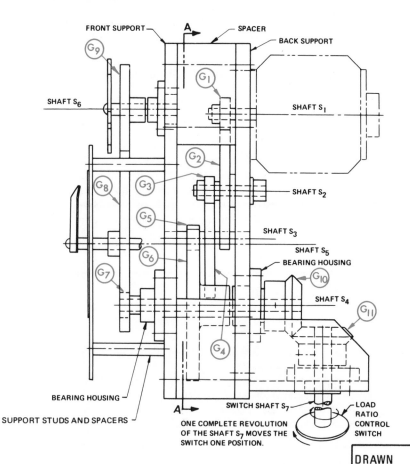

ONE COMPLETE REVOLUTION OF THE SHAFT S_7 MOVES THE SWITCH ONE POSITION.

METRIC

DRAWN	DATE
MOTOR DRIVE ASSEMBLY	A-74

267

ROLLER-ELEMENT BEARINGS

Roller-element bearings use a type of rolling element between the loaded members. Relative motion is accommodated by rotation of the elements. Roller-element bearings are usually housed in bearing races conforming to the element shapes. In addition, a cage or separator is often used to locate the elements within the bearings. These bearings are usually categorized by the form of the rolling element and in some instances by the load type they carry, figures 39-1 and 39-2. Roller-element bearings are generally classified as either ball or roller.

Ball Bearings

Ball bearings may be roughly divided into three categories: radial, angular contact, and thrust. Radial-contact ball bearings are designed for applications in which the load is primarily radial with only low-magnitude thrust loads. Angular-contact bearings are used where loads are combined radial and high thrust, and where precise shaft location is required. Thrust bearings handle loads which are primarily thrust.

Roller Bearings

Roller bearings have higher load capacities than ball bearings for a given envelope size. They are widely used in moderate-speed, heavy-duty applications. The four principal types of roller bearings are: cylindrical, needle, tapered, and spherical. Cylindrical roller bearings utilize cylinders with approximate length-diameter ratios ranging

SINGLE ROW, DEEP GROOVE BALL BEARINGS

The *Single Row, Deep Groove Ball Bearing* will sustain, in addition to radial load, a substantial thrust load in either direction . . . even at very high speeds. This advantage results from the intimate contact existing between the balls and the deep, continuous groove in each ring. When using this type of bearing, careful alignment between the shaft and housing is essential. This bearing is also available with seals, which serve to exclude dirt and retain lubricant.

ANGULAR CONTACT BALL BEARINGS

The *Angular Contact Ball Bearing* supports a heavy thrust load in one direction . . . sometimes combined with a moderate radial load. A steep contact angle, assuring the highest thrust capacity and axial rigidity, is obtained by a high thrust supporting shoulder on the inner ring and a similar high shoulder on the opposite side of the outer ring. These bearings can be mounted singly or, when the sides are flush ground, in tandem for constant thrust in one direction; mounted in pairs, also when sides are flush ground, for a combined load . . . either face-to-face or back-to-back.

Fig. 39-1A Roller-element bearings

CYLINDRICAL ROLLER BEARINGS

The *Cylindrical Roller Bearing* has high radial capacity and provides accurate guiding of the rollers, resulting in a close approach to true rolling. Consequent low friction permits operation at high speed. Those types which have flanges on one ring only, allow a limited free axial movement of the shaft in relation to the housing. They are easy to dismount even when both rings are mounted with a tight fit. The double row type assures maximum radial rigidity and is particularly suitable for machine tool spindles.

BALL THRUST BEARINGS

The *Ball Thrust Bearing* is designed for thrust load in one direction only. The load line through the balls is parallel to the axis of the shaft . . . resulting in high thrust capacity and minimum axial deflection. Flat seats are preferred . . . particularly where the load is heavy . . . or where close axial positioning of the shaft is essential; as for example, in machine tool spindles.

SPHERICAL ROLLER THRUST BEARINGS

The *Spherical Roller Thrust Bearing* is designed to carry heavy thrust loads, or combined loads which are predominantly thrust. This bearing has a single row of rollers which roll on a spherical outer race with full self-alignment. The cage, centered by an inner ring sleeve, is constructed so that lubricant is pumped directly against the inner ring's unusually high guide flange. This ensures good lubrication between the roller ends and the guide flange. The spherical roller thrust bearing operates best with relatively heavy oil lubrication.

Fig. 39-1B Roller-element bearings, continued

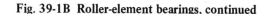

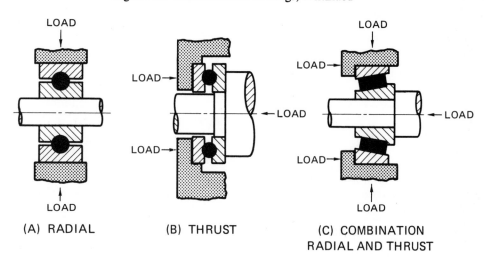

Fig. 39-2 Types of bearing loads

from 1:1 to 1:3 as rolling elements. Needle roller bearings utilize cylinders or needles of greater length-diameter ratios. Tapered and spherical roller bearings are capable of supporting combined radial and thrust loads.

The rolling elements of tapered roller bearings are truncated cones. Spherical roller bearings are available with both barrel and hourglass roller shapes. The primary advantage of spherical roller bearings is their self-aligning capability.

RETAINING RINGS

Retaining rings, or snap rings, are designed to provide a removable shoulder to locate, retain, or lock components accurately on shafts and in bores and housings, figures 39-3 and 39-4. They are easily installed and removed. Since they are usually made of spring steel, retaining rings have a high shear strength and impact capacity. In addition to fastening and positioning, a number of rings are designed for taking up end play caused by accumulated tolerances or wear in the parts being retained.

O-RING SEALS

O-rings are used as an axial mechanical seal (a seal which forms a running seal between a moving shaft and a housing) or a static seal (no moving parts). The advantage of using an O-ring as a gasket-type seal, figure 39-5, over conventional gaskets is that the nuts need not be tightened uniformly and sealing compounds are not required. A rectangular groove is the most common type of groove used for O-rings.

CLUTCHES

Clutches are used to start and stop machines or rotating elements without starting or stopping the prime mover. They are also used for automatic disconnection, quick starts and stops, and to permit shaft rotation

in one direction only such as the *overrunning* clutch shown in figure 39-6. A full complement of sprags between concentric inner and outer races transmits power from one race to the other by wedging action of the sprags when either race is rotated in the driving direction. Rotation in the opposite direction frees the sprags and the clutch is disengaged or *overruns.* This type of clutch is used in the power drive, drawing A-75.

BELT DRIVES

A *belt drive* consists of an endless flexible belt connecting two wheels or pulleys. Belt drives depend on friction between belt and pulley surfaces for transmission of power.

In a V-belt drive, the belt has a trapezoidal cross section, and runs in V-shaped grooves on the pulleys. These belts are made of cords or cables, impregnated and covered with rubber or other organic compound. The covering is formed to produce the required cross section. V-belts are usually manufactured as endless belts, although open-end and link types are available.

In the case of V-belts, the friction for the transmission of the driving force is increased by the wedging of the belt into the grooves on the pulley.

V-Belt Sizes

To facilitate interchangeability and to insure uniformity, V-belt manufacturers have developed industrial standards for the various types of V-belts, figure 39-7. Industrial V-belts are made in two types: heavy duty (conventional and narrow) and light duty. Conventional belts are available in A, B, C, D, and E sections. Narrow belts are made in 3V, 5V, and 8V sections. Light-duty belts come in 3L, 4L, and 5L sections.

AXIAL ASSEMBLY RINGS

INTERNAL — EXTERNAL

BASIC TYPES: Designed for axial assembly. Internal ring is compressed for insertion into bore or housing, external ring expanded for assembly over shaft. Both rings seat in deep grooves and are secure against heavy thrust loads and high rotational speeds.

INTERNAL — EXTERNAL

INVERTED RINGS: Same tapered construction as basic types, with lugs inverted to about bottom of groove. Section height increased to provide higher shoulder, uniformly concentric with housing or shaft. Rings provide better clearance, more attractive appearance than basic types.

END PLAY RINGS

INTERNAL — EXTERNAL

BOWED RINGS: For assemblies in which accumulated tolerances cause objectionable end play between ring and retained part. Bowed construction permits rings to provide resilient end-play takeup in axial direction while maintaining tight grip against groove bottom.

EXTERNAL — INTERNAL

RADIAL RINGS: Bowed E-rings are used for providing resilient end-play takeup in an assembly.

SELF-LOCKING RINGS

EXTERNAL — INTERNAL

CIRCULAR EXTERNAL RINGS: The push-on type of fastener with inclined prongs which bend from their initial position to grip the shaft. Ring at left has arched rim for increased strength and thrust load capacity; extra-long prongs accommodate wide shaft tolerances. Ring at right has flat rim, shorter locking prongs, smaller OD.

INTERNAL

CIRCULAR INTERNAL PINS: Designed for use in bores and housings. Functions in same manner as external types except that locking prongs are on the outside rim.

RADIAL LOCKING RINGS

EXTERNAL

CRESCENT RING: Has a tapered section similar to the basic axial types. Remains circular after installation on a shaft and provides a tight grip against the groove bottom.

INTERNAL

E-RINGS: Provide a large bearing shoulder on small-diameter shafts and is often used as a spring retainer. Three heavy prongs, spaced approximately 120 degrees apart, provide contact surface with groove bottom.

Fig. 39-3 Stamped retaining rings

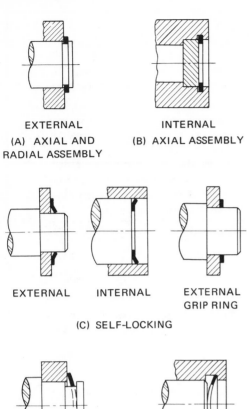

EXTERNAL

(A) AXIAL AND
RADIAL ASSEMBLY

INTERNAL

(B) AXIAL ASSEMBLY

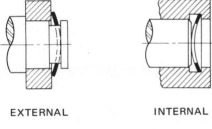

EXTERNAL   INTERNAL   EXTERNAL
                        GRIP RING

(C) SELF-LOCKING

EXTERNAL   INTERNAL

(D) END–PLAY TAKEUP

Fig. 39-4 Retaining ring application

Fig. 39-5 O-Ring Seal

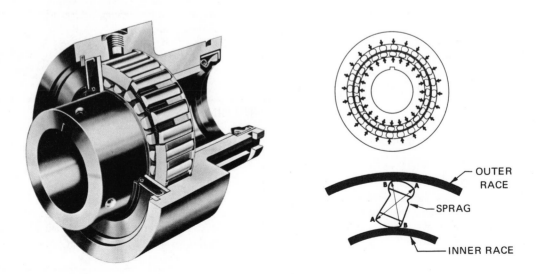

OUTER
RACE

SPRAG

INNER RACE

Fig. 39-6 Overrunning clutch

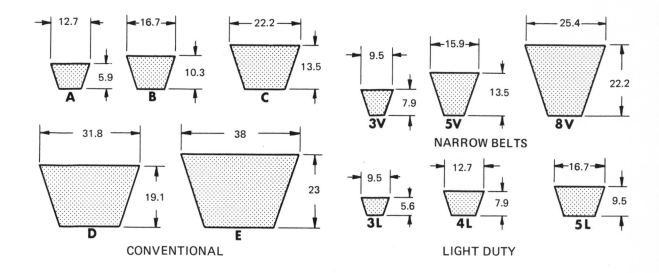

Fig. 39-7 V-belt sizes

Sheaves and Bushings

Sheaves (the grooved wheels of pulleys) are sometimes equipped with tapered bushings for ease of installation and removal, figure 39-8. They have extreme holding power, providing the equivalent of a shrink fit. The sheave and bushing used in the power drive, drawing A-75, have a six-hole drilling arrangement in both the bushing and sheave making it possible to insert the cap screw from either side. This is especially advantageous for applications where space is at a premium.

REFERENCES AND SOURCE MATERIAL

1. A.O. Dehayt, "Basic Bearing Types," *Machine Design* 40, No. 14 (1968).

2. *Machine Design* 39, No. 14 (1967).

3. American Chain Association.

Fig. 39-8 V-belt sheave and bushing

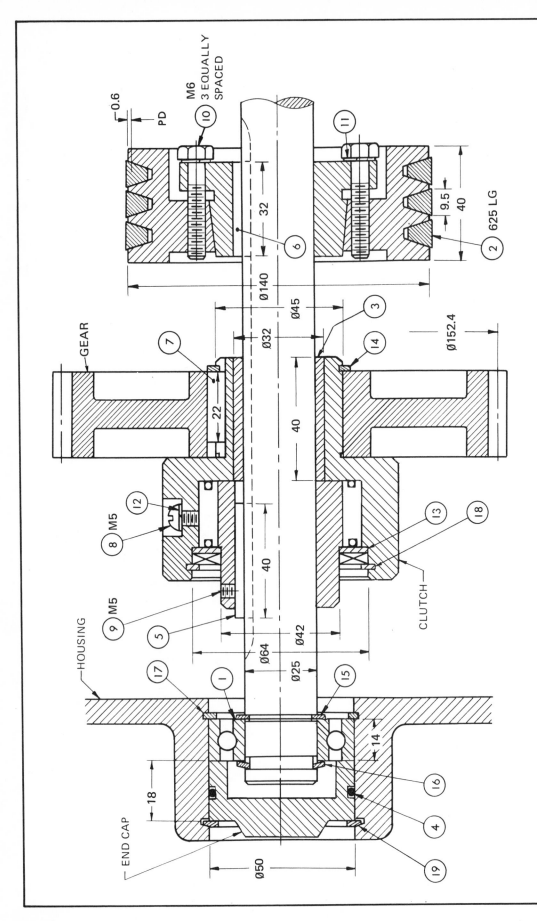

NOTE: ALL DIMENSIONS SHOWN ARE NOMINAL SIZE.

SKETCHING ASSIGNMENT

In the sketching area provided, make a one-view detail drawing of the end cap, using dimensions taken from O-ring catalogs for the groove. Use your judgment for dimensions not shown.

BILL OF MATERIAL

Prepare a bill of material for parts 1 to 19. Refer to manufacturers' catalogs and tables found in drafting manuals.

QUESTIONS

1. How many cap screws fasten the sheave to the bushing?

2. List five parts or methods that are used to lock or join parts together on this assembly.

3. How many V-belts are used?

4. How many keys are there?

5. How many retaining rings are used?

6. What type of bearing is part ③ ?

7. What type of bearing is part ① ?

8. What prevents the oil from leaking out between the housing and the end cap?

9. Can the gear be driven in both directions?

10. If the module on the gear is 3.175, what is the number of teeth?

11. What is the pitch diameter of the sheave?

12. What size V-belt is required?

MAKE A DETAIL DRAWING OF THE
END CAP – 5 mm SQUARES

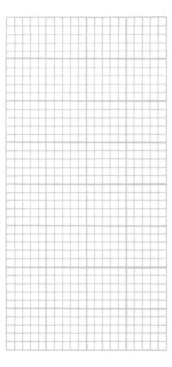

5 mm SQUARES

ANSWERS

1 _____

2 _____

_____

_____

_____

3 _____

4 _____

5 _____

6 _____

7 _____

8 _____

9 _____

10 _____

11 _____

12 _____

METRIC	
SCALE	NOT TO SCALE
DRAWN	DATE

POWER DRIVE

A-75

REVISIONS	1	MAR. 1/78	CAMPBELL
		PT 10 WAS M5	

RATCHET WHEELS

Ratchet wheels are used to transform reciprocating or oscillatory motion into intermittent motion, to transmit motion in one direction only, or to serve as an indexing device.

Common forms of ratchets and pawls are shown in figure 40-1. The teeth in the ratchet engage with the teeth in the pawl, permitting rotation in one direction only.

When designing a ratchet wheel and pawl, lay out points A, B, and C as shown in figure 40-1(A) on the same circle to insure that the smallest forces are acting on the system.

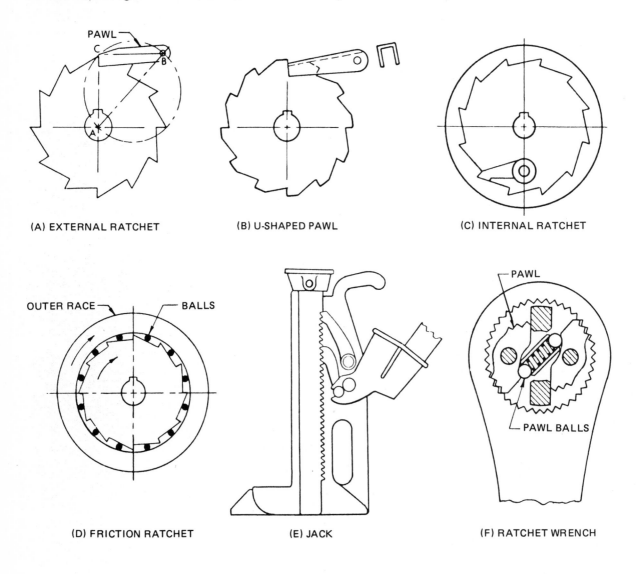

(A) EXTERNAL RATCHET (B) U-SHAPED PAWL (C) INTERNAL RATCHET

(D) FRICTION RATCHET (E) JACK (F) RATCHET WRENCH

Fig. 40-1 Rachets and pawls

Mechanical Advantage

Mechanical advantage occurs when a mass in one place lifts a heavier mass in another place, or a force applied at one point on a lever produces a greater force at another point. Examples of mechanical advantage are the teeter-totter, the winch, and the gears shown in figure 40-2.

The teeter-totter shows how mechanical advantage can be applied. If a 1 kilogram (kg) mass is placed on one end of a teeter-totter and a 2 kg mass is placed on the other end, the 1 kg mass goes up and the 2 kg mass goes down when both masses are placed equidistant from the fulcrum.

If the fulcrum is moved to make the distance from the 1 kg mass to the fulcrum twice that of the distance between the fulcrum and the 2 kg mass, the masses become balanced. If the fulcrum is moved still closer to the 2 kg mass, the 1 kg mass goes down and the 2 kg mass goes up. As the distance between the fulcrum and the 1 kg mass increases, the distance that the 1 kg mass moves must increase proportionately to maintain the same movement of the 2 kg mass. Thus, a light mass can move a heavier mass, but in doing so, the smaller mass must travel further than the larger one.

Mechanical advantage also occurs when a winch or different size gears are used. With the winch design shown in figure 40-2(B), a mechanical advantage of 10 is obtained because the handle is 10 times the distance from the fulcrum as compared to the center of the rope.

The mechanical advantage of gears can easily be determined by obtaining the ratio between the number of teeth on the gears or the ratio between the pitch diameters.

In the winch for example, the center of the handle bar is 256 mm from the center of the shaft to which the pinion gear is attached. Half the pitch diameter of the pinion is 16 mm, Drawing A-76. This produces a mechan-

ical advantage of 256:16 or 16:1. Further mechanical advantage is gained through the gear attached to the Ø25 mm shaft on which the rope revolves.

The hand will move a distance of approximately 1.5 metres when turning the handle one complete revolution. This will turn the rope drum one fifth of a revolution (50:10 teeth ratio), winding an 08 mm rope up approximately 20 mm. A greater force can be exerted using the winch than by simply pulling the rope.

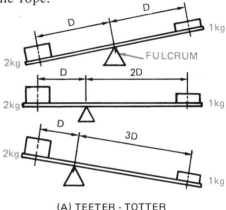

(A) TEETER - TOTTER

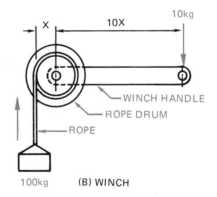

(B) WINCH

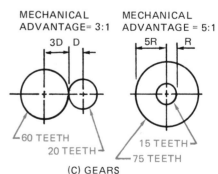

(C) GEARS

Fig. 40-2 Mechanical advantage

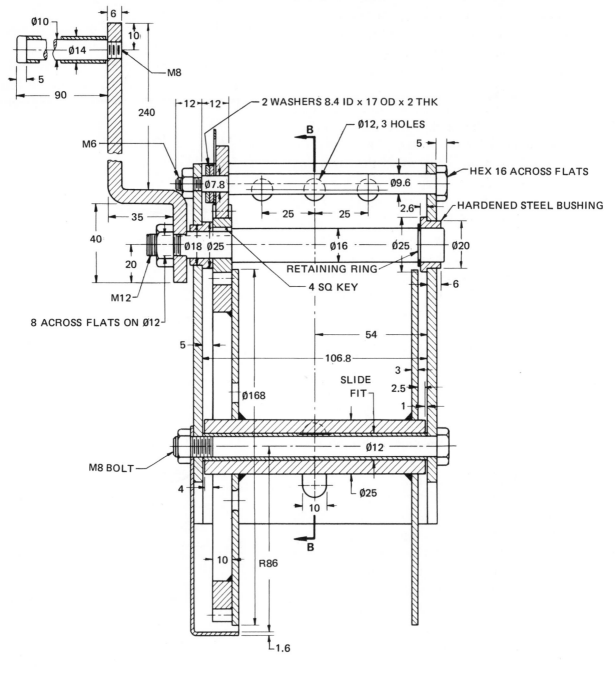

ASSIGNMENT: MAKE DETAIL DRAWINGS OF PARTS. NOTE, ALL DIMENSIONS SHOWN ARE NOMINAL SIZES. ALLOWANCES AND TOLERANCES ARE TO BE DETERMINED.

Ø10
Ø14
6
10
M8
5
90
240
M6
12 12
2 WASHERS 8.4 ID x 17 OD x 2 THK
B
Ø12, 3 HOLES
5
HEX 16 ACROSS FLATS
Ø7.8
Ø9.6
HARDENED STEEL BUSHING
35
25 25
2.6
40
Ø18 Ø25
Ø16 Ø25
Ø20
20
RETAINING RING
M12
4 SQ KEY
8 ACROSS FLATS ON Ø12
6
54
5
106.8
3
Ø168
SLIDE
FIT
2.5
1
M8 BOLT
Ø12
4
Ø25
10
10
R86
B
1.6

SECTION A–A

278

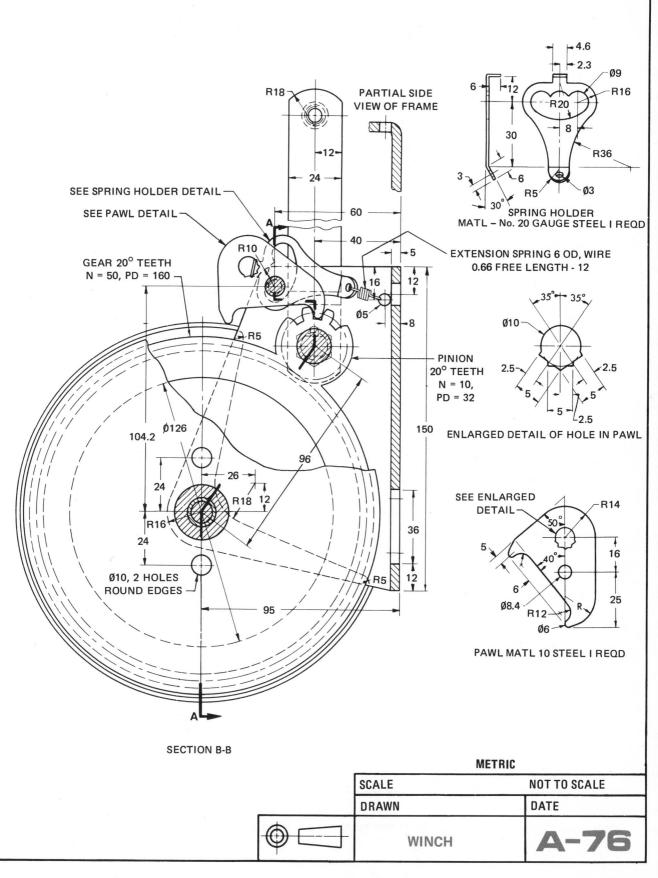

PARTIAL SIDE VIEW OF FRAME

R18

—12—
—24—

SEE SPRING HOLDER DETAIL
SEE PAWL DETAIL

GEAR 20° TEETH
N = 50, PD = 160

R10

R5

104.2 Ø126

26

24

R18 12

R16

24

Ø10, 2 HOLES
ROUND EDGES

95

96

60

40

5

16

Ø5

150

36

12

8

R5

SECTION B-B

A

PINION
20° TEETH
N = 10,
PD = 32

EXTENSION SPRING 6 OD, WIRE
0.66 FREE LENGTH - 12

SPRING HOLDER
MATL – No. 20 GAUGE STEEL I REQD

4.6
2.3
Ø9
R16
R20
8
6 12
30
3 6
R5 Ø3
30°
R36

35° 35°
Ø10
2.5 2.5
5 5
5
2.5

ENLARGED DETAIL OF HOLE IN PAWL

SEE ENLARGED
DETAIL
R14
50°
5
40°
16
6
Ø8.4
R12 R
Ø6
25

PAWL MATL 10 STEEL I REQD

METRIC	
SCALE	NOT TO SCALE
DRAWN	DATE

WINCH

A-76

UNIT

41

MODERN ENGINEERING TOLERANCING

An engineering drawing of a manufactured part conveys information from the designer to the manufacturer and inspector. It must contain all information necessary for the part to be correctly manufactured. It must also enable the inspector to determine precisely whether the finished parts are acceptable.

Therefore, each drawing must convey three essential items of information: the material to be used, the size or dimensions of the part, and the shape or geometrical characteristics of the part. The drawing must also indicate the permissible variation of size and form.

The actual size of a feature must be within the size limits specified on the draw-ing. Each measurement made at any cross section of the feature must not be greater than the maximum limit of size, nor smaller than the minimum limit of size, figure 41-1. Although each part is within the prescribed tolerance zones, the parts may not be usable because of their deviation from their form.

Formerly, tolerances for which there were no precise interpretations were often shown. While tolerancing of geometrical characteristics was sometimes limited to notes such as,

PARALLEL WITH SURFACE B WITHIN .001

or

STRAIGHT WITHIN .005

it did not precisely express permissible variations.

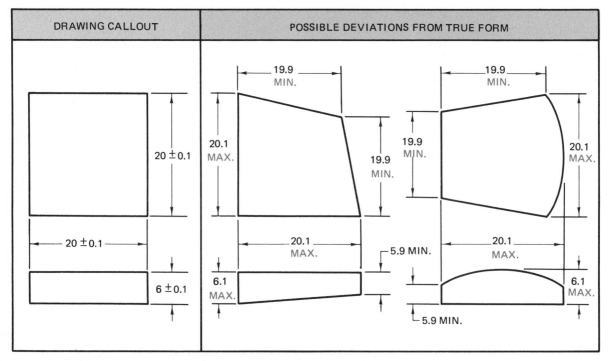

(A) SQUARE FEATURES

Fig. 41-1 Deviations permitted by toleranced dimensions

In order to meet functional requirements, it is often necessary to control errors of form including: squareness, roundness, and flatness, as well as deviation from true size. In the case of mating parts, such as holes and shafts, it is usually necessary to ensure that they do not cross the boundary of perfect form at their maximum material size (the smallest hole or the largest shaft) because of being bent or otherwise deformed. This condition is shown in figure 41-2 (page 282), where features are not permitted to cross the boundary of perfect form at the least material size (the largest hole or the smallest shaft).

The system of *geometrical tolerancing* offers a precise interpretation of drawing requirements. Geometrical tolerancing controls geometrical characteristics of parts. These characteristics include: flatness, roundness, angularity, profile, and position. Other techniques, such as datum systems, datum targets, and projected tolerance zones were developed in order to facilitate this precise interpretation.

Geometrical tolerances need not be used for every feature of a part. Generally, if each feature meets all dimensional tolerances, form variations will be adequately controlled by the accuracy of the manufacturing process and equipment used. A geometrical tolerance is used when geometrical errors must be limited more closely than might ordinarily be expected from the manufacturing process. A geometrical tolerance is also used to meet functional or interchangeability requirements.

Since some of the symbols used by the United States and other standardization bodies such as ISO, Canada, and Great Britain, vary slightly, both symbols are shown in this text, figure 41-3 (page 282).

GEOMETRICAL TOLERANCING

A geometrical tolerance is the maximum permissible variation of form, orientation, or location of a feature from that indicated or specified on a drawing. The tolerance value represents the width or diameter of the

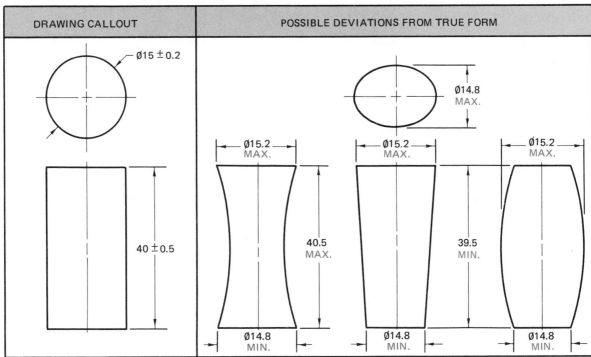

Fig. 41-1 Deviations permitted by toleranced dimensions, continued

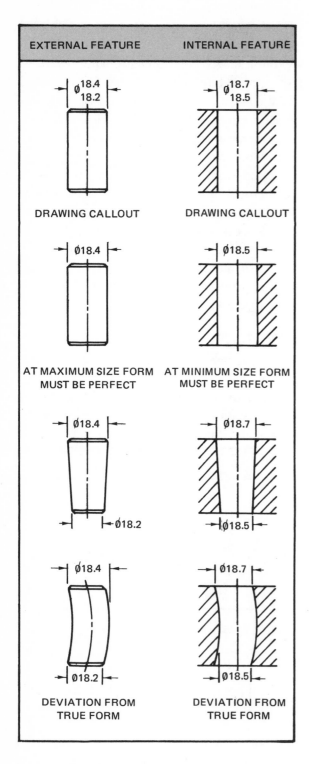

EXTERNAL FEATURE	INTERNAL FEATURE
⌀18.4 / 18.2 — DRAWING CALLOUT	⌀18.7 / 18.5 — DRAWING CALLOUT
⌀18.4 — AT MAXIMUM SIZE FORM MUST BE PERFECT	⌀18.5 — AT MINIMUM SIZE FORM MUST BE PERFECT
⌀18.4 / ⌀18.2	⌀18.7 / ⌀18.5
⌀18.4 / ⌀18.2 — DEVIATION FROM TRUE FORM	⌀18.7 / ⌀18.5 — DEVIATION FROM TRUE FORM

Fig. 41-2 Examples of deviation of form when perfect form at the maximum material size is required

GEOMETRICAL CHARACTERISTIC SYMBOLS

CHARACTERISTIC			SYMBOL
FORM TOLERANCES	FORM OF A LINE	STRAIGHTNESS	—
		ROUNDNESS	◯
		PROFILE OF A LINE	⌒
	FORM OF A SURFACE	FLATNESS	▱
		CYLINDRICITY	⌀
		PROFILE OF A SURFACE	◠
	ORIENTATION OF RELATED FEATURES	ANGULARITY	∠
		PARALLELISM	//
		PERPENDICULARITY	⊥
TOLERANCE FOR LOCATION		POSITION	⊕
		CONCENTRICITY	◎
		SYMMETRY	≡
	RUN-OUT –	CIRCULAR	↗
		TOTAL	⫽
SUPPLEMENTARY SYMBOLS		MAXIMUM MATERIAL SIZE	Ⓜ
		TRUE POSITION	[XX]
		DATUM IDENTIFICATION	ANSI [-A-] ISO Ⓐ
		DATUM TARGETS	⌀5 / A2
		REGARDLESS OF FEATURE SIZE	Ⓢ
		PROJECTED TOLERANCE ZONE	Ⓟ

Fig. 41-3 Geometrical characteristic symbols

tolerance zone, within which the point, line, or surface of the feature should lie.

Feature Control Symbol

The geometrical tolerance is shown on the drawing by a *feature control symbol,* figure 41-4. A feature control symbol consists of a rectangular frame divided into two or more compartments. The first compartment (starting from the left) contains the geometric characteristic. The second compartment contains the allowable tolerance. Where applicable, the tolerance is preceded by the diameter symbol and followed by the symbol for maximum material. Other compartments are added when datums must be specified. The feature control symbol joins the feature being controlled by a leader which is usually perpendicular to the surface being controlled. However, ANSI permits the leader to touch the surface at an angle other than 90 degrees and also permits the feature control symbol to be attached to the extension in figure 41-5.

Figures 41-6 and 41-7 illustrate the application and preferred location of a feature control symbol.

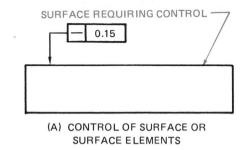

(A) CONTROL OF SURFACE OR SURFACE ELEMENTS

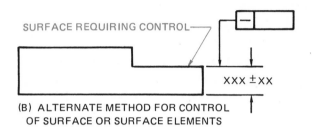

(B) ALTERNATE METHOD FOR CONTROL OF SURFACE OR SURFACE ELEMENTS

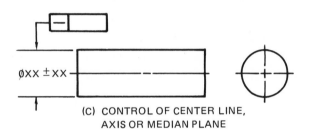

(C) CONTROL OF CENTER LINE, AXIS OR MEDIAN PLANE

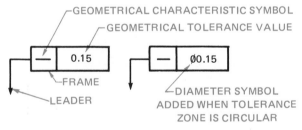

Fig. 41-4 Feature control symbol

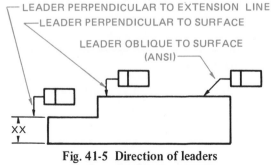

Fig. 41-5 Direction of leaders for feature control symbols

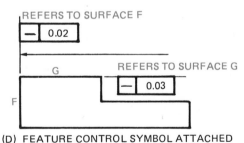

(D) FEATURE CONTROL SYMBOL ATTACHED TO EXTENSION LINES (ANSI ONLY)

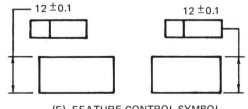

(E) FEATURE CONTROL SYMBOL ASSOCIATED WITH SIZE DIMENSIONS

Fig. 41-6 Application of feature control symbols

FORM TOLERANCE – STRAIGHTNESS

Lines and Surfaces

Straightness is fundamentally a characteristic of a line, such as the edge of a part or a line scribed on a surface. Figure 41-8(A) states in symbolic form that the line shall be straight within 0.15 mm. This means the line shall be contained within a tolerance zone consisting of an area between two parallel straight lines in the same plane, separated by the specified tolerance.

When a feature control symbol is intended to apply to a surface, the leader touches the surface or an extension line as shown in figure 41-5.

Application to Center Lines

When a feature control symbol is intended to apply to an axis, center line, or median plane of a feature, the leader is directed to the feature size dimension as shown in figure

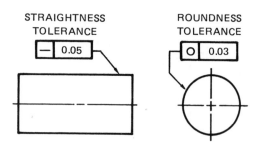

Fig. 41-7 Preferred location of feature control symbol

41-9(A) and (B). Alternately, a common leader may be used with the size dimension and feature control symbol, as in figure 41-6(E).

When controlling one or all of the features sharing the same center line, one of the methods shown in figure 41-9(B) or (C) is used.

When the resulting tolerance zone is circular or cylindrical, such as when controlling the straightness of a center line, figure 41-9,

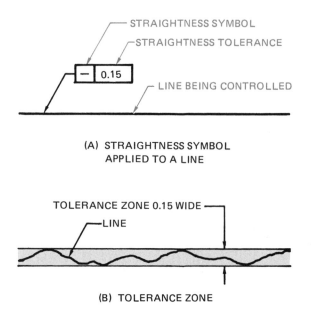

(A) STRAIGHTNESS SYMBOL APPLIED TO A LINE

(B) TOLERANCE ZONE

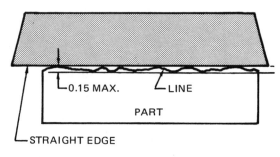

(C) CHECKING WITH STRAIGHTEDGE

Fig. 41-8 Straightness symbol and application (applying to a surface)

a diameter symbol precedes the tolerance value in the feature control symbol.

Straightness in a Specified Length

It is often desirable on long parts to specify a straightness tolerance over a specific length, either with or without a maximum overall tolerance. This requirement is specified on a drawing by including the specified length with the tolerance as shown in figures 41-10 and 41-11 (page 286).

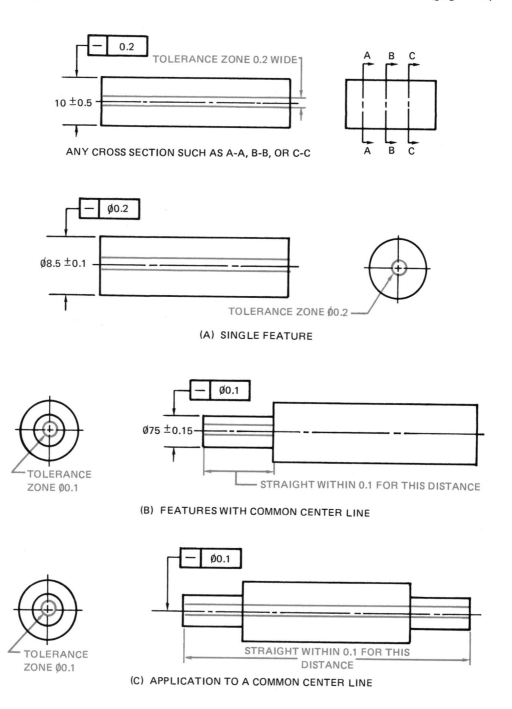

(A) SINGLE FEATURE

(B) FEATURES WITH COMMON CENTER LINE

(C) APPLICATION TO A COMMON CENTER LINE

Fig. 41-9 Application of straightness symbol to center lines

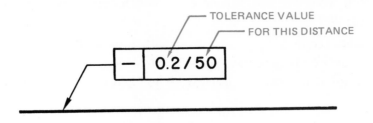

Fig. 41-10 Tolerance in a specified length

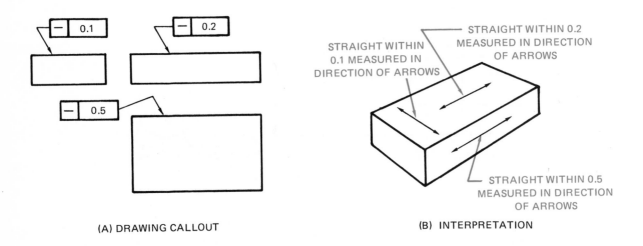

(A) DRAWING CALLOUT

(B) INTERPRETATION

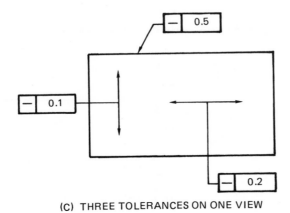

(C) THREE TOLERANCES ON ONE VIEW

Fig. 41-11 Straightness in several directions

1.

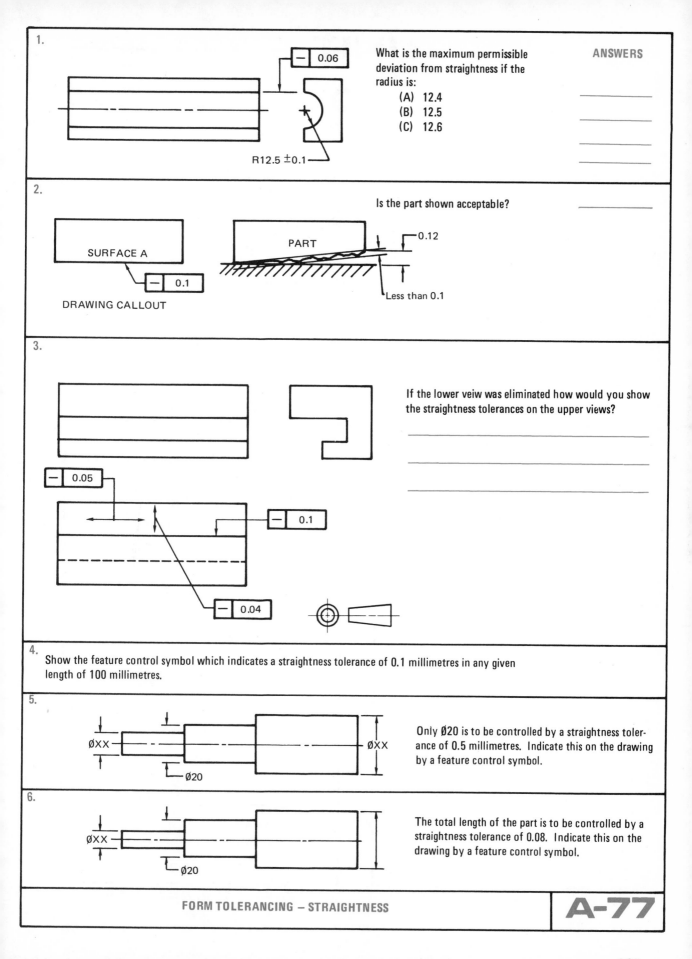

What is the maximum permissible deviation from straightness if the radius is:

   (A) 12.4
   (B) 12.5
   (C) 12.6

ANSWERS

_____

_____

_____

_____

R12.5 ±0.1

—| 0.06

2.

Is the part shown acceptable? _____

SURFACE A

—| 0.1

DRAWING CALLOUT

PART

0.12

Less than 0.1

3.

If the lower veiw was eliminated how would you show the straightness tolerances on the upper views?

_____

_____

_____

—| 0.05

—| 0.1

—| 0.04

4.
Show the feature control symbol which indicates a straightness tolerance of 0.1 millimetres in any given length of 100 millimetres.

5.

øXX

ø20

øXX

Only ø20 is to be controlled by a straightness tolerance of 0.5 millimetres. Indicate this on the drawing by a feature control symbol.

6.

øXX

ø20

The total length of the part is to be controlled by a straightness tolerance of 0.08. Indicate this on the drawing by a feature control symbol.

FORM TOLERANCING — STRAIGHTNESS

A-77

287

MAXIMUM MATERIAL CONDITION

Maximum material condition (MMC) is a state in which a feature contains the maximum amount of material within the stated limits of size. An example is the minimum hole diameter, maximum shaft diameter, figure 42-1. *Virtual Condition* is the boundary generated by the collective effects of the MMC limit of a feature and any applicable form or positional tolerance. For an internal feature, such as a hole, it is the minimum measured size minus any form variation. For an external feature, such as a shaft, it is the maximum measured size plus any form variation, figure 42-2(C).

Advantage of the MMC Concept

Parts are generally toleranced so they will assemble when mating features are at MMC. Additional tolerance on form or location is permitted when features depart from their MMC size.

APPLYING MMC TO A STRAIGHTNESS TOLERANCE

When specifying a straightness tolerance as shown in figure 42-3(B), the maximum

DRAWING CALLOUT	MAXIMUM MATERIAL CONDITION

Ø25 ±0.1   Ø24.9

SMALLEST HOLE (MMC)

(A) HOLE

Ø24.8 ±0.1   Ø24.9

LARGEST SHAFT (MMC)

(B) SHAFT

Fig. 42-1 Maximum material condition (MMC)

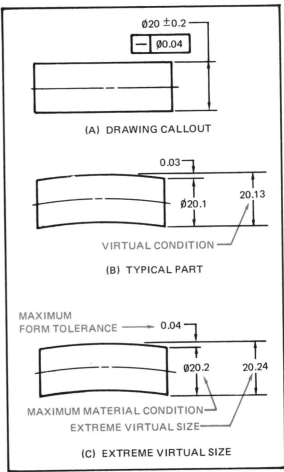

Ø20 ±0.2

Ø0.04

(A) DRAWING CALLOUT

0.03

Ø20.1   20.13

VIRTUAL CONDITION

(B) TYPICAL PART

MAXIMUM FORM TOLERANCE   0.04

Ø20.2   20.24

MAXIMUM MATERIAL CONDITION
EXTREME VIRTUAL SIZE

(C) EXTREME VIRTUAL SIZE

Fig. 42-2 Virtual size and condition

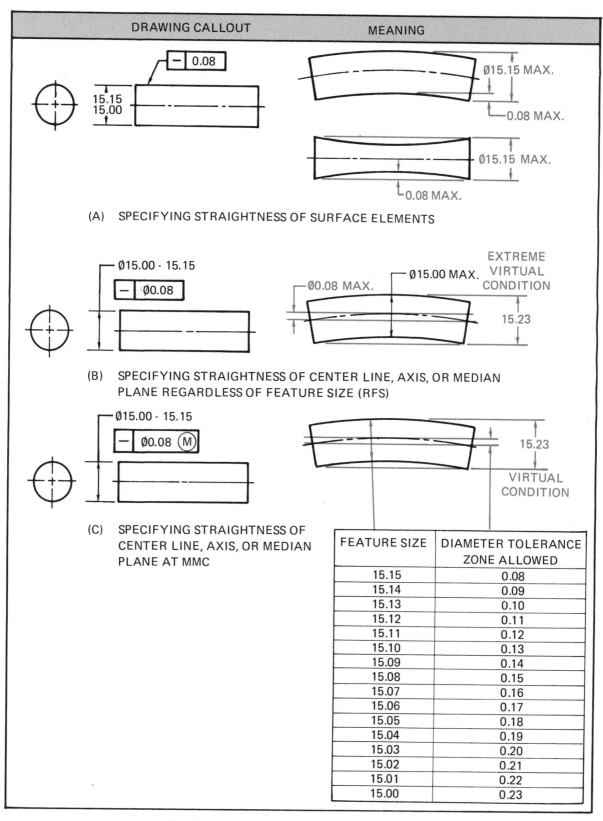

DRAWING CALLOUT	MEANING

(A) SPECIFYING STRAIGHTNESS OF SURFACE ELEMENTS

(B) SPECIFYING STRAIGHTNESS OF CENTER LINE, AXIS, OR MEDIAN PLANE REGARDLESS OF FEATURE SIZE (RFS)

(C) SPECIFYING STRAIGHTNESS OF CENTER LINE, AXIS, OR MEDIAN PLANE AT MMC

FEATURE SIZE	DIAMETER TOLERANCE ZONE ALLOWED
15.15	0.08
15.14	0.09
15.13	0.10
15.12	0.11
15.11	0.12
15.10	0.13
15.09	0.14
15.08	0.15
15.07	0.16
15.06	0.17
15.05	0.18
15.04	0.19
15.03	0.20
15.02	0.21
15.01	0.22
15.00	0.23

Fig. 42-3 A comparison of straightness tolerancing

allowable straightness tolerance is 0.08 mm regardless of the feature size. If, however, the straightness tolerance of 0.08 mm is required only at MMC, further geometric variation can be permitted without jeopardizing assembly, as the features approach their least material size, figure 42-4.

If the tolerance can be modified on an MMC basis, this is specified on the drawing by including the symbol (M) immediately after the tolerance value in the feature control symbol, figures 42-3(C) and 42-5.

If the virtual size must be kept within the maximum material boundary, the form tolerance must be specified as zero at MMC, as shown in figure 42-6.

It is sometimes necessary to ensure that the geometrical tolerance does not vary over

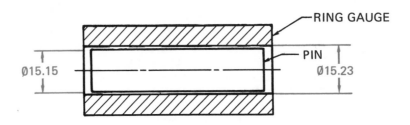

(A) MAXIMUM DIAMETER OF PIN WITH PERFECT FORM WITH GAUGE HOLE OF Ø15.23

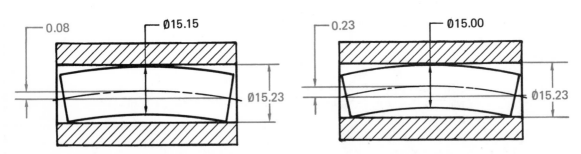

(B) WITH THE PIN AT MAXIMUM DIAMETER OF 15.15 THE GAUGE WILL ACCEPT THE PIN WITH UP TO A MAXIMUM OF 0.08 VARIATION IN STRAIGHTNESS.

(C) WITH THE PIN AT MINIMUM DIAMETER OF 15.00 THE GAUGE WILL ACCEPT THE PIN WITH UP TO A MAXIMUM OF 0.23 VARIATION IN STRAIGHTNESS AND THE PIN IS ACCEPTABLE.

Fig. 42-4 Size and tolerance variations for pin shown in figure 42-1(B)

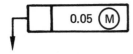

Fig. 42-5 Application of MMC symbol to tolerance

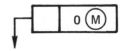

Fig. 42-6 MMC symbol with zero tolerance

the full range permitted by size variations. For such applications a maximum limit may be applied to the geometrical tolerance, in addition to the tolerance permitted at the maximum material limit, as shown in figure 42-7.

Before the 1973 publication of ANSI Y14.5, the ANSI standard stated that if a positional tolerance was used on a regardless-of-feature-size-basis, the symbol (S) was added after the tolerance, figure 42-8. In Y14.5 in 1973 the rule was changed, making the symbol (S) no longer necessary. However the symbol (S) appears on many existing drawings and is used when modification to these existing drawings is necessary.

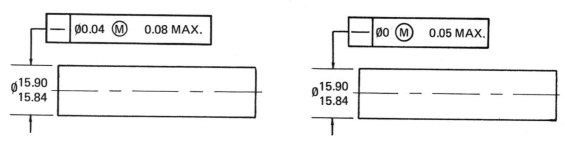

(A) DRAWING CALLOUT

FEATURE SIZE	PERMISSIBLE STRAIGHTNESS ERROR
15.90	0.04
15.89	0.05
15.88	0.06
15.87	0.07
15.86	0.08
15.85	0.08
15.84	0.08

FEATURE SIZE	PERMISSIBLE STRAIGHTNESS ERROR
15.90	0
15.89	0.01
15.88	0.02
15.87	0.03
15.86	0.04
15.85	0.05
15.84	0.05

(B) PERMISSIBLE VARIATIONS

Fig. 42-7 Tolerance with a maximum value

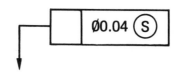

Fig. 42-8 Regardless-of-feature-size (RFS) symbol used on drawings

1. What is the extreme virtual size for each of the features shown?

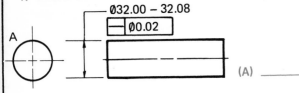

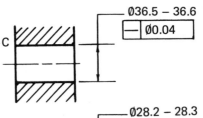

(A) _____

(C) _____

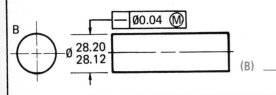

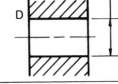

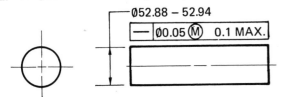

(B) _____

(D) _____

2. Complete the charts showing the largest permissible straightness error for the feature sizes shown.

Ø52.88 – 52.94

⏤ Ø0.05 Ⓜ 0.1 MAX.

Ø36.50 – 36.56

⏤ Ø0.08

FEATURE SIZE	PERMISSIBLE STRAIGHTNESS ERROR
52.94	
52.93	
52.92	
52.91	
52.90	
52.89	
52.88	

FEATURE SIZE	PERMISSIBLE STRAIGHTNESS ERROR
36.56	
36.55	
36.54	
36.53	
36.52	
36.51	
36.50	

3. Design a ring gauge to check the pins shown below.

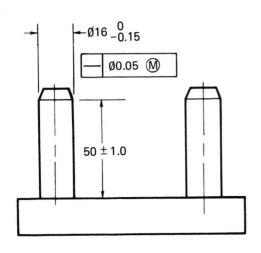

4. With reference to the drawing, are the following parts acceptable?

PART	FEATURE SIZE	STRAIGHTNESS DEVIATION
A	29.14	0.02
B	29.06	0.14
C	29.00	0.20
D	28.92	0.08
E	28.94	0.26

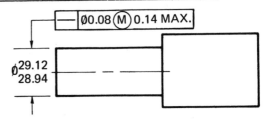

UNIT
43

FORM TOLERANCING

Form tolerancing is applied to the surface of an object to control its shape in terms of flatness and roundness.

Flatness

Flatness is a condition in which all surface elements are in one plane. On such a surface, all line elements in two or more directions are straight. A flatness symbol and tolerance are applied to a line representing the surface of a part by a feature control symbol, figure 43-1.

If the same control is wanted on two or more surfaces, a note may be added instead of repeating the symbol, figure 43-2.

Flatness on an MMC basis is very useful for controlling relatively thin parts which may be subject to bending or dishing. It is also preferred to a straightness tolerance for hexagons, squares, and other relatively long parts having opposing flat surfaces. The symbol is the same as for the flatness of a surface, except the modifier (M) is placed after the tolerance value, and the feature control symbol is directed toward the thickness dimension, figure 43-3.

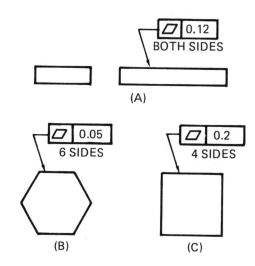

Fig. 43-2 Controlling flatness on two or more surfaces

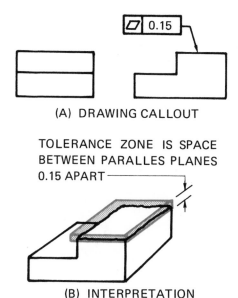

(A) DRAWING CALLOUT

TOLERANCE ZONE IS SPACE BETWEEN PARALLES PLANES 0.15 APART

(B) INTERPRETATION

Fig. 43-1 Flatness of a surface

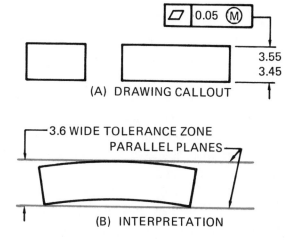

(A) DRAWING CALLOUT

3.6 WIDE TOLERANCE ZONE
PARALLEL PLANES

(B) INTERPRETATION

Fig. 43-3 Flatness on an MMC basis

To insure that the flatness tolerance remains stable over the entire length, a maximum limit may be applied to the geometrical tolerance, in addition to the tolerance permitted by the maximum material limit, figure 43-4.

Roundness

Roundness is a condition of a circular line or the surface of a circular feature in which all points on the line are equidistant from a common center point.

Errors of roundness (out-of-roundness) in a circular feature may occur as ovality, as lobing, or as random irregularities from a true circle. These errors are illustrated in figure 43-5.

The geometric characteristic symbol for roundness is a circle. A roundness tolerance may be specified by using this symbol in the feature control symbol, figure 43-6(A). It may be expressed on an MMC basis, but when not specified as such, applies to regardless-of-feature-size.

Roundness Tolerance – RFS. A roundness tolerance specifies the width of an annular tolerance zone, bounded by two concentric circles in the same plane, within which the circular line of the feature in that plane should be, figures 43-6 and 43-7.

Roundness on an MMC Basis. It is often advisable to insure that errors of roundness do not cause the outline of the feature to cross the maximum material boundary or to control the amount it does cross, in order to guarantee that the part will join satisfactorily with its mating parts. This is accomplished by specifying the roundness on an MMC basis. A tolerance on this basis is generally directed to the diametral dimension, such as the zero MMC tolerance shown in figure 43-8. This type of tolerancing does not prevent the part from crossing the least material boundary of Ø49.88.

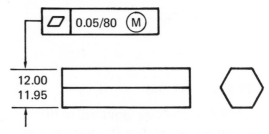

Fig. 43-4 Controlling flatness for a specified length

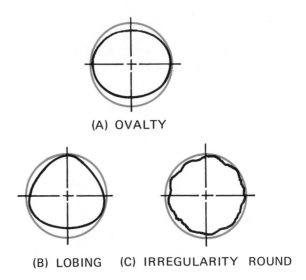

(A) OVALTY

(B) LOBING   (C) IRREGULARITY ROUND

Fig. 43-5 Common roundness errors

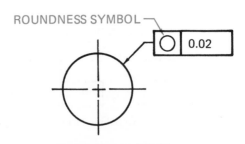

ROUNDNESS SYMBOL

0.02

(A) DRAWING CALLOUT

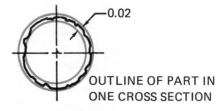

0.02

OUTLINE OF PART IN
ONE CROSS SECTION

(B) INTERPRETATION

Fig. 43-6 Roundness tolerance

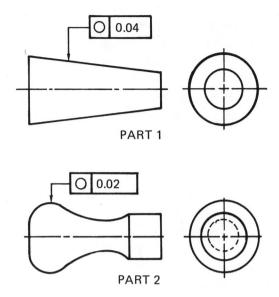

PART 1

PART 2

Fig. 43-7 Roundness tolerance of noncylindrical parts

Cylindricity

Cylindricity is the condition of a surface forming a cylinder where the surface elements in cross sections parallel to the axis are straight and parallel and in cross sections perpendicular to the axis are round. Cylindricity is a combination of geometrical form tolerances for roundness, straightness, and parallelism of the surface elements.

Cylindricity tolerances can only be applied to cylindrical surfaces, such as round holes and shafts. A conical surface must be controlled by a combination of tolerances for roundness, straightness, and angularity.

Errors of cylindricity may be caused by out-of-roundness, such as ovality or lobing; by errors of straightness caused by bending or diametral variation; by errors of parallelism

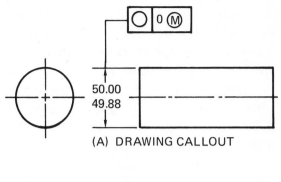

(A) DRAWING CALLOUT

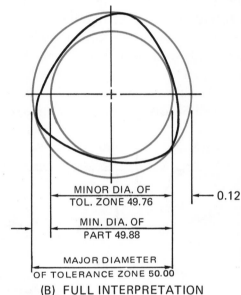

(B) FULL INTERPRETATION

Fig. 43-8 Roundness tolerance on an MMC basis

such as conicity or taper; and by random irregularities from a true cylindrical form.

The geometric characteristic symbol for cylindricity consists of a circle with two tangent lines at 60 degrees, figure 43-9. It is used in a feature control symbol and is directed toward the cylindrical surface, in either the side or end view, figures 43-10 and 43-11. When

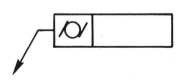

Fig. 43-9 Cylindricity symbol

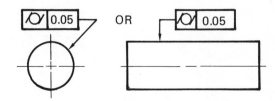

Fig. 43-10 Cylindricity tolerance directed toward either view

not modified on an MMC basis, a cylindricity tolerance applies regardless of feature size. The feature or part must be within its diameter limits at all points of measurement.

The *tolerance zone* is the annular space between two coaxial cylinders having a difference in radii equal to the specified tolerance, within which the entire surface of the feature must lie. The axis of the annular tolerance zone does not always coincide with the center line of the part. The diameters of the tolerance zone do not necessarily fall within the diameter limits of the part.

Because the measurement of cylindricity on an RFS basis is a tedious and time-consum-

ing procedure, a cylindricity tolerance should be on an MMC basis, figure 43-12. Whenever a fit between mating cylindrical features is required, a cylindricity tolerance of zero MMC is usually the preferred condition because it controls straightness and parallelism as well as roundness.

However, a cylindricity tolerance larger than zero may be specified. It is also satisfactory to specify a larger diameter tolerance with a zero geometrical tolerance, figure 43-12. Instead of specifying a maximum diameter of 29.97 mm with a cylindricity tolerance of 0.03 mm, it would be better to specify a maximum diameter of 30.00 mm with a cylindricity tolerance of zero MMC.

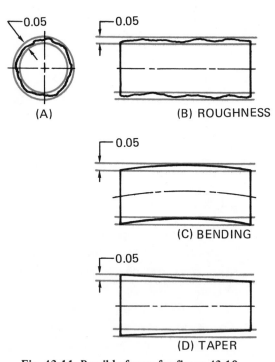

Fig. 43-11 Possible forms for figure 43-10

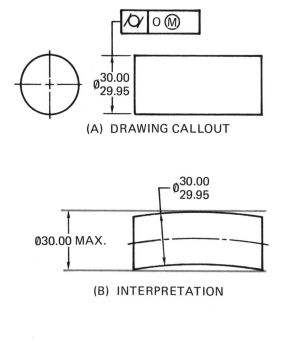

Fig. 43-12 Cylindricity tolerance on MMC basis

1. Show the tolerance zones and widths for the parts illustrated.

2. What are the extreme virtual sizes for the parts shown at (B) and (C)?

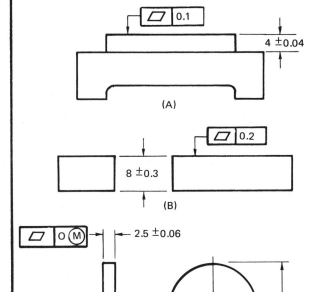

(A)

(B)

(C)

3. If measurements made at cross sections A-A, B-B and C-C indicate that all points on the circumference fall within the two annular rings shown, does the part meet the specified roundness tolerances?

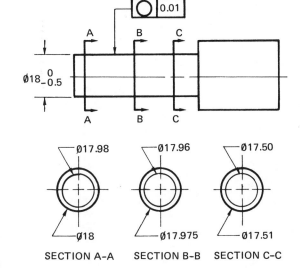

SECTION A-A   SECTION B-B   SECTION C-C

4. Sketch the tolerance zone for the cylindricity tolerance shown indicating its size and shape.

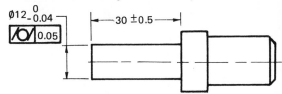

5. The parts shown below must assemble without interference. Add the largest cylindricity tolerances which will ensure this condition.

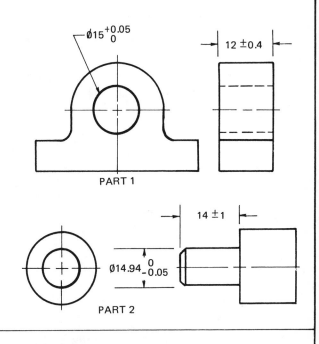

PART 1

PART 2

6. Add roundness tolerances to the 19 millimetre and 12 millimetre diameter features so that the features will not cross the boundary of perfect form at the maximum material size.

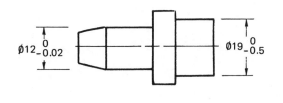

FORM TOLERANCING – FLATNESS, ROUNDNESS, CYLINDRICITY

A-79

UNIT 44

DATUMS AND THE THREE-PLANE METHOD OF TOLERANCING

A *datum* is a point, line, plane, or other geometrical surface from which dimensions are measured, or to which geometrical tolerances are referenced. A datum has an exact form and represents an exact or fixed location for purposes of manufacture or measurement.

A *datum feature* is a feature of a part, such as an edge, surface, or hole, which forms the basis for a datum or is used to establish the location of a datum.

DATUMS FOR GEOMETRICAL TOLERANCING

Datums are exact geometrical points, lines, or surfaces, each based on one or more datum features of the part. Surfaces are usually either flat or cylindrical, but other shapes are used when necessary. Since the datum features are physical surfaces of the part, they are subject to manufacturing errors and variations. For example, a flat surface of a part, if greatly magnified, will show some irregularity. If brought into contact with a perfect plane, this flat surface will touch only at the highest points, figure 44-1. The true datums exist only in theory but are considered to be in the form of locating surfaces of machines, fixtures, and gauging equipment on which the part rests or with which it makes contact during manufacture and measurement.

Although these surfaces are not perfect geometrical surfaces, they are intended to be of a sufficiently high quality to render any errors they introduce insignificant in comparison with the great tolerances.

Usually only one datum is required for orientation purposes, but positional relationships may require a datum system consisting of two or three datums. These datums are designated as primary, secondary, and tertiary. When these datums are mutually perpendicular plane surfaces, they are referred to as a three-plane system or a datum reference frame.

Primary Datum

If the primary datum feature is a flat surface, it could be laid on a suitable plane surface, such as the surface of a gauge, which would then become a primary datum, figure 44-2. Theoretically, there will be a minimum of three high spots on the flat surface coming in contact with the gauge surface.

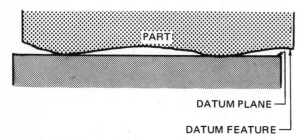

DATUM PLANE ⌐
DATUM FEATURE ⌐

Fig. 44-1 Magnified section of a flat surface

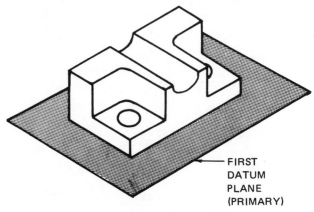

FIRST
DATUM
PLANE
(PRIMARY)

Fig. 44-2 Primary datum

Secondary Datum

If the part is brought into contact with a secondary plane while lying on the primary plane, it will theoretically touch at a minimum of two points, figure 44-3.

Tertiary Datum

The part can be slid along while maintaining contact with both the primary and secondary planes until it contacts a third plane, figure 44-4. This plane then becomes the tertiary datum and the part will, in theory, touch it at only one point.

These three planes constitute a datum system from which measurements can be taken. They will appear on the drawing as shown in figure 44-5 (page 300), except that the datum features should be identified in their correct sequence by the methods described later in the unit.

DATUM IDENTIFYING SYMBOL

Datum symbols have two functions. They indicate the datum surface or feature on the drawing and identify the datum

feature so it can be easily referred to in other requirements.

There are two methods of datum symbolization for such purposes: the American method used in ANSI standards, and the ISO method which is used in most other countries, figure 44-6, page 301.

ANSI Symbol

In the ANSI system every datum feature is identified by a capital letter enclosed in a rectangular box. A dash is placed before and after the letter to indicate it applies to a datum feature.

ISO Symbol

The ISO method is used in Canadian, British, and most other national standards. The datum symbol is a triangle figure 44-6 (B).

When identification of a datum is necessary to permit reference to it in another requirement, each datum feature is identified by a capital letter placed in a square frame and connected to the datum indicator symbol.

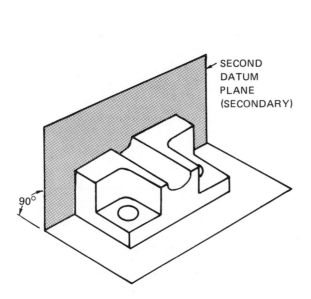

Fig. 44-3 Secondary datum

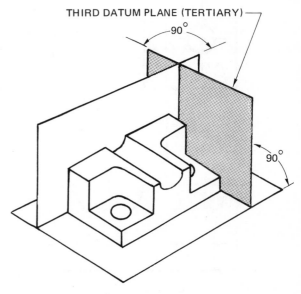

Fig. 44-4 Tertiary datum

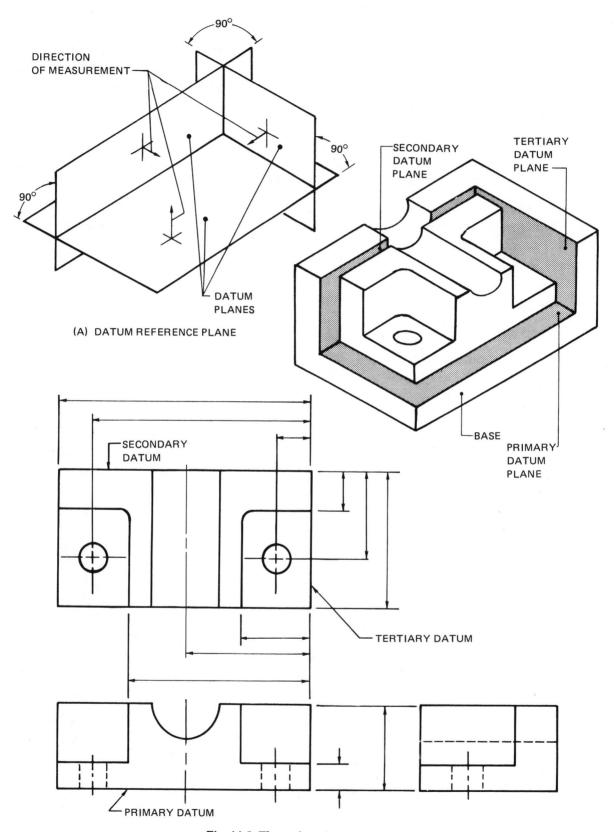

(A) DATUM REFERENCE PLANE

DIRECTION OF MEASUREMENT

DATUM PLANES

SECONDARY DATUM PLANE

TERTIARY DATUM PLANE

BASE

PRIMARY DATUM PLANE

SECONDARY DATUM

TERTIARY DATUM

PRIMARY DATUM

Fig. 44-5 Three-plane datum system

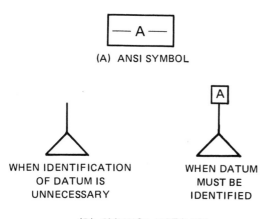

(A) ANSI SYMBOL

WHEN IDENTIFICATION
OF DATUM IS
UNNECESSARY

WHEN DATUM
MUST BE
IDENTIFIED

(B) SYMBOL USED BY
ISO AND CSA

Fig. 44-6 Datum identifying symbol

This identifying symbol may be directed to the datum feature in any of the following methods: by attaching a side, end, or corner of the symbol frame to an extension line from the feature, figure 44-7; by running a leader with arrowheads from the symbol frame to the feature, figure 44-8; by adding the symbol to a dimension or a feature control symbol pertaining to the feature, figures 44-8 and 44-9 (page 302).

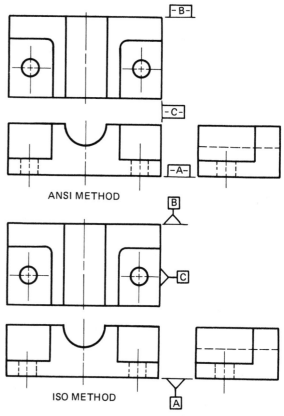

ANSI METHOD

ISO METHOD

**Fig. 44-7 Identifying datums for
part shown in figure 44-5**

SYMBOL ATTACHED TO AN EXTENSION LINE OR LEADER	SYMBOL COMBINED WITH A FEATURE CONTROL SYMBOL	SYMBOL ATTACHED TO A DIMENSION LINE
-A- -B- -D- -C-	XXX^{+XX}_{-XX} -A-	-A- XX
ANSI DATUM IDENTIFYING SYMBOL		
B A D C	XXX^{+XX}_{-XX} A	A XX
ISO DATUM IDENTIFYING SYMBOL		

Fig. 44-8 Common methods of connecting datum identifying symbol to datum

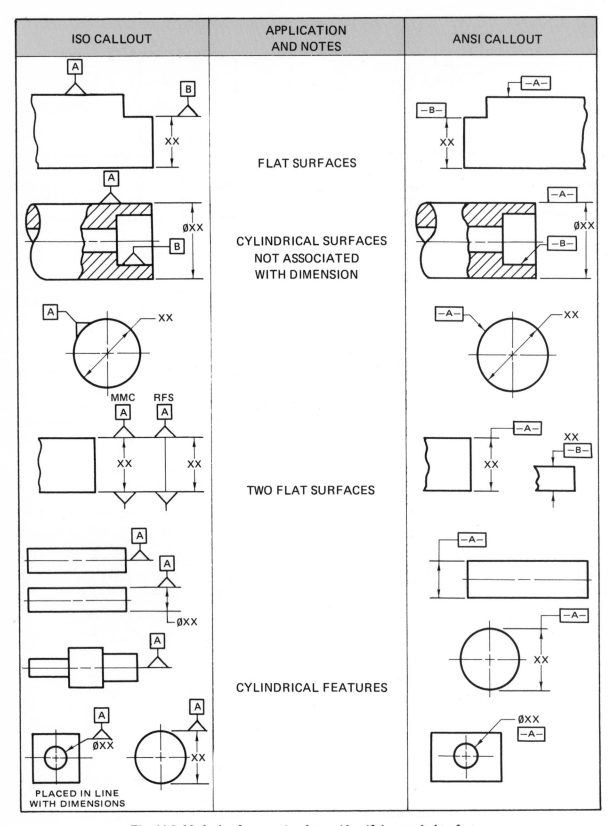

Fig. 44-9 Methods of connecting datum identifying symbol to feature

Multiple Datum Features

If a single datum is established by two datum features, such as two ends of a shaft, figure 44-10, the features are identified by separate letters. Both letters are then placed in the same compartment of the feature control symbol, separated by a dash. The datum in this case is the common line between the two datum features.

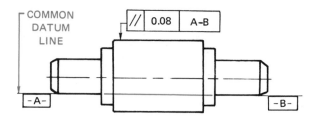

Fig. 44-10 Two datum features for one datum

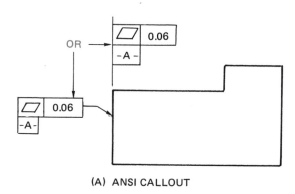

(A) ANSI CALLOUT

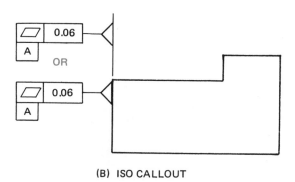

(B) ISO CALLOUT

Fig. 44-11 Datum surface also controlled by a feature control symbol

Combined Feature Control and Datum Identifying Symbol

When a feature is controlled by a positional or form tolerance and serves as a datum, the feature control and datum identifying symbol are combined, figures 44-11 through 44-14. Applications of combined feature control and datum identifying symbols are shown in units 45 to 49.

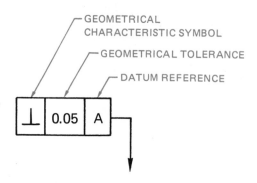

Fig. 44-12 Feature control symbol referenced to a datum

Fig. 44-13 Former ANSI location of datum in feature control symbol

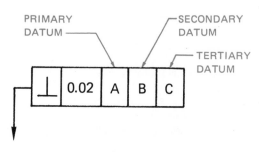

Fig. 44-14 Feature referenced to more than one datum

ASSIGNMENT

1. Make a three-view drawing of the part illustrated and show the following information:
   - Surface A is datum A and is to be straight within 0.2 for the 100 millimetre length; the straightness error should not exceed 0.05 for any 25 millimetre length.
   - Surface B is datum B and is to be flat within 0.1. Surface B is parallel to datum C-D within 0.1.
   - Surface C and D are datum features C and D respectively which form a single datum. Surfaces C and D should be flat within 0.05.

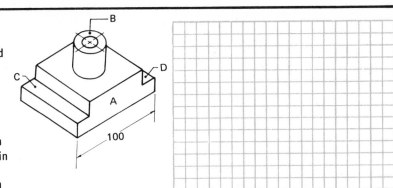

2. A. Pins 1, 2, and 3 are used to establish the secondary and tertiary datums for the part shown. What is used for the primary datum?
   B. Make a two-view drawing of the part and identify the primary, secondary and tertiary datum planes as A, B, and C respectively.
   C. How far is the center of the hole from (1) tertiary datum, (2) secondary datum?
   D. The back of the slot is to be flat within 0.3 millimetres. Place the form tolerance on the drawing.

(A) _____
(C) _____

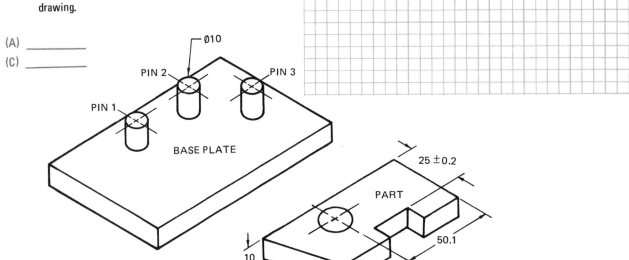

BASE PLATE

Ø10

PIN 1   PIN 2   PIN 3

PART

25 ± 0.2

50.1

10

3. What is the minimum number of contact points in a three-plane datum system for (A) primary datum, (B) secondary datum, (C) tertiary datum?   (A) _____   (B) _____   (C) _____

4. A. The bottom surface of the part shown is to be identified as datum B and should be flat within 0.1 millimetre. Specify both the datum and the tolerance on the drawing.
   B. What is the extreme virtual height of the part?   (B) _____

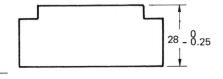

$28 {}_{-0.25}^{0}$

DATUMS AND THREE-PLANE TOLERANCING

A-80

304

UNIT

45

ORIENTATION TOLERANCING

Orientation is the angular relationship existing between two or more lines, surfaces, or other features. *Angularity* is the general geometric characteristic for orientation. This term describes relationships of any angle, between straight lines or surfaces such as flat or cylindrical surfaces. Special terms are used for two particular types of angularity. These are *perpendicularity,* or squareness, for features related to each other by a 90-degree angle; and *parallelism* for features related to one another by an angle of zero, figure 45-1.

An orientation tolerance indicates a relationship between two or more features. Whenever possible, the feature to which the controlled feature is related should be designated as a datum.

ANGULARITY   PERPENDIC-   PARALLELISM
              LARITY

Fig. 45-1 Orientation symbols

Tolerance of Flat Surfaces – Regardless of Feature Size

Figure 45-2 shows three simple parts in which one flat surface is designated as a datum.

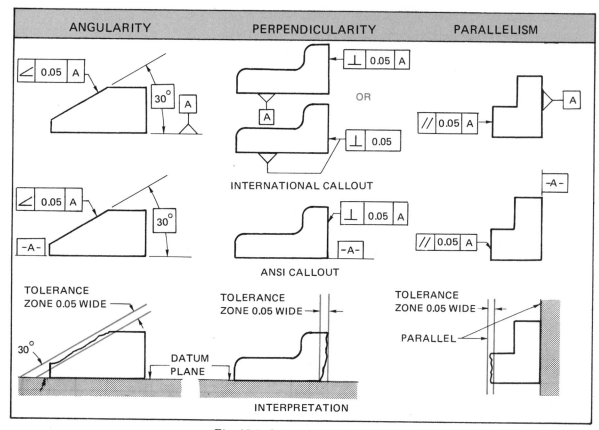

Fig. 45-2 Orientation tolerancing

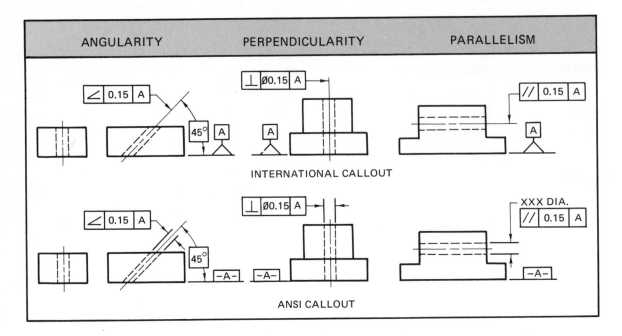

Fig. 45-3 Orientation of lines and surfaces

feature and another flat surface is related to it by one of the orientation tolerances.

Each of these tolerances means that the designated surface shall be contained within a tolerance zone consisting of the space between two parallel planes separated by the specified tolerance (0.05 mm) and related to the datum by the specified angle (30 degrees, 90 degrees, or 0 degrees).

Tolerancing of Lines Related to Surfaces — Regardless of Feature Size

Figure 45-3 shows some simple parts in which the axis or center line of a hole is related by an orientation tolerance to a flat surface. The flat surface is designated as the datum feature.

The center line of the hole must be contained within a tolerance zone consisting of the space between two parallel planes. These planes are separated by a specified tolerance of 0.15 mm and are related to the datum by one of the basic angles, 45 degrees, 90

degrees, or 0 degrees. Figure 45-4 clearly illustrates the tolerance zone for angularity.

When the tolerance is one of perpendicularity, the tolerance zone planes can be revolved around the feature axis without affecting the angle. The tolerance zone therefore becomes a cylinder. This cylindrical zone is perpendicular to the datum and has a diameter equal to the specified tolerance, figure 45-5.

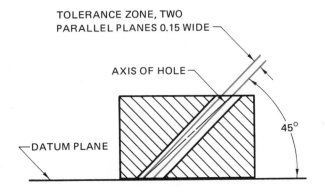

Fig. 45-4 Tolerance zone for angularity shown in fig. 45-3

Control in Two Directions

The feature control symbol shown in figure 45-3 controls angularity and parallelism with the base (datum A) only. If control with a side is also required, the side should be designated as the secondary datum, figure 45-6. The tolerance zone will be shaped like a prism whose bases are parallelograms (a parallelepiped).

Control of Center Lines

Tolerances intended to control orientation of the center line of a feature are applied to drawings as shown in figure 45-7 (page 308). The resultant tolerance zone is illustrated in figure 45-8. Perpendicularity is considered to apply in all directions, unless a note

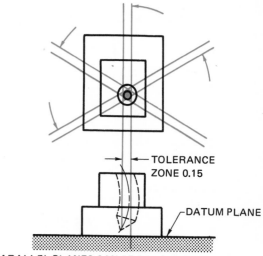

PARALLEL PLANES CAN BE REVOLVED MAKING THE TOLERANCE ZONE A CYLINDER

Fig. 45-5 Tolerance zone for perpendicularity shown in figure 45-3

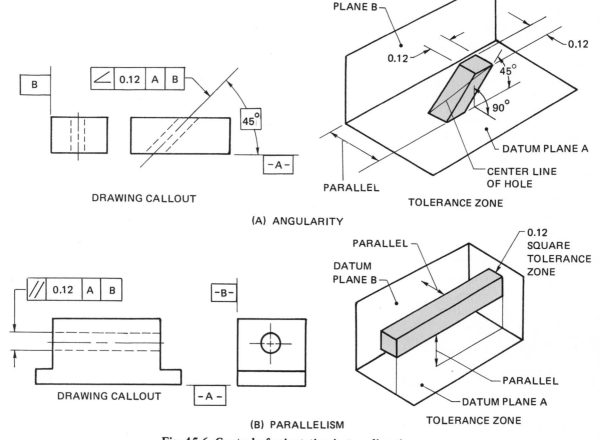

(A) ANGULARITY

(B) PARALLELISM

Fig. 45-6 Control of orientation in two directions

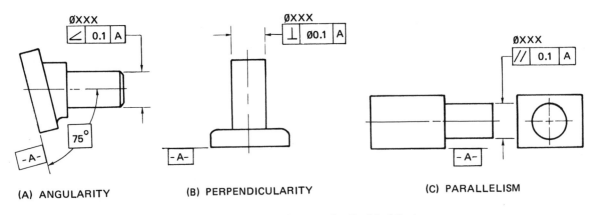

(A) ANGULARITY (B) PERPENDICULARITY (C) PARALLELISM

Fig. 45-7 Orientation of external cylindrical features

such as THIS DIRECTION ONLY or DIRECTION OF ARROW ONLY is added.

In some cases it is more important to control elements of the cylindrical surface than of its axis or center line. To control elements of the cylindrical surface, the tolerance is directed to the surface with a single arrowhead, figure 45-9.

Because the cylindrical features represent size, orientation symbols may be applied on an MMC basis. This is indicated by adding the modifier symbol after the tolerance, figure 45-10. This method provides additional permissible variation in orientation as the size of the feature decreases to its least material size.

A comparison between the basic geometrical tolerancing methods is shown in figure 45-11.

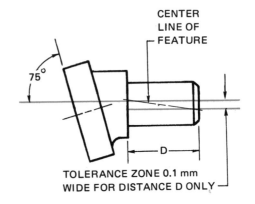

Fig. 45-8 Tolerance zone for angularity shown in figure 45-7

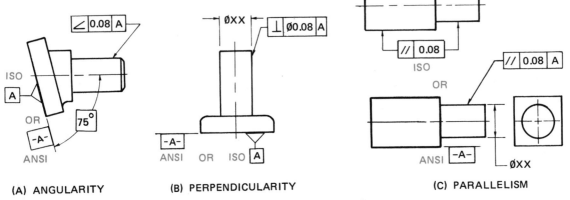

(A) ANGULARITY (B) PERPENDICULARITY (C) PARALLELISM

Fig. 45-9 Orientation of surface elements

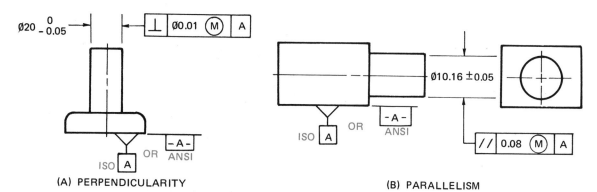

(A) PERPENDICULARITY

(B) PARALLELISM

Fig. 45-10 Perpendicularity and parallelism on MMC basis

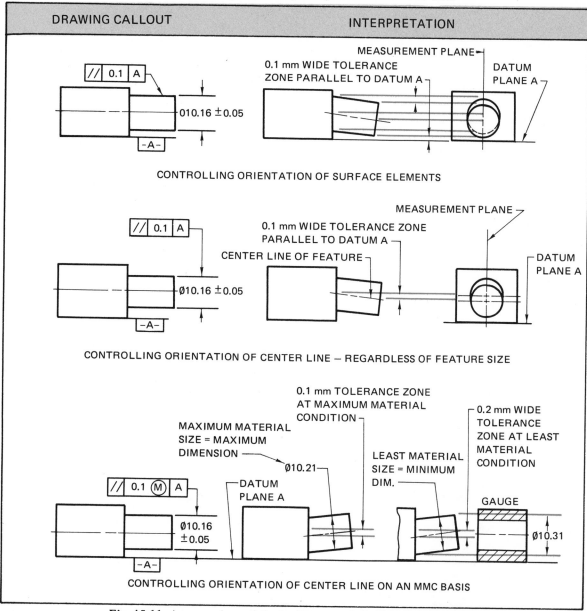

DRAWING CALLOUT	INTERPRETATION

CONTROLLING ORIENTATION OF SURFACE ELEMENTS

CONTROLLING ORIENTATION OF CENTER LINE — REGARDLESS OF FEATURE SIZE

CONTROLLING ORIENTATION OF CENTER LINE ON AN MMC BASIS

Fig. 45-11 A comparison of the basic geometrical tolerancing methods

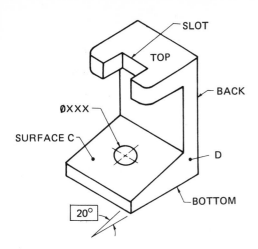

1. From the information given below prepare a three-view sketch of the part showing the datums and geometrical tolerances:
   - Bottom to be datum **A**
   - Back to be datum **B**
   - Hole to be perpendicular to bottom within 0.08
   - Back to be perpendicular to bottom within 0.1
   - Top to be parallel with bottom within 0.12
   - Surface **C** to have an angularity tolerance of 0.16 with the bottom. Surface **D** to be the secondary datum for this feature.
   - The sides of the slot to be parallel to each other within 0.05 and be perpendicular to back within 0.1.

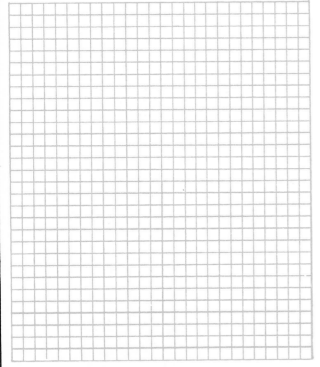

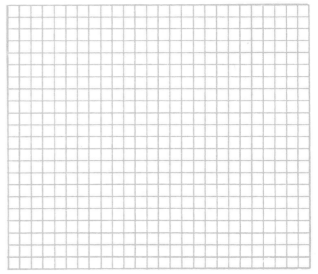

2. Prepare a three-view drawing showing the datums and feature control symbols from the information supplied:
   - Surfaces **A**, **B**, **C**, and **D** are datums **A**, **B**, **C**, and **D** respectively.
   - Surface **D** of the dovetail must have an angularity tolerance of 0.05 millimetres with datum **A**.
   - Surface **C** should be perpendicular to datum **B** within 0.03.
   - Surface **E** must be perpendicular to datums **A** and **D** within 0.02.

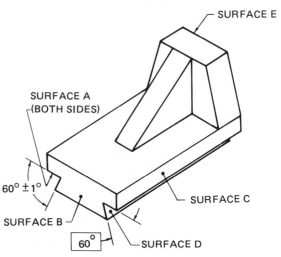

ORIENTATION TOLERANCING — ANGULARITY, PERPENDICULARITY, PARALLELISM

A-81

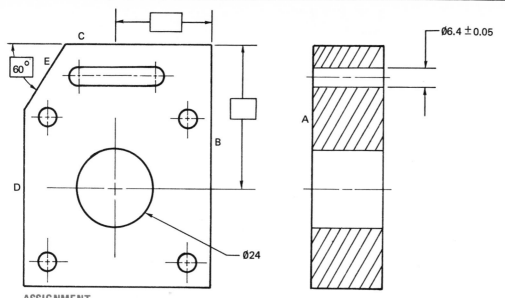

ASSIGNMENT

1. Add the following information to the drawing:
   - Surfaces marked **A**, **B**, and **C** are datums **A**, **B**, and **C**, respectively.
   - Surface **A** is perpendicular to datums **B** and **C** within 0.25.
   - Surface **D** is parallel to datum **B** within 0.1.
   - The slot is parallel to datum **C** within 0.02 and perpendicular to datum **A** within 0.01 at MMC.
   - The Ø24 hole is perpendicular to datum **A** within 0.03 at MMC.
   - Surface **E** has an angularity tolerance of 0.15.

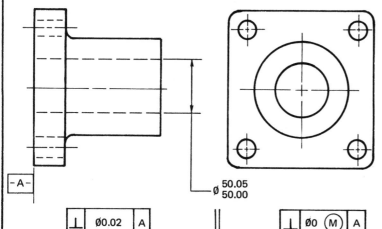

2. Calculate the maximum diameter tolerance zone allowed for the diameters shown in the tables below. The feature control symbol is associated with Ø $\frac{50.05}{50.00}$.

⊥	Ø0.02	A

FEATURE SIZE	DIAMETER TOLERANCE ZONE ALLOWED
50.05	
50.04	
50.03	
50.02	
50.01	
50.00	

⊥	Ø0 Ⓜ	A

FEATURE SIZE	DIAMETER TOLERANCE ZONE ALLOWED
50.05	
50.04	
50.03	
50.02	
50.01	
50.00	

⊥	Ø0 Ⓜ	Ø0.03 MAX.	A

FEATURE SIZE	DIAMETER TOLERANCE ZONE ALLOWED
50.05	
50.04	
50.03	
50.02	
50.01	
50.00	

ORIENTATION TOLERANCING — ANGULARITY, PERPENDICULARITY, PARALLELISM

A-82

TOLERANCING OF FEATURES BY POSITION

The location of features is one of the most frequent applications of dimensions on technical drawings. Tolerancing is accomplished either by applying coordinate tolerances to the dimensions or by geometrical (positional) tolerancing.

Positional tolerancing is especially useful when applied on an MMC basis to groups or patterns of holes. This method meets functional requirements in most cases and at the same time permits inspection with simple gauge procedures.

Most examples in these units will be devoted to the principles involved in the location of small holes, because they represent the most commonly used applications. The same principles apply to the location of other features such as slots, tabs, bosses, and noncircular holes.

Tolerancing Methods

The location of a single hole is usually indicated by rectangular coordinate dimensions, extending from suitable edges or other features of the part to the axis of the hole. Other dimensioning methods, such as that of polar coordinates, may be used when circumstances warrant. There are two standard methods of tolerancing the location of holes, *coordinate tolerancing* and *positional tolerancing*.

- Coordinate tolerancing refers to the tolerances applied directly to the coordinate dimensions or to applicable tolerances specified in a general tolerance note, figure 46-1(A).
- Positional tolerancing encompasses:
  (A) Positional Tolerancing, Regardless of Feature Size (RFS), figure 46-1(B).
  (B) Positional Tolerancing, Maximum Material Condition Basis (MMC), figure 46-1(C).

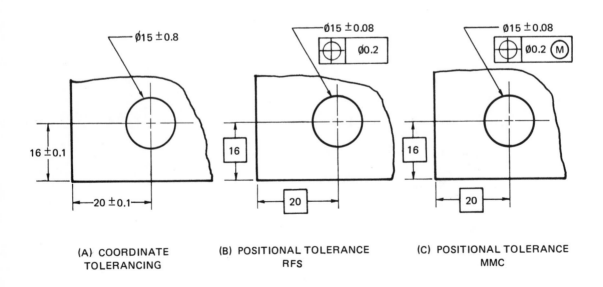

(A) COORDINATE TOLERANCING

(B) POSITIONAL TOLERANCE RFS

(C) POSITIONAL TOLERANCE MMC

Fig. 46-1 Comparison of tolerancing methods

These positional tolerancing methods are part of the system of geometrical tolerancing. To explain and understand the advantages and disadvantages of the positional tolerancing methods, the widely used method of coordinate tolerancing must be analyzed.

COORDINATE TOLERANCING

Coordinate dimensions and *tolerances* may be applied to the location of a single hole, figure 46-2.

If the two coordinate tolerances are equal, the tolerance zone formed will be a square. Unequal tolerances result in a rectangular tolerance zone. Polar dimensioning where one of the locating dimensions is a radius gives a circular ring section tolerance zone. For simplicity and because it is most frequently used, square tolerance zones are utilized in the analysis of most of the examples, figure 46-3.

The tolerance zone extends for the full depth of the hole. In most of the illustrations, tolerances will be analyzed as they apply at the surface of the part, where the axis is represented as a point.

The actual position of the center of the hole may be within the rectangular tolerance zone. For square tolerance zones, the maximum allowable variation from the theoretical exact position occurs in a direction of 45 degrees from the direction of the coordinate dimensions.

For the examples shown in figure 46-2, the tolerance zones are shown with their maximum tolerance values in figure 46-4 (page 314).

A quick and easy method of finding the maximum positional error permitted with coordinate tolerancing is by using a chart similar to that shown in figure 46-5.

In the first example shown in figure 46-2, the tolerance in both directions is 0.2 mm. The extension of the vertical and hor-

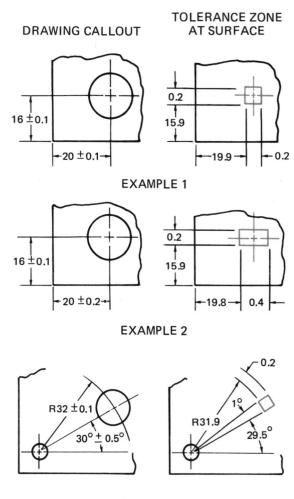

DRAWING CALLOUT   TOLERANCE ZONE AT SURFACE

EXAMPLE 1

EXAMPLE 2

EXAMPLE 3

Fig. 46-2 Tolerance zones for coordinate tolerances

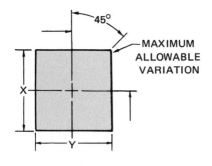

Fig. 46-3 Maximum permissible error for square tolerance zone

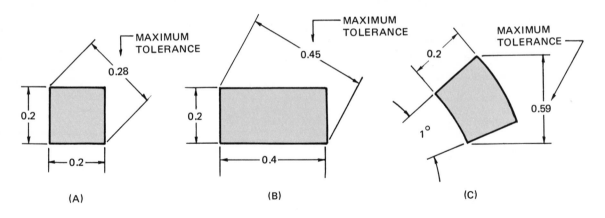

Fig. 46-4 Tolerance zones for examples in figure 46-2

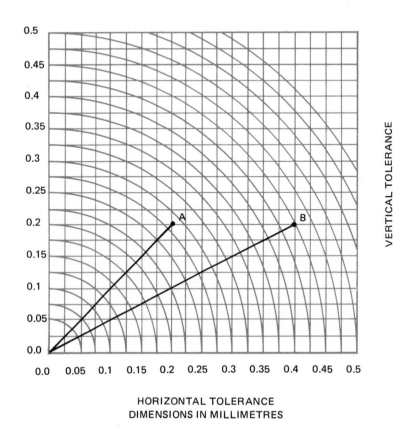

HORIZONTAL TOLERANCE
DIMENSIONS IN MILLIMETRES

Fig. 46-5 Chart for calculating maximum tolerance using coordinate tolerancing

izontal bars in the chart intersect at point A which lies between the radii of 0.275 and 0.3 mm. When rounded off to two decimal places it indicates a maximum permissible variation of position of 0.28 mm. In the second example shown in figure 46-2, the extension of the vertical and horizontal lines at 0.2 and 0.4 respectively in the chart intersect at point **B** indicating a maximum variation of position of 0.45 mm.

POSITIONAL TOLERANCING

Positional tolerancing is part of the system of geometrical tolerancing. In this system, the location of features is shown by coordinate dimensions or by polar and angular dimensions. The dimensions are shown, however, without direct tolerances. These dimensions represent the basic sizes and are commonly known as *true-position* dimensions. Each true-position dimension is enclosed in a rectangular frame to indicate that it represents an exact value to which tolerances shown in the general note do not apply, figure 46-6.

Symbol for Position

The geometric characteristic symbol for position is a circle and two solid center lines shown in the feature control symbol in the same way as for other geometrical symbols. Positional tolerances may be specified on a maximum material condition basis, but when MMC is not specified they apply to the position of the axis regardless of the size of the feature. The ANSI standard formerly showed the symbol Ⓢ , meaning regardless of feature size, for all positional tolerances not modified by Ⓜ , maximum material condition, figure 46-7.

Positional Tolerancing – Regardless of Feature Size

Positional tolerancing on this basis controls the axis of the hole. For this reason, the

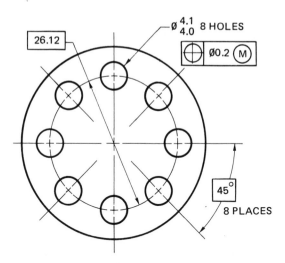

Fig. 46-6 Identifying true position dimensions

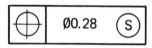

Fig. 46-7 Former ANSI symbol for regardless of feature size shown with positional tolerance

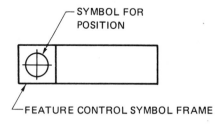

Fig. 46-8 Position symbol

control symbol is directed to the hole dimension, figure 46-8 and 46-9 (page 316). The positional tolerance represents the diameter of a cylindrical tolerance zone, located at true position as determined by the true position dimensions on the drawing.

It has already been shown that the rectangular coordinate tolerancing the maximum

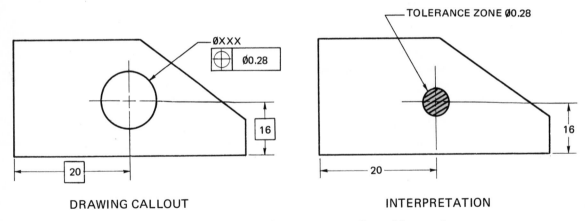

ØXXX

⌖ Ø0.28

16

20

DRAWING CALLOUT

TOLERANCE ZONE Ø0.28

16

20

INTERPRETATION

Fig. 46-9 Positional tolerancing – regardless of feature size

permissible error in location is not the value indicated by the horizontal and vertical tolerances, but is instead equivalent to the length of the diagonal between the two tolerances. For square tolerance zones this amount is 1.4 times the specified tolerance values. The specified tolerance can therefore be increased to an amount equal to the diagonal of the coordinate tolerance zone without affecting the clearance between the hole and its mating part.

It is practical to replace coordinate tolerances with a positional tolerance having a value equal to the diagonal of the coordinate tolerance zone. This provides 57 percent more

tolerance area, figure 46-10 and would probably result in the rejection of fewer parts for positional errors.

A simple method for checking positional tolerance errors is to take coordinate measurements and evaluate them on a chart as shown in figure 46-11. For example, the four parts shown in figure 46-12 (page 318) were rejected when the coordinate tolerances were applied to them.

If the part had been toleranced using the positional tolerance method and given a tolerance of Ø0.28 (equal to the diagonal of the coordinate tolerance zone), three of the parts would not have been rejected.

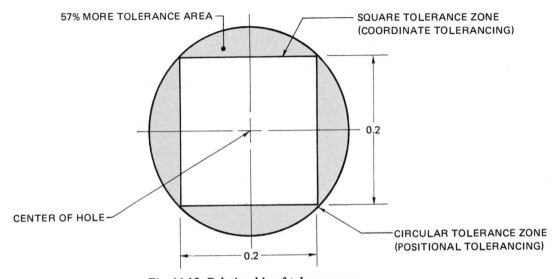

57% MORE TOLERANCE AREA

SQUARE TOLERANCE ZONE
(COORDINATE TOLERANCING)

0.2

CENTER OF HOLE

CIRCULAR TOLERANCE ZONE
(POSITIONAL TOLERANCING)

0.2

Fig. 46-10 Relationship of tolerance zones

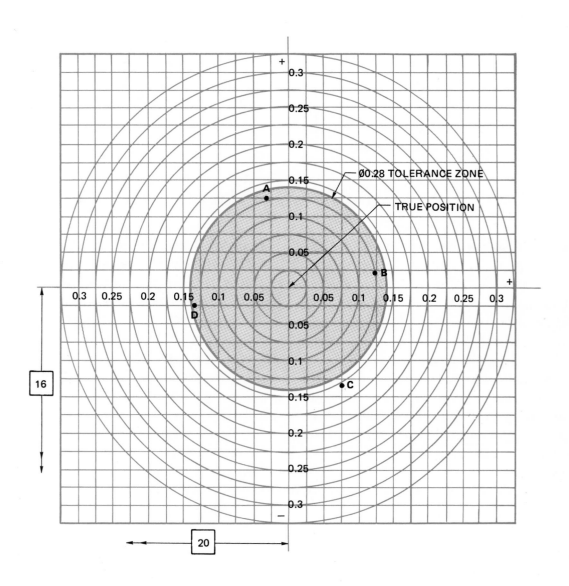

Fig. 46-11 Chart for evaluating positional tolerancing on an RFS basis

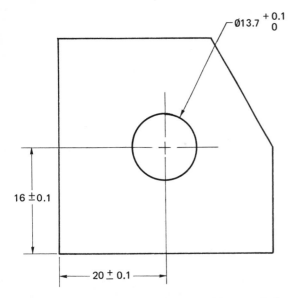

Ø13.7 $^{+0.1}_{0}$

16 ±0.1

20 ± 0.1

Part	Hole Location		Comment
A	16.13	19.97	Rejected
B	16.03	20.13	,,
C	15.87	20.8	,,
D	15.98	19.87	,,

Fig. 46-12 Parts A to D rejected because hole centers do not lie within coordinate tolerance zone

Positional Tolerancing – Maximum Material Condition

The problems of tolerancing for the position of holes are simplified when positional tolerancing is applied on an MMC basis. This permits an increase in positional variations as the size departs from the maximum material size and also permits the use of functional GO, NO-GO gauges.

To apply a positional tolerance on an MMC basis, the symbol (M) is added in the feature control symbol after the tolerance and the feature control symbol are associated with the size dimension.

Consider the part shown in figure 46-13. A positional tolerance applied to a hole on an MMC basis means that the boundary of the hole must fall outside a perfect cylinder having a diameter equal to the maximum material size of the feature (Ø13.7) minus the positional tolerance (Ø0.28). This cylinder (Ø13.42) is located with its axis at true position. The hole must be within the limits of Ø13.7 and Ø13.8. A simple gauge to check the part is shown in figure 46-14.

Should the hole be at its least material size of Ø13.8, as shown in figure 46-15, the positional tolerance can be increased to Ø0.38 without jeopardizing the function of the part.

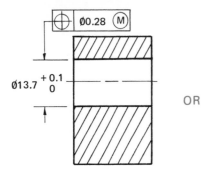

⊕ Ø0.28 (M)

Ø13.7 $^{+0.1}_{0}$

OR

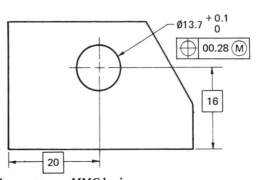

Ø13.7 $^{+0.1}_{0}$

⊕ 00.28 (M)

16

20

Fig. 46-13 Positional tolerance on an MMC basis

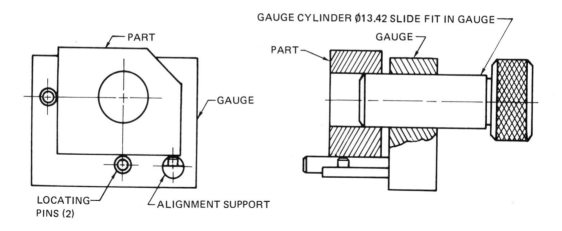

Fig. 46-14 Gauge for part shown in figure 46-13

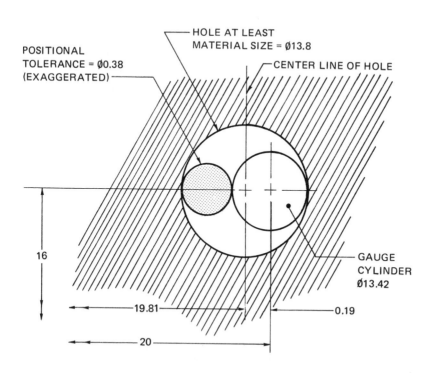

HOLE AT LEAST MATERIAL SIZE

Fig. 46-15 Positional variations for tolerancing on MMC basis

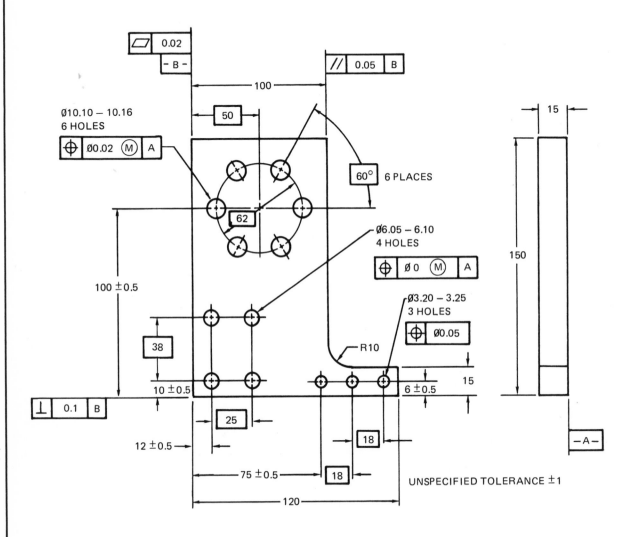

UNSPECIFIED TOLERANCE ±1

ASSIGNMENT

1. How many datum surfaces are there on the drawing?

2. How many true position dimensions are shown on the drawing?

3. What is the maximum flatness deviation permitted on datum surface B?

4. What is the thickness tolerance permitted on the part?

5. How many form tolerances are shown?

6. How many locational tolerances are indicated?

7. The part shown below was rejected because it did not meet the requirements specified on the drawing. List the reasons why it was not acceptable. Lines and surfaces are labelled for identification purposes.

8. Show the maximum positional tolerance allowed for each of the feature sizes in the table below.

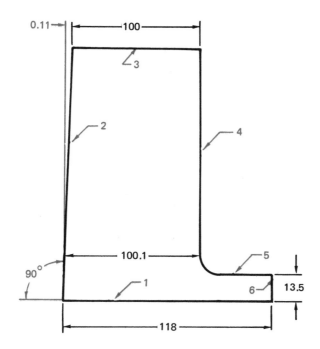

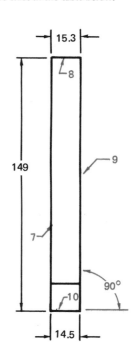

Ø3.20 - 3.25 HOLES		Ø6.05 - 6.10 HOLES		Ø10.10 - 10.16 HOLES	
FEATURE SIZE	MAX. POS. TOLERANCE ALLOWED	FEATURE SIZE	MAX. POS. TOLERANCE ALLOWED	FEATURE SIZE	MAX. POS. TOLERANCE ALLOWED
3.20		6.05		10.10	
3.21		6.06		10.11	
3.22		6.07		10.12	
3.23		6.08		10.13	
3.24		6.09		10.14	
3.25		6.10		10.15	
				10.16	

METRIC　　**POSITIONAL TOLERANCING**　　**A-83**

321

DATUMS FOR TOLERANCING BY POSITION

Occasionally it is preferable to have a hole related to surfaces or features (datums) other than the outside edges of the part. The first consideration in such applications is determination of the primary datum feature. Usually, the surface on which the hole is made is specified as the primary datum. This ensures that the true position of the axis is either perpendicular to this surface or at the basic angle, if it is other than 90 degrees. This surface is rested on the gauging plane or surface plate for measuring purposes. Secondary and tertiary datum features are then selected and identified if required.

Figure 47-1 shows a part having three datums specified, while figure 47-2 illustrates a gauge which could be used to check such a part.

DATUM TARGETS

It may be inadvisable to use a complete surface of a feature as a datum for the following reasons:

- The surface of a feature may be so large that a gauge designed to make contact with the full surface may be too expensive or too cumbersome to use.

- Functional requirements of a part may necessitate using only part of the surface as a datum feature.

- A surface selected as a datum feature may not be sufficiently accurate.

The datum target method is a useful technique for overcoming such problems. In this method, certain points, lines, or small areas on the surfaces are selected as the basis for establishing datums, figure 47-3. For flat

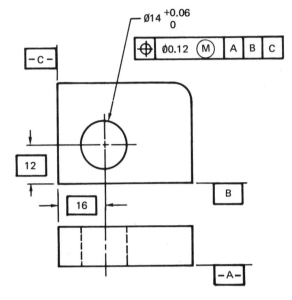

(A) DRAWING CALLOUT

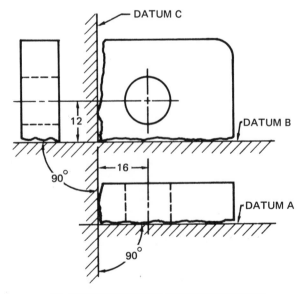

(B) INTERPRETATION OF TRUE POSITION

Fig. 47-1 Part with three datum features specified

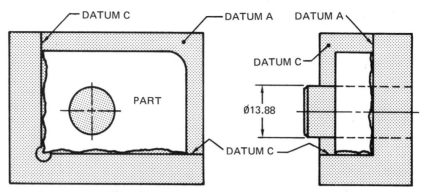

Fig. 47-2 Gauge for part in figure 47-1

surfaces, this usually requires three target points or areas for a primary datum; two for a secondary datum; and one for a tertiary datum. Often the use of such target areas eliminates the need for costly machining which might otherwise be required to produce surfaces suitable for use as datum features.

Targets need not be used for all datums. It is logical, for example, to use targets for the primary datum and other features or surfaces for secondary and tertiary datums. Datum targets should be spaced as far away as possible to provide maximum rigidity for making measurements.

Each datum target is shown on a view of the part in its desired location by a datum target symbol, figure 47-4 (page 324). The datum target symbol is a circle divided into two parts. The upper part contains the size of the datum area. The lower part contains a letter and number which identifies that particular target in the datum system. Targets are numbered consecutively. For example, in a three-plane, six-point datum system, if the datums are A, B, and C, the datum targets would be A1, A2, A3, B4, B5, and C6.

Each datum target symbol is connected to its corresponding datum by a leader ending with a dot or arrowhead, except when the symbol is attached to an extension line, figure 47-5. The extension line may be used as the leader line for the datum target symbols. When a datum target area is used, the size

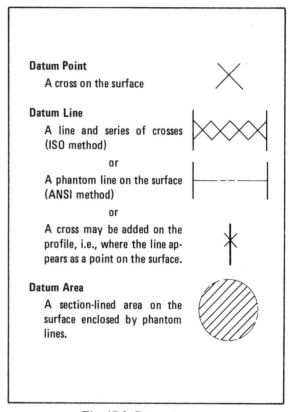

Datum Point

A cross on the surface

Datum Line

A line and series of crosses (ISO method)

or

A phantom line on the surface (ANSI method)

or

A cross may be added on the profile, i.e., where the line appears as a point on the surface.

Datum Area

A section-lined area on the surface enclosed by phantom lines.

Fig. 47-3 Datum targets

of the area is shown besides the datum target symbol. It is enclosed in a rectangular frame indicating that the general tolerance does not apply, figure 47-6.

Dimensions locating a set of datum targets should be either dimensionally related to one another or have a common origin.

The application and use of a surface and three lines used as datum features are shown in figure 47-7.

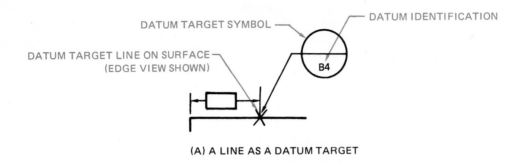

(A) A LINE AS A DATUM TARGET

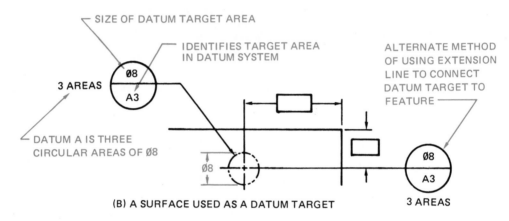

(B) A SURFACE USED AS A DATUM TARGET

Fig. 47-4 Datum target symbols

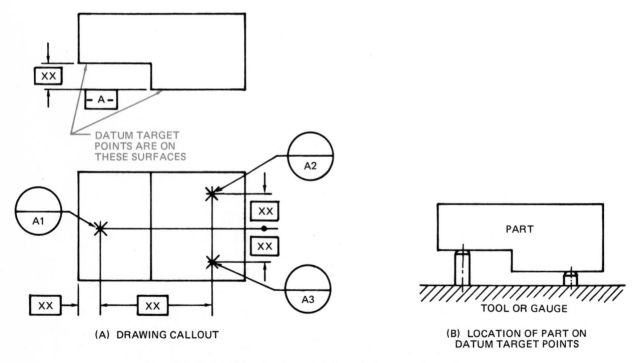

(A) DRAWING CALLOUT

(B) LOCATION OF PART ON
DATUM TARGET POINTS

Fig. 47-5 Datum target points on different planes used as a datum

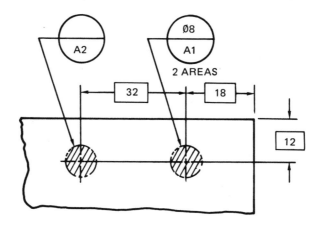

Fig. 47-6 Datum target areas

(A) DRAWING CALLOUT

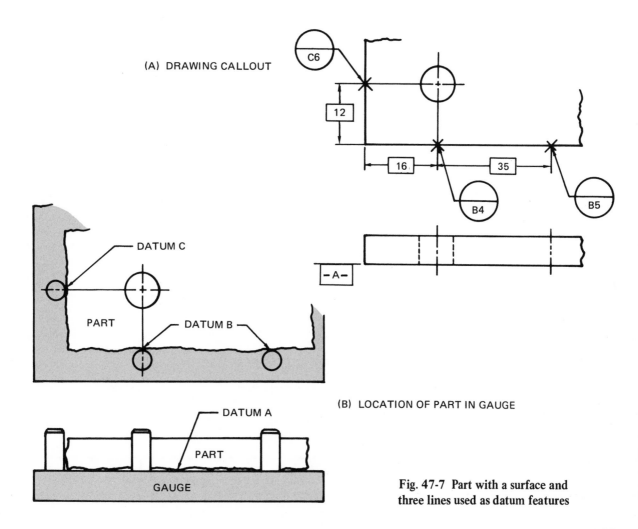

(B) LOCATION OF PART IN GAUGE

Fig. 47-7 Part with a surface and three lines used as datum features

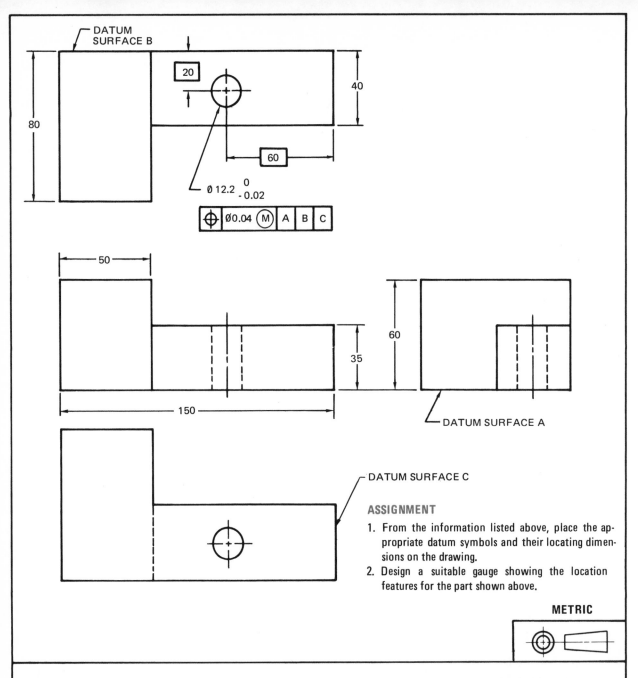

ASSIGNMENT

1. From the information listed above, place the appropriate datum symbols and their locating dimensions on the drawing.
2. Design a suitable gauge showing the location features for the part shown above.

METRIC

DATUM FEATURE	LOCATION OF DATUM FEATURES		
	FROM DATUM SURFACE A	FROM DATUM SURFACE B	FROM DATUM SURFACE C
DATUM TARGET A1 (10 mm)		15	135
DATUM TARGET A2 (10 mm)		65	135
DATUM TARGET A3 (10 mm)		15	15
DATUM LINE B2			125
DATUM LINE B3			25
DATUM POINT C1	15	20	

DATUMS FOR POSITIONAL TOLERANCING

A-84

PROFILE OF A LINE

A *profile* is the outline form or shape of a line or surface. A line profile may be the outline of a part or feature as depicted in a view on a drawing. A line profile may represent the edge of a part or it may refer to line elements of a surface in a single direction.

The symbol for *profile of a line* is a semicircle, figure 48-1. This is shown in the usual manner by including the symbol and tolerance in the feature control symbol directed toward the line to be controlled, as shown in figure 48-2.

Bilateral and Unilateral Tolerances

The profile tolerance zone is usually equally arranged around the basic profile in a form known as a *bilateral tolerance zone.* The width of this zone is always measured perpendicular to the profile surface.

Occasionally, it is advisable to have the tolerance zone on one side of the basic profile instead of being equally divided on both sides. Such zones are called *unilateral tolerance zones* and are indicated with a phantom line close to the profile surface. The tolerance is directed to this line, as shown in figure 48-3.

Extent of Controlled Profile

The profile is generally intended to extend to the first abrupt change or sharp corner. If the extent is not clearly identified by sharp corners or by basic profile dimensions, it must be indicated by a note under the feature control symbol, figure 48-4 (page 328).

Dual Tolerances

If different profile tolerances must be applied to different parts of a profile, the

Fig. 48-1 Profile-of-a-line tolerance

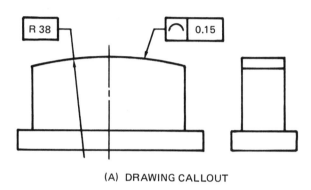

(A) DRAWING CALLOUT

TOLERANCE ZONE 0.15

(B) INTERPRETATION
TAKEN AT ANY SECTION, SUCH AS A-A OR B-B

Fig. 48-2 Simple profile with profile-of-a-line tolerance

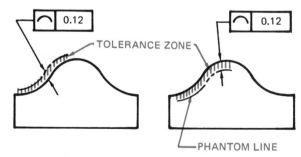

Fig. 48-3 Unilateral tolerance zones

tolerances are shown in a combined feature control symbol. However, the extent of each should be clearly stated, figure 48-5.

PROFILE OF A SURFACE

The symbol for *profile-of-a-surface* is a semicircle enclosed by a straight line at the bottom, figure 48-6.

The tolerance zone established by the profile-of-a-surface tolerance is a three-dimensional zone extending the length and width (or circumference) of the feature. Usually,

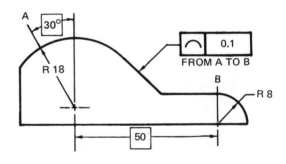

Fig. 48-4 Specifying extent of profile tolerance

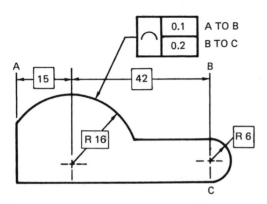

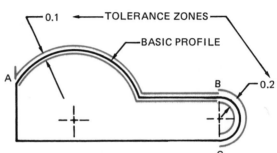

Fig. 48-5 Dual profile tolerance zones

profile-of-a-surface tolerancing requires reference to datums in order to provide proper orientation of a profile, figure 48-7.

Functional GO, NO-GO gauges are used to check parts which are toleranced on an MMC basis. Figure 48-8 shows two identical parts. One is toleranced with a profile-of-a-line tolerance which applies to the circumference of the part. This is often referred to as *all around tolerancing*. The second part has a profile-of-a-surface tolerance.

For figure 48-8(A), a suitable gauge would be made to the maximum material size plus the profile tolerance. If the profile tolerance is zero, the profile gauge will also check the maximum limit of size. The

Fig. 48-6 Profile-of-a-surface symbol

part must be capable of passing through the gauge.

When a profile-of-a-surface tolerance is applied on an MMC basis, as shown in figure 48-8(B), a functional GO, NO-GO gauge is required, similar to that shown in figure 48-8(A); except that it must be of such length as to completely encompass the part.

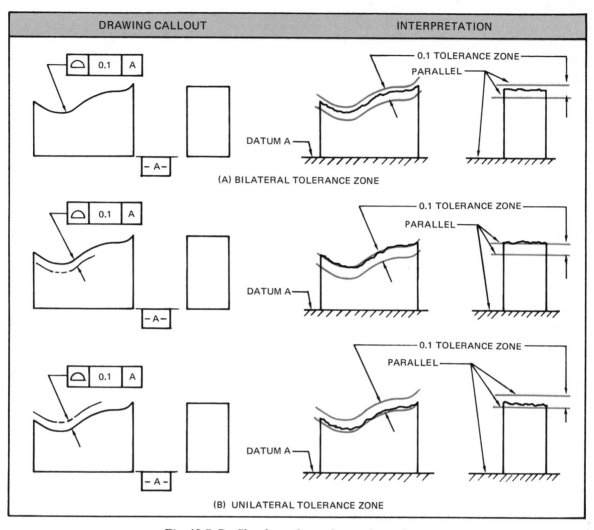

Fig. 48-7 Profile of a surface referenced to a datum

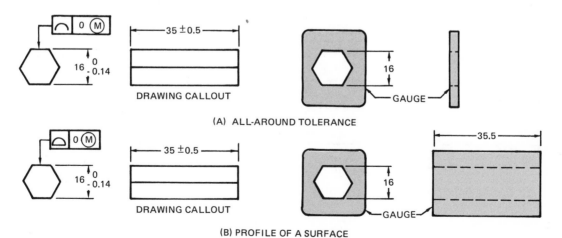

Fig. 48-8 A comparison of profile tolerances

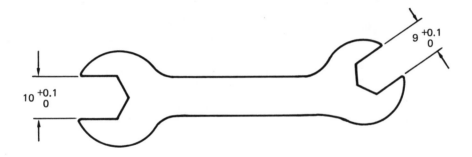

ASSIGNMENT

1. A profile tolerance is required for the wrench. The wrench openings must be held to zero tolerance at MMC; the remainder of the wrench profile tolerance must be held to 0.5 millimetres. Add the feature control symbol to the drawing.

2. Make a sketch showing the tolerance zone for the profile of the wrench.

3. Design suitable gauges to check the following parts.

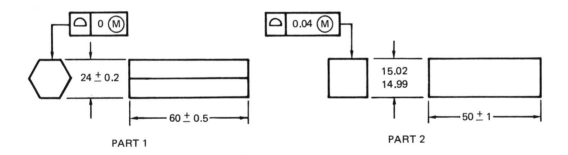

PART 1   PART 2

4. The cross-sectional shape of the part shown below must be controlled to ensure that it does not cross the boundary of perfect form at the maximum material size. If all dimensions are to be held to +0, -0.4, add a suitable profile tolerance to the part.

5. The profile must be controlled from A to B within a tolerance of 0.08 except that the 14 millimetre straight portion can be permitted to vary vertically by ± 0.1. Show how this would be specified on the drawing and sketch the resulting tolerance zone.

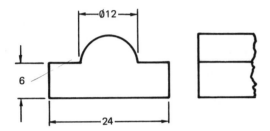

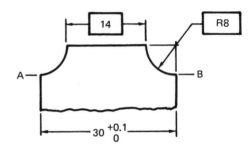

METRIC

PROFILE OF LINES AND SURFACES

A-85

UNIT

49

CORRELATIVE TOLERANCES

Correlative geometrical tolerancing refers to tolerancing for the control of two or more features which should be correlated in position or attitude. Examples of correlated tolerancing include: *coplanarity,* for controlling two or more flat surfaces; *symmetry,* for controlling features equally disposed around a center line; *concentricity* and *coaxiality,* for controlling features having common axes or center lines; and *runout,* for controlling surfaces related to an axis. These are all tolerances of location for which positional symbol and positional tolerances could be used. Special symbols, however, have been provided for some of them to clarify and simplify drawing callout requirements.

Coplanarity

Coplanarity refers to the relative position of two or more flat surfaces which are intended to lie in the same geometrical plane. There is no special symbol for coplanarity. There are several methods of tolerancing. Either the symbol for position or the symbol

for flatness may be used, depending on the type of control required.

Figure 49-1 depicts a case where coplanar surfaces must be accurately located and parallel to another surface of the part, which is then designated as the datum feature.

If coplanarity of two or more surfaces must be held within close limits while permitting a greater variation in the position of the surfaces in relation to other features of the part, two methods of tolerancing giving the identical results may be used. The surfaces may be positioned relative to one another using the symbol for position, figure 49-2 (page 332). Or, they may be treated as one flat surface, using a flatness tolerance. The word SIMULTANEOUS must be added beneath the feature control symbol.

Symmetry

Symmetry is a condition in which a feature or features correspond in size, shape and relative position around a center line or center plane of another feature. The center line or center plane of the second feature is usually specified as a datum.

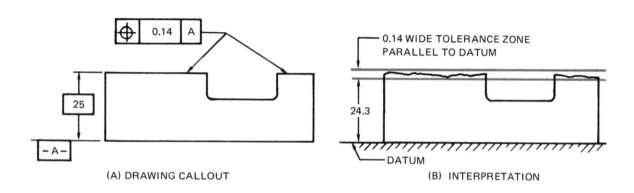

(A) DRAWING CALLOUT   (B) INTERPRETATION

Fig. 49-1 Coplanar surfaces parallel to a datum

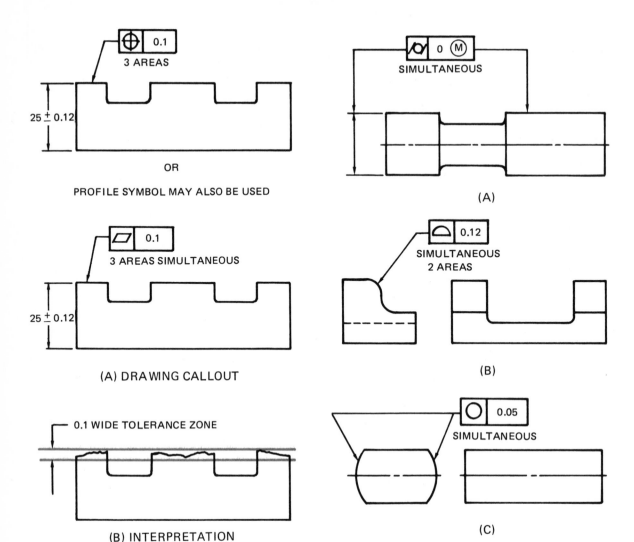

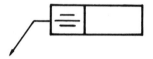

OR

PROFILE SYMBOL MAY ALSO BE USED

(A) DRAWING CALLOUT

0.1 WIDE TOLERANCE ZONE

(B) INTERPRETATION

Fig. 49-2 Coplanarity controlled
by position or flatness

(A)

SIMULTANEOUS
2 AREAS

(B)

SIMULTANEOUS

(C)

Fig. 49-3 Tolerancing correlated surfaces

Fig. 49-4 Symmetry tolerance

A symmetry tolerance specifies the width of a tolerance zone. This width is the area between two parallel lines or the space between two parallel planes equally arranged around the axis or median plane.

Symmetry is therefore a special case of position. The advantage of using the symmetry symbol instead of the position symbol is that it indicates that the true position is symmetrical.

The geometric characteristic symbol for symmetry consists of three horizontal lines, figure 49-4.

Figure 49-5 shows a simple example in which a slot is intended to be symmetrical with the overall width of the part, which is specified as the datum. The interpretation

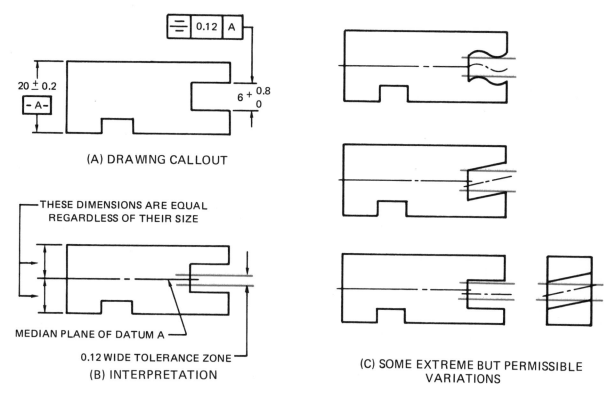

(A) DRAWING CALLOUT

THESE DIMENSIONS ARE EQUAL
REGARDLESS OF THEIR SIZE

MEDIAN PLANE OF DATUM A

0.12 WIDE TOLERANCE ZONE
(B) INTERPRETATION

(C) SOME EXTREME BUT PERMISSIBLE
VARIATIONS

Fig. 49-5 Symmetry tolerancing

shows the tolerance zone equally arranged around the center line or median plane of the datum feature. The width of the tolerance zone is the same for each part regardless of the actual size of the feature or datum feature.

Whenever possible, symmetry should be specified on an MMC basis. Such specifications solve most measurement problems by permitting use of suitable functional GO, NO-GO gauges. Some parts toleranced on this basis, with suitable gauging principles, are shown in figure 49-6.

Concentricity and Coaxiality

Concentricity is a condition in which two or more features, such as circles, spheres, cylinders, cones, or hexagons, share a common center or axis. An example would be a round hole through the center of a cylindrical part.

Coaxiality is a very similar condition in which two or more circular or similar features are arranged with their axes in the same straight line. Examples are a counterbored hole or a shaft having parts along its length turned to different diameters.

Both these terms are often used interchangeably. For geometrical tolerancing, the same symbol is used for both conditions. The geometric characteristic symbol used for both concentricity and coaxiality consists of two concentric circles, figure 49-7 (page 334).

Concentricity on an RFS basis is often difficult to evaluate. Every consideration should be given to alternate methods such as concentricity on an MMC basis or a runout tolerance.

When concentricity is specified on an MMC basis and referenced to datums which are also modified on an MMC basis, this enables concentricity to be evaluated by functional receiver type GO, NO-GO gauges.

Figure 49-8 shows a part modified on an MMC basis, together with a suitable gauge

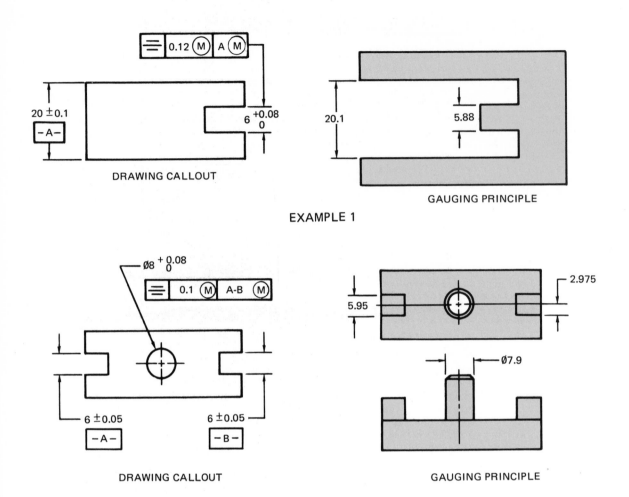

EXAMPLE 1

EXAMPLE 2

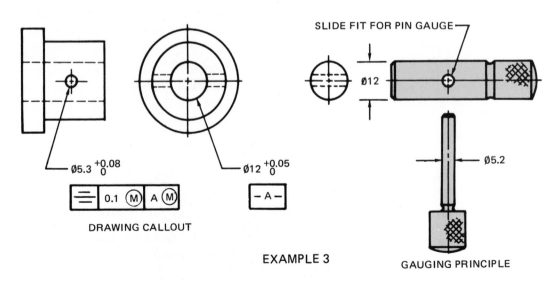

EXAMPLE 3

Fig. 49-6 Examples of parts toleranced on an MMC basis

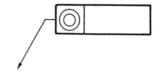

Fig. 49-7 Concentricity tolerancing

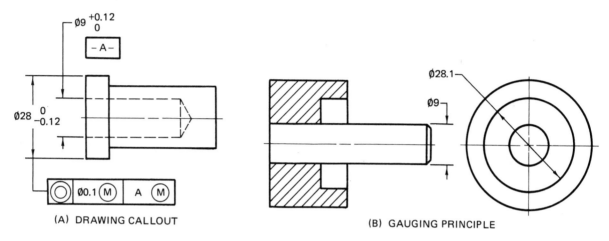

(A) DRAWING CALLOUT

(B) GAUGING PRINCIPLE

Fig. 49-8 Concentricity on MMC basis

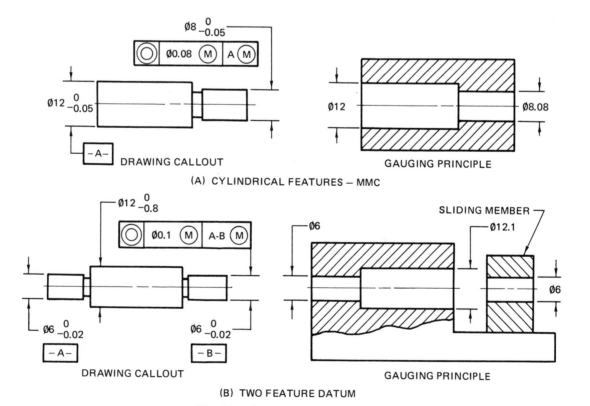

DRAWING CALLOUT

GAUGING PRINCIPLE

(A) CYLINDRICAL FEATURES – MMC

DRAWING CALLOUT

GAUGING PRINCIPLE

(B) TWO FEATURE DATUM

Fig. 49-9 Concentricity tolerancing

which makes the complete check for concentricity and also checks the maximum material size (minimum diameter) of the center hole.

Runout

Runout is the deviation in position of a surface of revolution as a part is revolved around a datum axis.

A *runout tolerance* represents the maximum permissible variation of position of a surface, measured at a fixed point, when the part is revolved without axial movement through 360 degrees around the datum axis. The tolerance zone is the area, or space, of uniform thickness between two shapes coaxial with the datum axis and parallel to the true profile of the controlled surface.

There are two concepts in runout tolerancing. One is known as *circular runout* and concerns runout of each circular element or cross section. The other is called *total runout* and provides composite control of all surface elements simultaneously. Circular runout is simpler than total runout and is generally easier to measure.

There are two geometrical characteristic symbols for runout, one for circular and one for total runout, figure 49-10.

Circular Runout. Circular runout means that the runout tolerance must be met in each circular element, and a sufficient number of measurements must be made at different positions along the surface to guarantee that the tolerance is met at all cross sections of the part.

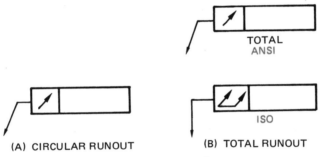

(A) CIRCULAR RUNOUT   (B) TOTAL RUNOUT

Fig. 49-10 Runout tolerance

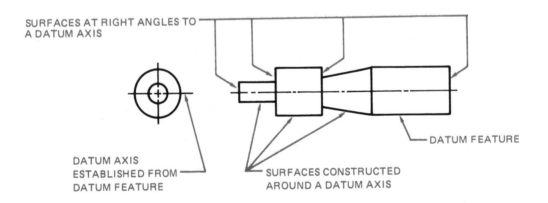

Fig. 49-11 Features applicable to runout tolerancing

Thus, in figure 49-12, the surface is measured at several positions along the surface. At each position the indicator movement during one revolution of the part must not exceed the specified tolerance, in this case 0.1 millimetre. Circular runout does not control the profile elements of the surface.

Total Runout. Total runout concerns the runout of a complete surface, not merely the runout of each circular element. For measurement purposes, the checking indicator must traverse the full length or extent of the surface while the part is revolved about its datum axis. Total runout is the difference between the lowest indicator reading in any position and the highest reading in that or any other position on the same surface, figures 49-13 and 49-14.

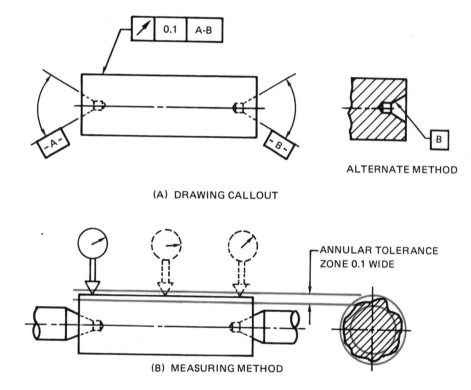

Fig. 49-12 Circular runout of cylindrical features

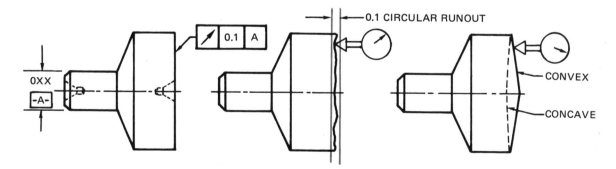

Fig. 49-13 Runout perpendicular to axis

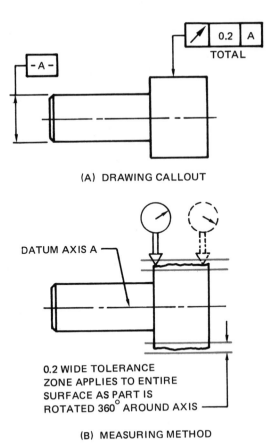

(A) DRAWING CALLOUT

DATUM AXIS A

0.2 WIDE TOLERANCE
ZONE APPLIES TO ENTIRE
SURFACE AS PART IS
ROTATED 360° AROUND AXIS

(B) MEASURING METHOD

Fig. 49-14 Specifying total runout relative to a datum diameter

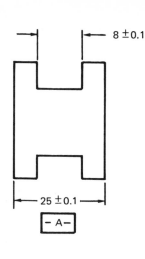

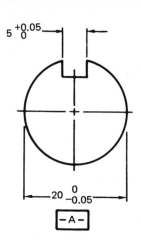

8 ±0.1

5 +0.05 / 0

25 ±0.1

−A−

20 0 / −0.05

−A−

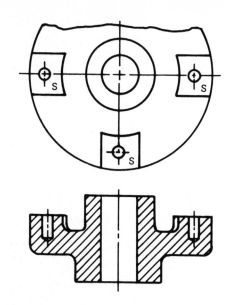

ASSIGNMENT

1. The two 8 slots must be simultaneously symmetrical with the 25 width within zero tolerance. Add a suitable geometrical tolerance to the drawing.

2. The 5 keyway is to be symmetrically located on the Ø20 shaft within 0.05 MMC when the diameter is at its maximum permissible size. Add a suitable geometrical tolerance to the drawing.

5. The four surfaces marked S must be coplanar with one another within 0.01 and perpendicular with the axis of the center hole within 0.03. Add the geometrical tolerance to the drawing.

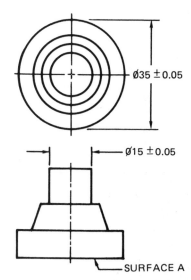

Ø35 ±0.05

Ø15 ±0.05

SURFACE A

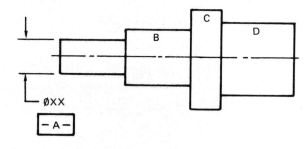

C

B

D

ØXX

−A−

6. Diameters **B** and **C** are to have a circular runout of 0.1 with reference to datum **A**. Diameter **D** is to have a total runout of 0.05 with reference to datum **A**. Add the geometrical tolerances to the drawing.

3. The Ø15 must be concentric with the Ø35 when resting on surface A within 0.03. Specify this on the drawing on a RFS basis.

4. If the tolerance and the diametral datum were both specified on an MMC basis, show the elements and sizes of a suitable gauge.

METRIC

CORRELATIVE TOLERANCES — COPLANARITY, SYMMETRY, CONCENTRICITY, RUNOUT

A-86

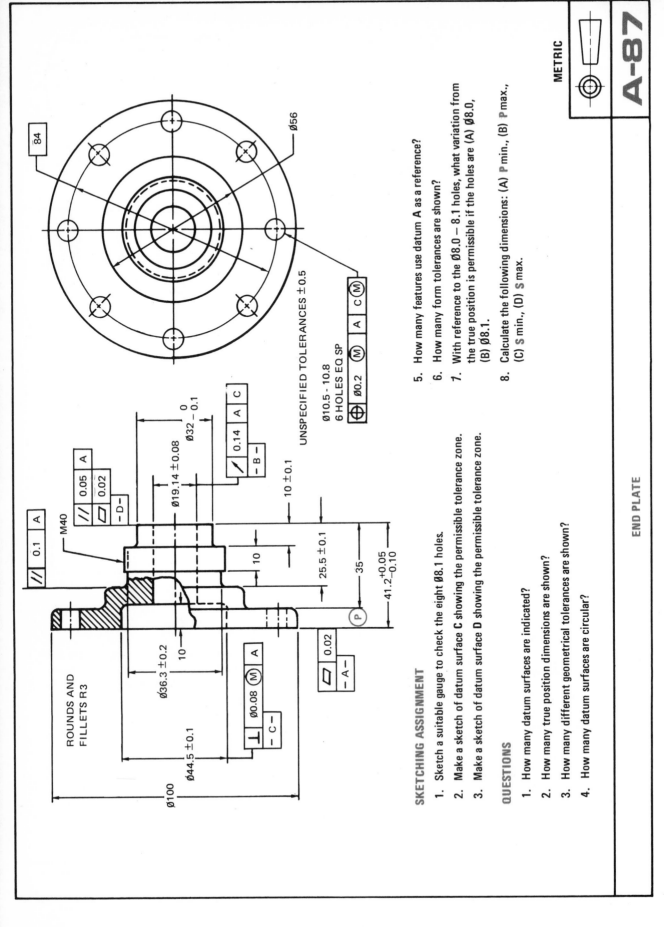

SKETCHING ASSIGNMENT

1. Sketch a suitable gauge to check the eight Ø8.1 holes.

2. Make a sketch of datum surface **C** showing the permissible tolerance zone.

3. Make a sketch of datum surface **D** showing the permissible tolerance zone.

QUESTIONS

1. How many datum surfaces are indicated?

2. How many true position dimensions are shown?

3. How many different geometrical tolerances are shown?

4. How many datum surfaces are circular?

5. How many features use datum **A** as a reference?

6. How many form tolerances are shown?

7. With reference to the Ø8.0 – 8.1 holes, what variation from the true position is permissible if the holes are (A) Ø8.0, (B) Ø8.1.

8. Calculate the following dimensions: (A) **P** min., (B) **P** max., (C) **S** min., (D) **S** max.

END PLATE

340

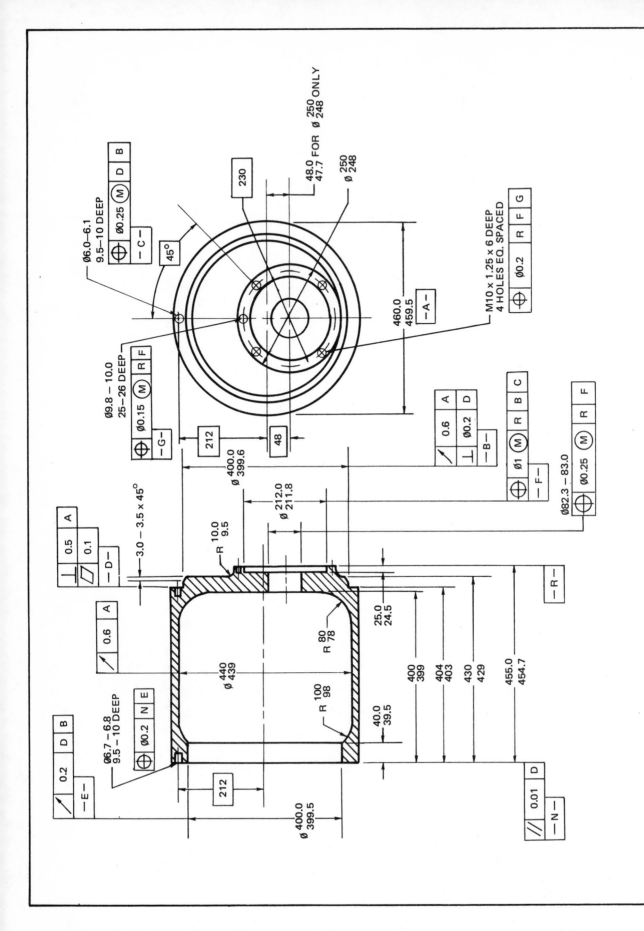

342

SKETCHING ASSIGNMENT

1. Sketch a suitable gauge to check the Ø82.3 – 83.0 hole.

2. Make a sketch of datum surface **D** showing the permissible tolerance zone.

3. Make a sketch of datum surface **N** showing the permissible tolerance zone.

QUESTIONS

1. How many datum surfaces or points are indicated?

2. How many true position dimensions are indicated?

3. How many datum surfaces are flat?

4. How many datum surfaces are circular?

5. How many dimensions show a true position tolerancing?

6. How many form tolerances are required?

7. How many orientation tolerances are shown?

8. How many features use datum **A** as a reference?

9. If the diameter of datum **F** was Ø211.8, what would be the maximum permissible positional tolerance?

10. What is the tertiary datum for the positional tolerance of datum **F**?

11. The geometrical tolerances placed on datum surface **D** controls _____ and _____ .

12. With reference to the Ø6.7 – 6.8 hole, what variation from the true position dimension 48 is permitted when the hole is (A) Ø82.3, (B) Ø83.0?

13. With reference to the Ø6.7 – 6.8 hole, what is the maximum permissible variation from true position if the hole was (A) Ø6.7, (B) Ø6.8?

14. What is (A) maximum, (B) minimum wall thickness permitted at the Ø438 – 440?

15. With reference to the Ø6.0 – 6.1 hole, what variation from the true position 212 is permitted when the hole is (A) 6.0, (B) 6.05, (C) 6.1?

ANSWERS

1. _____
2. _____
3. _____
4. _____
5. _____
6. _____
7. _____
8. _____
9. _____
10. _____
11. _____
12. A. _____
    B. _____
13. A. _____
    B. _____
14. A. _____
    B. _____
15. A. _____
    B. _____
    C. _____

HOUSING

Common Prefixes			
Name	Symbol	Meaning	Multiplier
Mega	M	one million	1 000 000
Kilo	k	one thousand	1 000
hecto	h	one hundred	100
deca	da	ten	10
deci	d	one tenth	0.1
centi	c	one hundredth	0.01
Milli	m	one thousandth	0.001
Micro	μ	one millionth	0.000 001

Units to be Adopted		
	Metric	Discontinued English Unit
Length	millimetre metre kilometre	inch foot yard mile
Area	sq. centimetre sq. metre sq. kilometre	sq. inch sq. foot acre sq. mile
Volume	cubic centimetre cubic metre millilitre litre	cubic inch cubic foot cubic yard pint, quart, gallon bushel fluid ounce
Mass	gram kilogram tonne	ounce pound ton
Force	newton kilonewton	pound force
Pressure	pascal kilopascal	pounds per sq. inch pounds per sq. foot

Applications of Length Units		
Name	Equal to	Applications
Kilometre (km)	1000 metres length of a brisk 10 minute walk	Highway distances
Metre (m)	1000 millimetres approximately half the height of a bedroom door	Dimensions of a building lot
Centimetre (cm)	10 millimetres radius of a nickel	Not generally used in engineering applications
Millimetre (mm)	One thousandth of a metre thickness of a dime	Most common small unit in engineering for part sizes etc.

Table 1 Conversion factors — English to metric

Inch	.00	.01	.02	.03	.04	.05	.06	.07	.08	.09
.00	0.00	0.25	0.51	0.76	1.02	1.27	1.52	1.78	2.03	2.29
.10	2.54	2.79	3.05	3.30	3.56	3.81	4.06	4.32	4.57	4.83
.20	5.08	5.33	5.59	5.84	6.10	6.35	6.60	6.86	7.11	7.37
.30	7.62	7.87	8.13	8.38	8.64	8.89	9.14	9.40	9.65	9.91
.40	10.16	10.41	10.67	10.92	11.18	11.43	11.68	11.94	12.19	12.45
.50	12.70	12.95	13.21	13.46	13.72	13.97	14.22	14.48	14.73	14.99
.60	15.24	15.49	15.75	16.00	16.26	16.51	16.76	17.02	17.27	17.53
.70	17.78	18.03	18.29	18.54	18.80	19.05	19.30	19.56	19.81	20.07
.80	20.32	20.57	20.83	21.08	21.34	21.59	21.84	22.10	22.35	22.61
.90	22.86	23.11	23.37	23.62	23.88	24.13	24.38	24.64	24.89	25.15

ONE HUNDREDTH OF AN INCH INCREMENTS TO 0.99 INCH

Inches	.0	.10	.20	.30	.40	.50	.60	.70	.80	.90
0	0.0	2.5	5.1	7.6	10.2	12.7	15.2	17.8	20.3	22.9
1	25.4	27.9	30.5	33.0	35.6	38.1	40.6	43.2	45.7	48.3
2	50.8	53.3	55.9	58.4	61.0	63.5	66.0	68.6	71.1	73.7
3	76.2	78.7	81.3	83.8	86.4	88.9	91.4	94.0	96.5	99.1
4	101.6	104.1	106.7	109.2	111.8	114.3	116.8	119.4	121.9	124.5
5	127.0	129.5	132.1	134.6	137.2	139.7	142.2	144.8	147.3	149.9
6	152.4	154.9	157.5	160.0	162.6	165.1	167.6	170.2	172.7	175.3
7	177.8	180.3	182.9	185.4	188.0	190.5	193.0	195.6	198.1	200.7
8	203.2	205.7	208.3	210.8	213.4	215.9	218.4	221.0	223.5	226.1
9	228.6	231.1	233.7	236.2	238.8	241.3	243.8	246.4	248.9	251.5
10	254.0	256.5	259.1	261.6	264.2	266.7	269.2	271.8	274.3	276.9
11	279.4	281.9	284.5	287.0	289.6	292.1	294.6	297.2	299.7	302.3
12	304.8	307.3	309.9	312.4	315.0	317.5	320.0	322.6	325.1	327.7
13	330.2	332.7	335.3	337.8	340.4	342.9	345.4	348.0	350.5	353.1
14	355.6	358.1	360.7	363.2	365.8	368.3	370.8	373.4	375.9	378.5
15	381.0	383.5	386.1	388.6	391.2	393.7	396.2	398.8	401.3	403.9
16	406.4	408.9	411.5	414.0	416.6	419.1	421.6	424.2	426.7	429.3
17	431.8	434.3	436.9	439.4	442.0	444.5	447.0	449.6	452.1	454.7
18	457.2	459.7	462.3	464.8	467.4	469.9	472.4	475.0	477.5	480.1
19	482.6	485.1	487.7	490.2	492.8	495.3	497.8	500.4	502.9	505.5
20	508.0	510.5	513.1	515.6	518.2	520.7	523.2	525.8	528.3	530.9

ONE TENTH OF AN INCH INCREMENTS TO 20.9 INCHES

Table 2 Conversion of decimals of an inch to millimetres

IN.	0	1/16	1/8	3/16	1/4	5/16	3/8	7/16	1/2	9/16	5/8	11/16	3/4	13/16	7/8	15/16
0	.0	1.6	3.2	4.8	6.4	7.9	9.5	11.1	12.7	14.3	15.9	17.5	19.1	20.6	22.2	23.8
1	25.4	27.0	28.6	30.2	31.8	33.3	34.9	36.5	38.1	39.7	41.3	42.9	44.5	46.0	47.6	49.2
2	50.8	52.4	54.0	55.6	57.2	58.7	60.3	61.9	63.5	65.1	66.7	68.3	69.9	71.4	73.0	74.6
3	76.2	77.8	79.4	81.0	82.6	84.1	85.7	87.3	88.9	90.5	92.1	93.7	95.3	96.8	98.4	100.0
4	101.6	103.2	104.8	106.4	108.0	109.5	111.1	112.7	114.3	115.9	117.5	119.1	120.7	122.2	123.8	125.4
5	127.0	128.6	130.2	131.8	133.4	134.9	136.5	138.1	139.7	141.3	142.9	144.5	146.1	147.6	149.2	150.8
6	152.4	154.0	155.6	157.2	158.8	160.3	161.9	163.5	165.1	166.7	168.3	169.9	171.5	173.0	174.6	176.2
7	177.8	179.4	181.0	182.6	184.2	185.7	187.3	188.9	190.5	192.1	193.7	195.3	196.9	198.4	200.0	201.6
8	203.2	204.8	206.4	208.0	209.6	211.1	212.7	214.3	215.9	217.5	219.1	220.7	222.3	223.8	225.4	227.0
9	228.6	230.2	231.8	233.4	235.0	236.5	238.1	239.7	241.3	242.9	244.5	246.1	247.7	249.2	250.8	252.4
10	254.0	255.6	257.2	258.8	260.4	261.9	263.5	265.1	266.7	268.3	269.9	271.5	273.1	274.6	276.2	277.8
11	279.4	281.0	282.6	284.2	285.8	287.3	288.9	290.5	292.1	293.7	295.3	296.9	298.5	300.0	301.6	303.2
12	304.8	306.4	308.0	309.6	311.2	312.7	314.3	315.9	317.5	319.1	320.7	322.3	323.9	325.4	327.0	328.6
13	330.2	331.8	333.4	335.0	336.6	338.1	339.7	341.3	342.9	344.5	346.1	347.7	349.3	350.8	352.4	354.0
14	355.6	357.2	358.8	360.4	362.0	363.5	365.1	366.7	368.3	369.9	371.5	373.1	374.7	376.2	377.8	379.4

Table 3 Conversion of fractions of an inch to millimetres (1 inch = 25.4 millimetres)

Table 4 ISO metric threads

Nominal Size DIA (mm) Preferred	Coarse Thread Pitch	Coarse Tap Drill Size	Fine Thread Pitch	Fine Tap Drill Size	4 Thread Pitch	4 Tap Drill Size	3 Thread Pitch	3 Tap Drill Size	2 Thread Pitch	2 Tap Drill Size	1.5 Thread Pitch	1.5 Tap Drill Size	1.25 Thread Pitch	1.25 Tap Drill Size	1 Thread Pitch	1 Tap Drill Size	0.75 Thread Pitch	0.75 Tap Drill Size	0.5 Thread Pitch	0.5 Tap Drill Size	0.35 Thread Pitch	0.35 Tap Drill Size
1.6	0.35	1.25																				
1.8	0.35	1.45																				
2	0.4	1.6																				
2.2	0.45	1.75																				
2.5	0.45	2.05																			0.35	2.15
3	0.5	2.5																			0.35	2.65
3.5	0.6	2.9																			0.35	3.15
4	0.7	3.3																	0.5	3.5		
4.5	0.75	3.7																	0.5	4		
5	0.8	4.2																	0.5	4.5		
*6	1	5															0.75	5.2				
**6.3	1	5.3																				
8	1.25	6.7	1	7											1	7	0.75	7.2				
10	1.5	8.5	1.25	8.7									1.25	8.7	1	9	0.75	9.2				
12	1.75	10.2	1.25	10.8							1.5	10.5	1.25	10.7	1	11						
14	2	12	1.5	12.5							1.5	12.5	1.25	12.7	1	13						
16	2	14	1.5	14.5							1.5	14.5			1	15						
18	2.5	15.5	1.5	16.5					2	16	1.5	16.5			1	17						
20	2.5	17.5	1.5	18.5					2	18	1.5	18.5			1	19						
22	2.5	19.5	1.5	20.5					2	20	1.5	20.5			1	21						
24	3	21	2	22					2	22	1.5	22.5			1	23						
27	3	24	2	25					2	25	1.5	25.5			1	26						
30	3.5	26.5	2	28					2	28	1.5	28.5			1	29						
33	3.5	29.5	2	31					2	31	1.5	31.5										
36	4	32	3	33					2	34	1.5	34.5										
39	4	35	3	36					2	37	1.5	37.5										
42	4.5	37.5	3	39	4	38	3	39	2	40	1.5	40.5										
45	4.5	39	3	42	4	41	3	42	2	43	1.5	43.5										
48	5	43	3	45	4	44	3	45	2	46	1.5	46.5										

Series with Graded Pitches · Series with Constant Pitches

* ISO thread size
** ANSI thread size (To be discontinued)

Metric Drill Sizes		Reference Decimal Equivalent (Inches)	Metric Drill Sizes		Reference Decimal Equivalent (Inches)	Metric Drill Sizes		Reference Decimal Equivalent (Inches)
Preferred	Available		Preferred	Available		Preferred	Available	
–	0.40	.0157	–	2.7	.1063	14	–	.5512
–	0.42	.0165	2.8	–	.1102	–	14.5	.5709
–	0.45	.0177	–	2.9	.1142	15	–	.5906
–	0.48	.0189	3	–	.1181	–	15.5	.6102
0.5	–	.0197	–	3.1	.1220	16	–	.6299
–	0.52	.0205	3.2	–	.1260	–	16.5	.6496
0.55	–	.0217	–	3.3	.1299	17	–	.6693
–	0.58	.0228	3.4	–	.1339	–	17.5	.6890
0.6	–	.0236	–	3.5	.1378	18	–	.7087
–	0.62	.0244	3.6	–	.1417	–	18.5	.7283
0.65	–	.0256	–	3.7	.1457	19	–	.7480
–	0.68	.0268	3.8	–	.1496	–	19.5	.7677
0.7	–	.0276	–	3.9	.1535	20	–	.7874
–	0.72	.0283	4	–	.1575	–	20.5	.8071
0.75	–	.0295	–	4.1	.1614	21	–	.8268
–	0.78	.0307	4.2	–	.1654	–	21.5	.8465
0.8	–	.0315	–	4.4	.1732	22	–	.8661
–	0.82	.0323	4.5	–	.1772	–	23	.9055
0.85	–	.0335	–	4.6	.1811	24	–	.9449
–	0.88	.0346	4.8	–	.1890	25	–	.9843
0.9	–	.0354	5	–	.1969	26	–	1.0236
–	0.92	.0362	–	5.2	.2047	–	27	1.0630
0.95	–	.0374	5.3	–	.2087	28	–	1.1024
–	0.98	.0386	–	5.4	.2126	–	29	1.1417
1	–	.0394	5.6	–	.2205	30	–	1.1811
–	1.03	.0406	–	5.8	.2283	–	31	1.2205
1.05	–	.0413	6	–	.2362	32	–	1.2598
–	1.08	.0425	–	6.2	.2441	–	33	1.2992
1.1	–	.0433	6.3	–	.2480	34	–	1.3386
–	1.15	.0453	–	6.5	.2559	–	35	1.3780
1.2	–	.0472	6.7	–	.2638	36	–	1.4173
1.25	–	.0492	–	6.8	.2677	–	37	1.4567
1.3	–	.0512	–	6.9	.2717	38	–	1.4961
–	1.35	.0531	7.1	–	.2795	–	39	1.5354
1.4	–	.0551	–	7.3	.2874	40	–	1.5748
–	1.45	.0571	7.5	–	.2953	–	41	1.6142
1.5	–	.0591	–	7.8	.3071	42	–	1.6535
–	1.55	.0610	8	–	.3150	–	43.5	1.7126
1.6	–	.0630	–	8.2	.3228	45	–	1.7717
–	1.65	.0650	8.5	–	.3346	–	46.5	1.8307
1.7	–	.0669	–	8.8	.3465	48	–	1.8898
–	1.75	.0689	9	–	.3543	50	–	1.9685
1.8	–	.0709	–	9.2	.3622	–	51.5	2.0276
–	1.85	.0728	9.5	–	.3740	53	–	2.0866
1.9	–	.0748	–	9.8	.3858	–	54	2.1260
–	1.95	.0768	10	–	.3937	56	–	2.2047
2	–	.0787	–	10.3	.4055	–	58	2.2835
–	2.05	.0807	10.5	–	.4134	60	–	2.3622
2.1	–	.0827	–	10.8	.4252			
–	2.15	.0846	11	–	.4331			
2.2	–	.0866	–	11.5	.4528			
–	2.3	.0906	12	–	.4724			
2.4	–	.0945	12.5	–	.4921			
2.5	–	.0984	13	–	.5118			
2.6	–	.1024	–	13.5	.5315			

Table 5 Twist drill sizes

Number or Letter Size Drill	Size		Number or Letter Size Drill	Size		Number or Letter Size Drill	Size		Number or Letter Size Drill	Size	
	mm	Inches		mm	Inches		mm	Inches		mm	Inches
80	0.343	.014	50	1.778	.070	20	4.089	.161	K	7.137	.821
79	0.368	.015	49	1.854	.073	19	4.216	.166	L	7.366	.290
78	0.406	.016	48	1.930	.076	18	4.305	.170	M	7.493	.295
77	0.457	.018	47	1.994	.079	17	4.394	.173	N	7.671	.302
76	0.508	.020	46	2.057	.081	16	4.496	.177	O	8.026	.316
75	0.533	.021	45	2.083	.082	15	4.572	.180	P	8.204	.323
74	0.572	.023	44	2.184	.086	14	4.623	.182	Q	8.433	.332
73	0.610	.024	43	2.261	.089	13	4.700	.185	R	8.611	.339
72	0.635	.025	42	2.375	.094	12	4.800	.189	S	8.839	.348
71	0.660	.026	41	2.438	.096	11	4.851	.191	T	9.093	.358
70	0.711	.028	40	2.489	.098	10	4.915	.194	U	9.347	.368
69	0.742	.029	39	2.527	.100	9	4.978	.196	V	9.576	.377
68	0.787	.031	38	2.578	.102	8	5.080	.199	W	9.804	.386
67	0.813	.032	37	2.642	.104	7	5.105	.201	X	10.084	.397
66	0.838	.033	36	2.705	.107	6	5.182	.204	Y	10.262	.404
65	0.889	.035	35	2.794	.110	5	5.220	.206	Z	10.490	.413
64	0.914	.036	34	2.819	.111	4	5.309	.209			
63	0.940	.037	33	2.870	.113	3	5.410	.213			
62	0.965	.038	32	2.946	.116	2	5.613	.221			
61	0.991	.039	31	3.048	.120	1	5.791	.228			
60	1.016	.040	30	3.264	.129	A	5.944	.234			
59	1.041	.041	29	3.354	.136	B	6.045	.238			
58	1.069	.042	28	3.569	.141	C	6.147	.242			
57	1.092	.043	27	3.658	.144	D	6.248	.246			
56	1.181	.047	26	3.734	.147	E	6.350	.250			
55	1.321	.052	25	3.797	.150	F	6.528	.257			
54	1.397	.055	24	3.861	.152	G	6.629	.261			
53	1.511	.060	23	3.912	.154	H	6.756	.266			
52	1.613	.064	22	3.988	.157	I	6.909	.272			
51	1.702	.067	21	4.039	.159	J	7.036	.277			

Table 6 Number and letter size of drills

U.S. Standard (USS)		U.S. Standard (Revised)		Birmingham (BWG)		New Birmingham (BG)		Browne And Sharpe (B & S)		Imperial Standard (SWG)		Electrical Steel	
Gauge	mm	Gauge	mm	Gauge	mm	Gauge	mm	Gauge	mm	Gauge	mm	Gauge	mm
		3	6.01					3	5.83				
4	5.95	4	5.70	4	6.05	4	6.35	4	5.19	4	5.89		
5	5.56	5	5.31	5	5.59	5	5.65	5	4.62	5	5.39		
6	5.16	6	4.94	6	5.16	6	5.03	6	4.12	6	4.88		
7	4.76	7	4.55	7	4.57	7	4.48	7	3.67	7	4.47		
8	4.37	8	4.18	8	4.19	8	3.99	8	3.26	8	4.06		
9	3.97	9	3.80	9	3.76	9	3.55	9	2.91	9	3.66		
10	3.57	10	3.42	10	3.40	10	3.18	10	2.59	10	3.25		
11	3.18	11	3.04	11	3.05	11	2.83	11	2.30	11	2.95	11	3.18
12	2.78	12	2.66	12	2.77	12	2.52	12	2.05	12	2.64	12	2.77
13	2.38	13	2.78	13	2.41	13	2.24	13	1.83	13	2.34	13	2.39
14	1.98	14	1.90	14	2.11	14	1.99	14	1.63	14	2.03	14	1.98
15	1.79	15	1.71	15	1.83	15	1.78	15	1.45	15	1.83	15	1.78
16	1.59	16	1.52	16	1.65	16	1.59	16	1.29	16	1.63	16	1.59
17	1.43	17	1.37	17	1.47	17	1.41	17	1.15	17	1.42	17	1.42
18	1.27	18	1.21	18	1.25	18	2.58	18	1.02	18	1.22	18	1.27
19	1.11	19	1.06	19	1.07	19	1.19	19	0.91	19	1.02	19	1.11
20	0.95	20	0.91	20	0.89	20	1.00	20	0.81	20	0.91	20	0.95
21	0.87	21	0.84	21	0.81	21	0.89	21	0.72	21	0.81		
22	0.79	22	0.76	22	0.71	22	0.79	22	0.65	22	0.71	22	0.79
23	0.71	23	0.68	23	0.64	23	0.71	23	0.57	23	0.61	23	0.71
24	0.64	24	0.61	24	0.56	24	0.63	24	0.51	24	0.56	24	0.64
25	0.56	25	0.53	25	0.51	25	0.56	25	0.46	25	0.51	25	0.56
26	0.48	26	0.46	26	0.46	26	0.50	26	0.40	26	0.46	26	0.47
27	0.44	27	0.42	27	0.41	27	0.44	27	0.36	27	0.42	27	0.43
28	0.40	28	0.38	28	0.36	28	0.40	28	0.32	28	0.38	28	0.39
29	0.36	29	0.34	29	0.33	29	0.35	29	0.29	29	0.35	29	
30	0.32	30	0.31	30	0.31	30	0.31	30	0.25	30	0.32	30	0.36
31	0.28	31	0.27	31	0.25	31	0.28	31	0.23				
32	0.26	32	0.25	32	0.23			32	0.20	32	0.27	32	0.32
33	0.24	33	0.23	33	0.20	33	0.22	33	0.18	33	0.25		
34	0.22	34	0.21	34	0.18	34	0.20	34	0.16	34	0.23		
				35	0.13	35	0.18			35	0.21		
36	0.18	36	0.17	36	0.10	36	0.16	36	0.13				
										37	0.17		
38	0.16	38	0.15									38	0.15
						38	0.12	38	0.10				
						40	0.10			40	0.12		
										42	0.10		

Metric standards governing gauge sizes were not available at the time of publication. The sizes given in the above chart are "soft conversion" from current inch standards, and are not meant to be representative of the precise metric gauge sizes which may be available in the future. Conversions are given only to allow the student to compare gauge sizes readily with metric drill sizes.

Table 7 Sheet metal gauges and thicknesses

Imperial Standard				American Standard				Steel			
Gauge	mm	Gauge	mm	Gauge	mm	Gauge	mm	Gauge	mm	Gauge	mm
7/0	12.70							7/0	12.45		
6/0	11.79							6/0	11.72		
5/0	10.97							5/0	10.94		
4/0	10.16			4/0	11.68			4/0	10.00		
3/0	9.45			3/0	10.40			3/0	9.21		
2/0	8.84			2/0	9.27			2/0	8.41		
1/0	8.23			1/0	8.25			1/0	7.79		
1	7.62	26	0.46	1	7.35	26	0.40	1	7.19	26	0.46
2	7.01	27	0.42	2	6.54	27	0.36	2	6.67	27	0.44
3	6.40	28	0.38	3	5.83	28	0.32	3	6.19	28	0.41
4	5.89	29	0.35	4	5.19	29	0.29	4	5.72	29	0.38
5	5.39	30	0.32	5	4.62	30	0.25	5	5.26	30	0.36
6	4.88	31	0.30	6	4.12	31	0.27	6	4.88	31	0.34
7	4.47	32	0.27	7	3.67	32	0.20	7	4.50	32	0.33
8	4.06	33	0.25	8	3.26	33	0.18	8	4.12	33	0.30
9	3.66	34	0.23	9	2.91	34	0.16	9	3.77	34	0.26
10	3.25	35	0.21	10	2.59	35	0.14	10	3.43	35	0.24
11	2.95	36	0.19	11	2.30	36	0.13	11	3.06	36	0.23
12	2.64	37	0.17	12	2.05	37	0.11	12	2.68	37	0.22
13	2.34	38	0.15	13	1.83	38	0.10	13	2.32	38	0.20
14	2.03	39	0.13	14	1.63	39	0.09	14	2.03	39	0.19
15	1.83	40	0.12	15	1.45	40	0.08	15	1.83	40	0.18
16	1.63	41	0.11	16	1.29	41	0.07	16	1.59		
17	1.42	42	0.10	17	1.15	42	0.06	17	1.37		
18	1.22	43	0.09	18	1.02	43	0.06	18	1.21		
19	1.02	44	0.08	19	0.91	44	0.05	19	1.04		
20	0.91	45	0.07	20	0.81	45	0.05	20	0.88		
21	0.81	46	0.06	21	0.72	46	0.04	21	0.81		
22	0.71	47	0.05	22	0.64	47	0.04	22	0.73		
23	0.61	48	0.04	23	0.57	48	0.03	23	0.66		
24	0.56	49	0.03	24	0.51	49	0.03	24	0.58		
25	0.51	50	0.02	25	0.46	50	0.02	25	0.52		

Metric standards governing gauge sizes were not available at the time of publication. The sizes given in the above chart are "soft conversions" from current inch standards and are not meant to be representative of the precise metric gauge sizes which may be available in the future. Conversions are given only to allow the student to compare gauge sizes readily with metric drill sizes.

Table 8 Wire gauges and diameters in millimetres

And	&	Malleable Iron	MI	
Across Flats	A/F	Material	MATL	
Angular	ANG	Maximum	MAX	
Approximate	APPROX	Maximum Material		
Assembly	ASSY	Condition	M or MMC	
Basic	BASIC	Metre	m	
Bill of Material	B/M	Metric Thread	M	
Bolt Circle	BC	Micrometre	μm	
Brass	BR	Mild Steel	MS	
Bronze	BRZ	Millimetre	mm	
Brown and Sharpe Gauge	B & S GA	Minimum	MIN	
Bushing	BUSH	Minute (Angle)	MIN	
Casting	CSTG	Nominal	NOM	
Cast Iron	CI	Not to Scale	—— or NTS	
Centimetre	cm	Number	NO	
Center Line	℄	Outside Diameter	OD	
Center to Center	C to C	Parallel	PAR	
Chamfered	CHAM	Perpendicular	PERP	
Circularity	CIR	Pitch	P	
Cold Rolled Steel	CRS	Pitch Circle Diameter	PCD	
Concentric	CONC	Pitch Diameter	PD	
Counterbore	CBORE	Plate	PL	
Countersink	CSK	Radian	rad	
Cubic Centimetre	cm^3	Radius	R	
Cubic Metre	m^3	Reference or		
Datum	DATUM .. $\boxed{A}$ .. $\boxed{\text{-A-}}$	Reference Dimension	REF	
Degree (Angle)	° or DEG	Regardless of Feature Size	RFS	
Diameter	DIA or Ø	Revolutions per Minute	REV/MIN	
Diametral Pitch	DP	Right Hand	RH	
Dimension	DIM	Second (Arc)	(″)	
Drawing	DWG	Second (Time)	SEC	
Eccentric	ECC	Section	SECT	
Figure	FIG	Slotted	SLOT	
Finish All Over	FAO	Socket	SOCK	
Gauge	GA	Spherical	SPHER	
Heat Treat	HT TR	Spotface	SFACE	
Head	HD	Square	SQ	
Heavy	HY	Square Centimetre	cm^2	
Hexagon	HEX	Square Metre	m^2	
Hydraulic	HYD	Steel	ST	
Inside Diameter	ID	Straight	STR	
International Organization		Symmetrical	SYM	
for Standardization	ISO	Thread	THD	
Iron Pipe Size	IPS	Through	THRU	
Kilogram	kg	Tolerance	TOL	
Kilometre	km	True Profile	TP	
Large End	LE	Undercut	UCUT	
Left Hand	LH	U.S. Sheet Metal Gauge	USS GA	
Machined	▽	Wrought Iron	WI	
Machine Steel	MST			

ISO   ANSI (above Datum row)

Table 9 Abbreviations and symbols

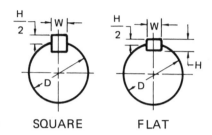

Diameter of Shaft (mm)		Square Key		Flat Key	
		Nominal Size		Nominal Size	
Over	Up To	W	H	W	H
6	8	2	2		
8	10	3	3		
10	12	4	4		
12	17	5	5		
1·7	22	6	6		
22	30	7	7	8	7
30	38	8	8	10	8
38	44	9	9	12	8
44	50	10	10	14	9
50	58	12	12	16	10

Table 10 Square and flat stock keys

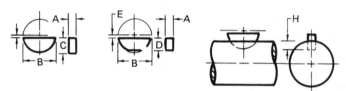

Key No.	Nominal (A x B)		Key			Key Seat
	Millimetres	Inches	E	C	D	H
204	1.6 x 6.4	0.062 x 0.250	0.5	2.8	2.8	4.3
304	2.4 x 12.7	0.094 x 0.500	1.3	5.1	4.8	3.8
305	2.4 x 15.9	0.094 x 0.625	1.5	6.4	6.1	5.1
404	3.2 x 12.7	0.125 x 0.500	1.3	5.1	4.8	3.6
405	3.2 x 15.9	0.125 x 0.625	1.5	6.4	6.1	4.6
406	3.2 x 19.1	0.125 x 0.750	1.5	7.9	7.6	6.4
505	4.0 x 15.9	0.156 x 0.625	1.5	6.4	6.1	4.3
506	4.0 x 19.1	0.156 x 0.750	1.5	7.9	7.6	5.8
507	4.0 x 22.2	0.156 x 0.875	1.5	9.7	9.1	7.4
606	4.8 x 19.1	0.188 x 0.750	1.5	7.9	7.6	5.3
607	4.8 x 22.2	0.188 x 0.875	1.5	9.7	9.1	7.1
608	4.8 x 25.4	0.188 x 1.000	1.5	11.2	10.9	8.6
609	4.8 x 28.6	0.188 x 1.250	2.0	12.2	11.9	9.9
807	6.4 x 22.2	0.250 x 0.875	1.5	9.7	9.1	6.4
808	6.4 x 25.4	0.250 x 1.000	1.5	11.2	10.9	7.9

NOTE: Metric key sizes were not available at the time of publication. Sizes shown are inch-designed key sizes soft converted to millimetres. Conversion was necessary to allow the student to compare keys with slot sizes given in millimetres.

Table 11 Woodruff keys

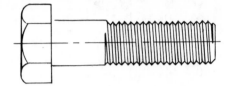

Nominal Size (Millimetres)	Width Across Flats	Thickness
1.6	3.2	1.1
2	4	1.4
2.5	5	1.7
3	5.5	2
4	7	2.8
5	8	3.5
6	10	4
8	13	5.5
10	17	7
12	19	8
14	22	9
16	24	10
18	27	12
20	30	13
22	32	14
24	36	15
27	41	17
30	46	19
33	50	21
36	55	23
39	60	25

Table 12 Hexagon head bolts, regular series

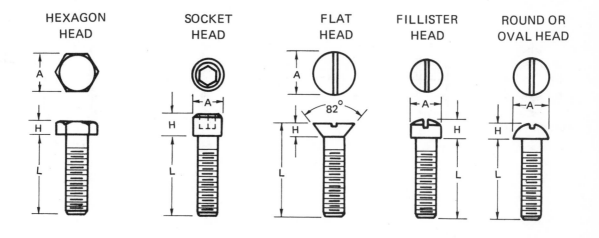

Nominal Size	Hexagon Head		Socket Head			Flat Head		Fillister Head		Round or Oval Head	
	A	**H**	**A**	**H**	**Key Size**	**A**	**H**	**A**	**H**	**A**	**H**
M3	5.5	2	5.5	3	2.5	5.6	1.6	6	2.4	5.6	
4	7	2.8	7	4	3	7.5	2.2	8	3.1	7.5	
5	8.5	3.5	9	5	4	9.2	2.5	10	3.8	9.2	
6	10	4	10	6	5	11	3	12	4.6	11	
8	13	5.5	13	8	6	14.5	4	16	6	14.5	
10	17	7	16	10	8	18	5	20	7.5	18	
12	19	8	18	12	10						
14	22	9	22	14	12						
16	24	10	24	16	14						
18	27	12	27	18	14						
20	30	13	30	20	17						
22	36	15	33	22	17						
24	36	15	36	24	19						
27	41	17	40	27	19						
30	46	19	45	30	22						

Table 13 Common cap screws

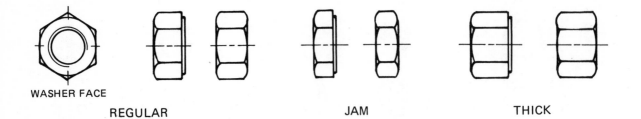

WASHER FACE

REGULAR   JAM   THICK

Nominal Size (Millimetres)	Distance Across Flats	Thickness		
		Regular	Jam	Thick
1.6	3.2	1.3		
2	4	1.6	1.2	
2.5	5	2		
3	5.5	2.4	1.6	4
4	7	3	2	5
5	8	4	2.5	5
6	10	5	3	6
8	13	6.5	5	8
10	17	8	6	10
12	19	10	7	12
14	22	11	8	14
16	24	13	8	16
18	27	15	9	18.5
20	30	16	9	20
22	32	18	10	22
24	36	19	10	24
27	41	22	12	27
30	46	24	12	30
33	50	26		
36	55	29		
39	60	31		

Table 14 Hexagon head nuts

FLAT WASHER   LOCKWASHER   SPRING LOCKWASHER

Bolt Size	Flat Washers			Lockwashers			Spring Lockwashers		
	ID	OD	Thickness	ID	OD	Thickness	ID	OD	Thickness
2	2.2	5.5	0.5	2.1	3.3	0.5			
3	3.2	7	0.5	3.1	5.7	0.8			
4	4.3	9	0.8	4.1	7.1	0.9	4.2	8	0.3 0.4
5	5.3	11	1	5.1	8.7	1.2	5.2	10	0.4 0.5
6	6.4	12	1.5	6.1	11.1	1.6	6.2	12.5	0.5 0.7
7	7.4	14	1.5	7.1	12.1	1.6	7.2	14	0.5 0.8
8	8.4	17	2	8.2	14.2	2	8.2	16	0.6 0.9
10	10.5	21	2.5	10.2	17.2	2.2	10.2	20	0.8 1.1
12	13	24	2.5	12.3	20.2	2.5	12.2	25	0.9 1.5
14	15	28	2.5	14.2	23.2	3	14.2	28	1 1.5
16	17	30	3	16.2	26.2	3.5	16.3	31.5	1.2 1.7
18	19	34	3	18.2	28.2	3.5	18.3	35.5	1.2 2
20	21	36	3	20.2	32.2	4	20.4	40	1.5 2.25
22	23	39	4	22.5	34.5	4	22.4	45	1.75 2.5
24	25	44	4	24.5	38.5	5			
27	28	50	4	27.5	41.5	5			
30	31	56	4	30.5	46.5	6			

Table 15 Common washer sizes

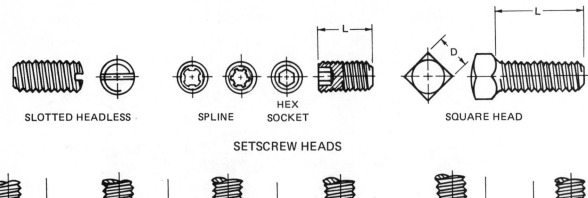

SETSCREW HEADS

SLOTTED HEADLESS   SPLINE   HEX SOCKET   SQUARE HEAD

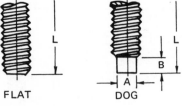
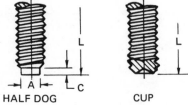
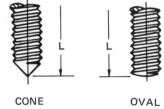

FLAT   DOG   HALF DOG   CUP   CONE   OVAL

SETSCREW POINTS

Nominal Size	Key Size
M 1.4	0.7
2	0.9
3	1.5
4	2
5	2.5
6	3
8	4
10	5
12	6
16	8

Table 16 Setscrews

Nominal Pipe Size (Inches)	Outside Diameter (Millimetres)	Wall Thickness (Millimetres)		
		Standard	Extra Strong	Double Extra Strong
.125 (1/8)	10.29	1.75	2.44	—
.250 (1/4)	13.72	2.29	3.15	—
.375 (3/8)	17.15	2.36	3.28	—
.500 (1/2)	21.34	2.82	3.84	7.80
.750 (3/4)	26.67	2.92	3.99	8.08
1.00	33.4	3.45	4.65	9.37
1.25	42.4	3.63	4.98	9.98
1.50	48.3	3.76	5.18	10.44
2.00	60.3	3.99	5.66	11.35
2.50	73.0	5.26	7.16	14.40
3.00	88.9	5.61	7.77	15.62
3.50	101.6	5.87	8.25	16.54
4.00	114.3	6.15	8.74	17.53
5.00	141.3	6.68	9.73	19.51
6.00	168.3	7.26	11.20	22.45
8.00	219.1	8.36	12.98	22.73
10.00	273.1	9.45	12.95	—
12.00	323.9	—	12.95	—
14.00	355.6	9.73	12.95	—
16.00	406.4	9.73	12.95	—

Nominal pipe sizes are specified in inches.
Outside diameter and wall thicknesses are specified in millimetres.

Table 17 Pipe sizes

VALUES IN MILLIMETRES

Nominal Size Range Millimetres		Class RC1 Precision Sliding			Class RC2 Sliding Fit			Class RC3 Precision Running			Class RC4 Close Running			Class RC5 Medium Running		
Over	To	Hole Tol. H5 (−0)	Clearance Minimum	Shaft Tol. g4 (+0)	Hole Tol. H6 (−0)	Clearance Minimum	Shaft Tol. g5 (+0)	Hole Tol. H7 (−0)	Clearance Minimum	Shaft Tol. f6 (+0)	Hole Tol. H8 (−0)	Clearance Minimum	Shaft Tol. f7 (+0)	Hole Tol. H8 (−0)	Clearance Minimum	Shaft Tol. e7 (+0)
0	3	+0.004	0.003	−0.003	+0.006	0.003	−0.004	+0.010	0.008	−0.006	+0.015	0.008	−0.010	+0.015	0.015	−0.010
3	6	+0.005	0.004	−0.004	+0.008	0.004	−0.005	+0.013	0.010	−0.008	+0.018	0.010	−0.013	+0.018	0.020	−0.013
6	10	+0.006	0.005	−0.004	+0.010	0.005	−0.006	+0.015	0.013	−0.010	+0.023	0.013	−0.015	+0.023	0.025	−0.015
10	18	+0.008	0.006	−0.005	+0.010	0.006	−0.008	+0.018	0.015	−0.010	+0.025	0.015	−0.018	+0.025	0.030	−0.018
18	30	+0.010	0.008	−0.006	+0.013	0.008	−0.010	+0.020	0.020	−0.013	+0.030	0.020	−0.020	+0.030	0.040	−0.020
30	50	+0.010	0.010	−0.008	+0.015	0.010	−0.010	+0.030	0.030	−0.015	+0.040	0.030	−0.030	+0.040	0.050	−0.030
50	80	+0.013	0.010	−0.008	+0.018	0.010	−0.013	+0.030	0.030	−0.020	+0.050	0.030	−0.030	+0.050	0.060	−0.030
80	120	+0.015	0.013	−0.010	+0.023	0.013	−0.015	+0.040	0.040	−0.020	+0.060	0.040	−0.040	+0.060	0.080	−0.040
120	180	+0.018	0.015	−0.013	+0.025	0.015	−0.018	+0.040	0.040	−0.030	+0.060	0.040	−0.040	+0.060	0.090	−0.040
180	250	+0.020	0.015	−0.015	+0.030	0.015	−0.020	+0.050	0.050	−0.030	+0.070	0.050	−0.050	+0.070	0.110	−0.050
250	315	+0.023	0.020	−0.015	+0.030	0.020	−0.023	+0.050	0.060	−0.030	+0.080	0.060	−0.050	+0.080	0.130	−0.050
315	400	+0.025	0.025	−0.018	+0.036	0.025	−0.025	+0.060	0.080	−0.040	+0.090	0.080	−0.060	+0.090	0.150	−0.060

Nominal Size Range Millimetres		Class RC6 Medium Running			Class RC7 Free Running			Class RC8 Loose Running			Class RC9 Loose Running		
Over	To	Hole Tol. H9 (−0)	Clearance Minimum	Shaft Tol. e8 (+0)	Hole Tol. H9 (−0)	Clearance Minimum	Shaft Tol. d8 (+0)	Hole Tol. H10 (−0)	Clearance Minimum	Shaft Tol. e9 (+0)	Hole Tol. GR11 (−0)	Clearance Minimum	Shaft Tol. gr10 (+0)
0	3	+0.025	0.015	−0.015	+0.025	0.025	−0.015	+0.041	0.064	−0.025	+0.060	0.100	−0.040
3	6	+0.030	0.015	−0.018	+0.030	0.030	−0.018	+0.046	0.071	−0.030	+0.080	0.110	−0.050
6	10	+0.036	0.025	−0.023	+0.036	0.040	−0.023	+0.056	0.076	−0.036	+0.070	0.130	−0.060
10	18	+0.040	0.030	−0.025	+0.040	0.050	−0.025	+0.070	0.090	−0.040	+0.100	0.150	−0.070
18	30	+0.050	0.040	−0.030	+0.050	0.060	−0.030	+0.090	0.110	−0.050	+0.130	0.180	−0.090
30	50	+0.060	0.050	−0.040	+0.060	0.080	−0.040	+0.100	0.130	−0.060	+0.150	0.200	−0.100
50	80	+0.080	0.060	−0.050	+0.080	0.100	−0.050	+0.110	0.150	−0.080	+0.180	0.230	−0.120
80	120	+0.090	0.080	−0.060	+0.090	0.130	−0.060	+0.130	0.180	−0.090	+0.230	0.250	−0.130
120	180	+0.100	0.090	−0.060	+0.100	0.150	−0.060	+0.150	0.200	−0.100	+0.250	0.300	−0.150
180	250	+0.110	0.100	−0.070	+0.110	0.180	−0.070	+0.180	0.250	−0.110	+0.300	0.380	−0.180
250	315	+0.130	0.130	−0.080	+0.130	0.200	−0.080	+0.200	0.300	−0.130	+0.300	0.460	−0.200
315	400	+0.150	0.150	−0.090	+0.150	0.250	−0.090	+0.230	0.360	−0.150	+0.360	0.560	−0.230

Table 18 Running and sliding fits

VALUES IN THOUSANDTHS OF AN INCH

Nominal Size Range Inches		Class RC1 Precision Sliding			Class RC2 Sliding Fit			Class RC3 Precision Running			Class RC4 Close Running			Class RC5 Medium Running		
Over	To	Hole Tol. GR5 −0	Clearance Minimum	Shaft Tol. GR4 +0	Hole Tol. GR6 −0	Clearance Minimum	Shaft Tol. GR5 +0	Hole Tol. GR7 −0	Clearance Minimum	Shaft Tol. GR6 +0	Hole Tol. GR8 −0	Clearance Minimum	Shaft Tol. GR7 +0	Hole Tol. GR8 −0	Clearance Minimum	Shaft Tol. GR7 +0
0	.12	+0.15	0.10	−0.12	+0.25	0.10	−0.15	+0.40	0.30	−0.25	+0.60	0.30	−0.40	+0.60	0.60	−0.40
.12	.24	+0.20	0.15	−0.15	+0.30	0.15	−0.20	+0.50	0.40	−0.30	+0.70	0.40	−0.50	+0.70	0.80	−0.50
.24	.40	+0.25	0.20	−0.15	+0.40	0.20	−0.25	+0.60	0.50	−0.40	+0.90	0.50	−0.60	+0.90	1.00	−0.60
.40	.71	+0.30	0.25	−0.20	+0.40	0.25	−0.30	+0.70	0.60	−0.40	+1.00	0.60	−0.70	+1.00	1.20	−0.70
.71	1.19	+0.40	0.30	−0.25	+0.50	0.30	−0.40	+0.80	0.80	−0.50	+1.20	0.80	−0.80	+1.20	1.60	−0.50
1.19	1.97	+0.40	0.40	−0.30	+0.60	0.40	−0.40	+1.00	1.00	−0.60	+1.60	1.00	−1.00	+1.60	2.00	−1.00
1.97	3.15	+0.50	0.40	−0.30	+0.70	0.40	−0.50	+1.20	1.20	−0.70	+1.80	1.20	−1.20	+1.80	2.50	−1.20
3.15	4.73	+0.60	0.50	−0.40	+0.90	0.50	−0.60	+1.40	1.40	−0.90	+2.20	1.40	−1.40	+2.20	3.00	−1.40
4.73	7.09	+0.70	0.60	−0.50	+1.00	0.60	−0.70	+1.60	1.60	−1.00	+2.50	1.60	−1.60	+2.50	3.50	−1.60
7.09	9.85	+0.80	0.60	−0.60	+1.20	0.60	−0.80	+1.80	2.00	−1.20	+2.80	2.00	−1.80	+2.80	4.50	−1.80
9.85	12.41	+0.90	0.80	−0.60	+1.20	0.80	−0.90	+2.00	2.50	−1.20	+3.00	2.50	−2.00	+3.00	5.00	−2.00
12.41	15.75	+1.00	1.00	−0.70	+1.40	1.00	−1.00	+2.20	3.00	−1.40	+3.50	3.00	−2.20	+3.50	6.00	−2.20

Nominal Size Range Inches		Class RC6 Medium Running			Class RC7 Free Running			Class RC8 Loose Running			Class RC9 Loose Running		
Over	To	Hole Tol. GR9 −0	Clearance Minimum	Shaft Tol. GR8 +0	Hole Tol. GR9 −0	Clearance Minimum	Shaft Tol. GR8 +0	Hole Tol. GR10 −0	Clearance Minimum	Shaft Tol. GR9 +0	Hole Tol. GR11 −0	Clearance Minimum	Shaft Tol. GR10 +0
0	.12	+1.00	0.60	−0.60	+1.00	1.00	−0.60	+1.60	2.50	−1.00	+2.50	4.00	−1.60
.12	.24	+1.20	0.80	−0.70	+1.20	1.20	−0.70	+1.80	2.80	−1.20	+3.00	4.50	−1.80
.24	.40	+1.40	1.00	−0.90	+1.40	1.60	−0.90	+2.20	3.00	−1.40	+3.50	6.00	−2.20
.40	.71	+1.60	1.20	−1.00	+1.60	2.00	−1.00	+2.80	3.50	−1.60	+4.00	6.00	−2.80
.71	1.19	+2.00	1.60	−1.20	+2.00	2.50	−1.20	+3.50	4.50	−2.00	+5.00	7.00	−3.50
1.19	1.97	+2.50	2.00	−1.60	+2.50	3.00	−1.60	+4.00	5.00	−2.50	+6.00	8.00	−4.00
1.97	3.15	+3.00	2.50	−1.80	+3.00	4.00	−1.80	+4.50	6.00	−3.00	+7.00	9.00	−4.50
3.15	4.73	+3.50	3.00	−2.20	+3.50	5.00	−2.20	+5.00	7.00	−3.50	+9.00	10.00	−5.00
4.73	7.09	+4.00	3.50	−2.50	+4.00	6.00	−2.50	+6.00	8.00	−4.00	+10.00	12.00	−6.00
7.09	9.85	+4.50	4.00	−2.80	+4.50	7.00	−2.80	+7.00	10.00	−4.50	+12.00	15.00	−7.00
9.85	12.41	+5.00	5.00	−3.00	+5.00	8.00	−3.00	+8.00	12.00	−5.00	+12.00	18.00	−8.00
12.41	15.75	+6.00	6.00	−3.50	+6.00	10.00	−3.50	+9.00	14.00	−6.00	+14.00	22.00	−9.00

Table 18 Running and sliding fits (continued)

VALUES IN MILLIMETRES

Nominal Size Range Millimetres		Class LC1			Class LC2			Class LC3			Class LC4			Class LC5			Class LC6		
Over	To	Hole Tol. H6 (-0)	Shaft Tol. h5 (+0)	Minimum Clearance	Hole Tol. H7 (-0)	Shaft Tol. h6 (+0)	Minimum Clearance	Hole Tol. H8 (-0)	Shaft Tol. h7 (+0)	Minimum Clearance	Hole Tol. H10 (-0)	Shaft Tol. h9 (+0)	Minimum Clearance	Hole Tol. H7 (-0)	Shaft Tol. g6 (+0)	Minimum Clearance	Hole Tol. H9 (-0)	Shaft Tol. f8 (+0)	Minimum Clearance
0	3	+0.006	-0.004	0	+0.010	-0.006	0	+0.015	-0.010	0	+0.041	-0.025	0	+0.010	-0.006	0.002	+0.025	-0.015	0.008
3	6	+0.008	-0.005	0	+0.013	-0.008	0	+0.018	-0.013	0	+0.046	-0.030	0	+0.013	-0.008	0.004	+0.030	-0.018	0.010
6	10	+0.010	-0.006	0	+0.015	-0.010	0	+0.023	-0.015	0	+0.056	-0.036	0	+0.015	-0.010	0.005	+0.036	-0.023	0.013
10	18	+0.010	-0.008	0	+0.018	-0.010	0	+0.025	-0.018	0	+0.070	-0.040	0	+0.018	-0.010	0.006	+0.041	-0.025	0.015
18	30	+0.013	-0.010	0	+0.020	-0.013	0	+0.030	-0.020	0	+0.090	-0.050	0	+0.020	-0.013	0.008	+0.050	-0.030	0.020
30	50	+0.015	-0.010	0	+0.025	-0.015	0	+0.041	-0.025	0	+0.100	-0.060	0	+0.025	-0.015	0.010	+0.060	-0.040	0.030
50	80	+0.018	-0.013	0	+0.030	-0.018	0	+0.046	-0.030	0	+0.110	-0.080	0	+0.030	-0.018	0.010	+0.080	-0.050	0.030
80	120	+0.023	-0.015	0	+0.036	-0.023	0	+0.056	-0.036	0	+0.130	-0.080	0	+0.036	-0.023	0.013	+0.090	-0.060	0.040
120	180	+0.025	-0.018	0	+0.041	-0.025	0	+0.064	-0.041	0	+0.150	-0.100	0	+0.041	-0.025	0.015	+0.100	-0.060	0.040
180	250	+0.030	-0.020	0	+0.046	-0.030	0	+0.071	-0.046	0	+0.180	-0.110	0	+0.046	-0.030	0.015	+0.110	-0.070	0.050
250	315	+0.020	-0.023	0	+0.051	-0.030	0	+0.076	-0.051	0	+0.200	-0.130	0	+0.051	-0.030	0.018	+0.130	-0.080	0.060
315	400	+0.036	-0.025	0	+0.056	-0.036	0	+0.089	-0.056	0	+0.230	-0.150	0	+0.056	-0.036	0.018	+0.150	-0.090	0.060

Nominal Size Range Millimetres		Class LC7			Class LC8			Class LC9			Class LC10			Class LC11		
Over	To	Hole Tol. H10 (-0)	Shaft Tol. e9 (+0)	Minimum Clearance	Hole Tol. H10 (-0)	Shaft Tol. d9 (+0)	Minimum Clearance	Hole Tol. H11 (-0)	Shaft Tol. c10 (+0)	Minimum Clearance	Hole Tol. GR12 (-0)	Shaft Tol. gr11 (+0)	Minimum Clearance	Hole Tol. GR13 (-0)	Shaft Tol. gr12 (+0)	Minimum Clearance
0	3	+0.041	-0.025	0.015	+0.041	-0.025	0.025	+0.064	-0.041	0.06	+0.10	-0.06	0.10	+0.15	-0.10	0.13
3	6	+0.046	-0.030	0.020	+0.046	-0.030	0.030	+0.076	-0.046	0.07	+0.13	-0.08	0.11	+0.18	-0.13	0.15
6	10	+0.056	-0.036	0.025	+0.056	-0.036	0.041	+0.089	-0.056	0.08	+0.15	-0.09	0.13	+0.23	-0.15	0.18
10	18	+0.070	-0.040	0.030	+0.070	-0.040	0.050	+0.100	-0.070	0.09	+0.18	-0.10	0.15	+0.25	-0.18	0.20
18	30	+0.090	-0.050	0.040	+0.090	-0.050	0.060	+0.130	-0.090	0.11	+0.20	-0.13	0.18	+0.31	-0.20	0.25
30	50	+0.100	-0.060	0.050	+0.100	-0.060	0.090	+0.150	-0.100	0.13	+0.25	-0.15	0.20	+0.41	-0.25	0.31
50	80	+0.110	-0.080	0.060	+0.110	-0.080	0.100	+0.180	-0.110	0.15	+0.31	-0.18	0.25	+0.46	-0.31	0.36
80	120	+0.130	-0.090	0.080	+0.130	-0.090	0.130	+0.230	-0.130	0.18	+0.36	-0.23	0.28	+0.56	-0.36	0.41
120	180	+0.150	-0.100	0.090	+0.150	-0.100	0.150	+0.250	-0.150	0.20	+0.41	-0.25	0.31	+0.64	-0.41	0.46
180	250	+0.180	-0.110	0.100	+0.180	-0.110	0.180	+0.310	-0.180	0.25	+0.46	-0.31	0.41	+0.71	-0.46	0.56
250	315	+0.200	-0.130	0.110	+0.200	-0.130	0.180	+0.310	-0.200	0.31	+0.51	-0.31	0.51	+0.76	-0.51	0.71
315	400	+0.230	-0.150	0.130	+0.230	-0.150	0.200	+0.360	-0.230	0.36	+0.56	-0.36	0.56	+0.89	-0.56	0.76

Table 19 Locational clearance fits

362

VALUES IN THOUSANDTHS OF AN INCH

Nominal Size Range Inches		Class LC1			Class LC2			Class LC3			Class LC4			Class LC5			Class LC6		
Over	To	Clearance Minimum	Hole Tol. GR6 (-0)	Shaft Tol. GR5 (+0)	Clearance Minimum	Hole Tol. GR8 (-0)	Shaft Tol. GR7 (+0)	Clearance Minimum	Hole Tol. GR10 (-0)	Shaft Tol. GR9 (+0)	Clearance Minimum	Hole Tol. GR7 (-0)	Shaft Tol. GR6 (+0)	Clearance Minimum	Hole Tol. GR9 (-0)	Shaft Tol. GR8 (+0)	Clearance Minimum	Hole Tol. GR9 (-0)	Shaft Tol. GR8 (+0)
0	.12	0	+0.25	-0.15	0	+0.4	-0.25	0	+0.6	-0.4	0	+1.6	-1.0	0.10	+0.4	-0.25	0.3	+1.0	-0.6
.12	.24	0	+0.30	-0.20	0	+0.5	-0.30	0	+0.7	-0.5	0	+1.8	-1.2	0.15	+0.5	-0.30	0.4	+1.2	-0.7
.24	.40	0	+0.40	-0.25	0	+0.6	-0.40	0	+0.9	-0.6	0	+2.2	-1.4	0.20	+0.6	-0.40	0.5	+1.4	-0.9
.40	.71	0	+0.40	-0.30	0	+0.7	-0.40	0	+1.0	-0.7	0	+2.8	-1.6	0.25	+0.7	-0.40	0.6	+1.6	-1.0
.71	1.19	0	+0.50	-0.40	0	+0.8	-0.50	0	+1.2	-0.8	0	+3.5	-2.0	0.30	+0.8	-0.50	0.8	+2.0	-1.2
1.19	1.97	0	+0.60	-0.40	0	+1.0	-0.60	0	+1.6	-1.0	0	+4.0	-2.5	0.40	+1.0	-0.60	1.0	+2.5	-1.6
1.97	3.15	0	+0.70	-0.50	0	+1.2	-0.70	0	+1.8	-1.2	0	+4.5	-3.0	0.40	+1.2	-0.70	1.2	+3.0	-1.8
3.15	4.73	0	+0.90	-0.60	0	+1.4	-0.90	0	+2.7	-1.4	0	+5.0	-3.5	0.50	+1.4	-0.90	1.4	+3.5	-2.2
4.73	7.09	0	+1.00	-0.70	0	+1.6	-1.00	0	+2.5	-1.6	0	+6.0	-4.0	0.60	+1.6	-1.00	1.6	+4.0	-2.5
7.09	9.85	0	+1.20	-0.80	0	+1.8	-1.20	0	+2.8	-1.8	0	+7.0	-4.5	0.60	+1.8	-1.20	2.0	+4.5	-2.8
9.85	12.41	0	+1.20	-0.90	0	+2.0	-1.20	0	+3.0	-2.0	0	+8.0	-5.0	0.70	+2.0	-1.20	2.2	+5.0	-3.0
12.41	15.75	0	+1.40	-1.00	0	+2.2	-1.40	0	+3.5	-2.2	0	+9.0	-6.0	0.70	+2.2	-1.40	2.5	+6.0	-3.5

Nominal Size Range Inches		Class LC7			Class LC8			Class LC9			Class LC10			Class LC11		
Over	To	Clearance Minimum	Hole Tol. GR10 (-0)	Shaft Tol. GR9 (+0)	Clearance Minimum	Hole Tol. GR10 (-0)	Shaft Tol. GR9 (+0)	Clearance Minimum	Hole Tol. GR11 (-0)	Shaft Tol. GR10 (+0)	Clearance Minimum	Hole Tol. GR12 (-0)	Shaft Tol. GR11 (+0)	Clearance Minimum	Hole Tol. GR13 (-0)	Shaft Tol. GR12 (+0)
0	.12	0.6	+1.6	-1.0	1.0	+1.6	-1.0	2.5	+2.5	-1.6	4.0	+1.0	-2.5	5.0	+6.0	-4.0
.12	.24	0.8	+1.8	-1.2	1.2	+1.8	-1.2	2.8	+3.0	-1.8	4.5	+5.0	-3.0	6.0	+7.0	-5.0
.24	.40	1.0	+2.2	-1.4	1.6	+2.2	-1.4	3.0	+3.5	-2.2	5.0	+6.0	-3.5	7.0	+9.0	-6.0
.40	.71	1.2	+2.8	-1.6	2.0	+2.8	-1.6	3.5	+4.0	-2.8	6.0	+7.0	-4.0	8.0	+10.0	-7.0
.71	1.19	1.6	+3.5	-2.0	2.5	+3.5	-2.0	4.5	+5.0	-3.5	7.0	+8.0	-5.0	10.0	+12.0	-8.0
1.19	1.97	2.0	+4.0	-2.5	3.6	+4.0	-2.5	5.0	+6.0	-4.0	8.0	+10.0	-6.0	12.0	+16.0	-10.0
1.97	3.15	2.5	+4.5	-3.0	4.0	+4.5	-3.0	6.0	+7.0	-4.5	10.0	+12.0	-7.0	14.0	+18.0	-12.0
3.15	4.73	3.0	+5.0	-3.5	5.0	+5.0	-3.5	7.0	+9.0	-5.0	11.0	+14.0	-9.0	16.0	+22.0	-14.0
4.73	7.09	3.5	+6.0	-4.0	6.0	+6.0	-4.0	8.0	+10.0	-6.0	12.0	+16.0	-10.0	18.0	+25.0	-16.0
7.09	9.85	4.0	+7.0	-4.5	7.0	+7.0	-4.5	10.0	+12.0	-7.0	16.0	+18.0	-12.0	22.0	+28.0	-18.0
9.85	12.41	4.5	+8.0	-5.0	7.0	+8.0	-5.0	12.0	+12.0	-8.0	20.0	+20.0	-12.0	28.0	+30.0	-20.0
12.41	15.75	5.0	+9.0	-6.0	8.0	+9.0	-6.0	14.0	+14.0	-9.0	22.0	+22.0	-14.0	30.0	+35.0	-22.0

Table 19 Locational clearance fits (continued)

VALUES IN MILLIMETRES

Nominal Size Range Millimetres		Class LT1			Class LT2			Class LT3			Class LT4			Class LT5			Class LT6		
		Hole Tol. H7	Maximum Interference	Shaft Tol. js6	Hole Tol. H8	Maximum Interference	Shaft Tol. js7	Hole Tol. H7	Maximum Interference	Shaft Tol. k6	Hole Tol. H8	Maximum Interference	Shaft Tol. k7	Hole Tol. H7	Maximum Interference	Shaft Tol. n6	Hole Tol. H8	Maximum Interference	Shaft Tol. n7
Over	To	-0		+0	-0		+0	-0		+0	-0		+0	-0		+0	-0		+0
0	3	+0.010	0.002	-0.006	+0.015	0.005	-0.010	+0.010	0.006	-0.006	+0.015	0.010	-0.010	+0.010	0.013	-0.006	+0.015	0.016	-0.010
3	6	+0.013	0.004	-0.008	+0.018	0.006	-0.013	+0.013	0.010	-0.008	+0.018	0.015	-0.013	+0.013	0.015	-0.008	+0.018	0.020	-0.013
6	10	+0.015	0.005	-0.010	+0.023	0.008	-0.015	+0.015	0.013	-0.010	+0.023	0.018	-0.015	+0.015	0.020	-0.010	+0.023	0.025	-0.015
10	18	+0.018	0.005	-0.010	+0.025	0.008	-0.018	+0.018	0.013	-0.010	+0.025	0.020	-0.018	+0.018	0.023	-0.010	+0.025	0.030	-0.018
18	30	+0.020	0.006	-0.013	+0.030	0.010	-0.020	+0.020	0.015	-0.013	+0.030	0.023	-0.020	+0.020	0.028	-0.013	+0.030	0.036	-0.020
30	50	+0.025	0.008	-0.015	+0.041	0.013	-0.025	+0.025	0.018	-0.015	+0.041	0.028	-0.025	+0.025	0.033	-0.015	+0.041	0.044	-0.025
50	80	+0.030	0.008	-0.018	+0.046	0.015	-0.030	+0.030	0.020	-0.018	+0.046	0.033	-0.030	+0.030	0.038	-0.018	+0.046	0.051	-0.030
80	120	+0.036	0.010	-0.023	+0.056	0.018	-0.036	+0.036	0.025	-0.023	+0.056	0.038	-0.036	+0.036	0.048	-0.023	+0.056	0.062	-0.036
120	180	+0.041	0.013	-0.025	+0.064	0.020	-0.041	+0.041	0.028	-0.025	+0.064	0.044	-0.041	+0.041	0.056	-0.025	+0.064	0.071	-0.041
180	250	+0.046	0.015	-0.030	+0.071	0.023	-0.046	+0.046	0.036	-0.030	+0.071	0.051	-0.046	+0.046	0.066	-0.030	+0.071	0.081	-0.046
250	315	+0.051	0.015	-0.030	+0.076	0.025	-0.051	+0.051	0.036	-0.030	+0.076	0.056	-0.051	+0.051	0.066	-0.030	+0.076	0.086	-0.051
315	400	+0.056	0.018	-0.036	+0.089	0.025	-0.056	+0.056	0.041	-0.036	+0.089	0.062	-0.056	+0.056	0.076	-0.036	+0.089	0.096	-0.056

VALUES IN THOUSANDTHS OF AN INCH

Nominal Size Range Inches		Class LT1			Class LT2			Class LT3			Class LT4			Class LT5			Class LT6		
		Hole Tol. GR7	Maximum Interference	Shaft Tol. GR6	Hole Tol. GR8	Maximum Interference	Shaft Tol. GR7	Hole Tol. GR7	Maximum Interference	Shaft Tol. GR6	Hole Tol. GR8	Maximum Interference	Shaft Tol. GR7	Hole Tol. GR7	Maximum Interference	Shaft Tol. GR6	Hole Tol. GR8	Maximum Interference	Shaft Tol. GR7
Over	To	-0		+0	-0		+0	-0		+0	-0		+0	-0		+0	-0		+0
0	.12	+0.4	0.10	-0.25	+0.6	0.20	-0.4	+0.4	0.25	-0.25	+0.6	0.4	-0.4	+0.4	0.5	-0.25	+0.6	0.65	-0.4
.12	.24	+0.5	0.15	-0.30	+0.7	0.25	-0.5	+0.5	0.40	-0.30	+0.7	0.6	-0.5	+0.5	0.6	-0.30	+0.7	0.80	-0.5
.24	.40	+0.6	0.20	-0.40	+0.9	0.30	-0.6	+0.6	0.50	-0.40	+0.9	0.7	-0.6	+0.6	0.8	-0.40	+0.9	1.00	-0.6
.40	.71	+0.7	0.20	-0.40	+1.0	0.30	-0.7	+0.7	0.50	-0.40	+1.0	0.8	-0.7	+0.7	0.9	-0.40	+1.0	1.20	-0.7
.71	1.19	+0.8	0.25	-0.50	+1.2	0.40	-0.8	+0.8	0.60	-0.50	+1.2	0.9	-0.8	+0.8	1.1	-0.50	+1.2	1.40	-0.8
1.19	1.97	+1.0	0.30	-0.60	+1.6	0.50	-1.0	+1.0	0.70	-0.60	+1.6	1.1	-1.0	+1.0	1.3	-0.60	+1.6	1.70	-1.0
1.97	3.15	+1.2	0.30	-0.70	+1.8	0.60	-1.2	+1.2	0.80	-0.70	+1.8	1.3	-1.2	+1.2	1.5	-0.70	+1.8	2.00	-1.2
3.15	4.73	+1.4	0.40	-0.90	+2.2	0.70	-1.4	+1.4	1.00	-0.90	+2.2	1.5	-1.4	+1.4	1.9	-0.90	+2.2	2.40	-1.4
4.73	7.09	+1.6	0.50	-1.00	+2.5	0.80	-1.6	+1.6	1.10	-1.00	+2.5	1.7	-1.6	+1.6	2.2	-1.00	+2.5	2.80	-1.6
7.09	9.85	+1.8	0.60	-1.20	+2.8	0.90	-1.8	+1.8	1.40	-1.20	+2.8	2.0	-1.8	+1.8	2.6	-1.20	+2.8	3.20	-1.8
9.85	12.41	+2.0	0.60	-1.20	+3.0	1.00	-2.0	+2.0	1.40	-1.20	+3.0	2.2	-2.0	+2.0	2.6	-1.20	+3.0	3.40	-2.0
12.41	15.75	+2.2	0.70	-1.40	+3.5	1.00	-2.2	+2.2	1.60	-1.40	+3.5	2.4	-2.2	+2.2	3.0	-1.40	+3.5	3.80	-2.2

Table 20 Locational transition fits

VALUES IN MILLIMETRES

Nominal Size Range Millimetres		Class LN1 Light Press Fit			Class LN2 Medium Press Fit			Class LN3 Heavy Press Fit			Class LN4			Class LN5			Class LN6		
Over	To	Hole Tol. GR6 -0	Maximum Interference	Shaft Tol. gr5 +0	Hole Tol. H7 -0	Maximum Interference	Shaft Tol. p6 +0	Hole Tol. H7 -0	Maximum Interference	Shaft Tol. t6 +0	Hole Tol. GR8 -0	Maximum Interference	Shaft Tol. gr7 +0	Hole Tol. GR9 -0	Maximum Interference	Shaft Tol. gr8 +0	Hole Tol. GR10 -0	Maximum Interference	Shaft Tol. gr9 +0
0	3	+0.006		-0.004	+0.010	0.016	-0.006	+0.010	0.019	-0.006	+0.015	0.030	-0.010	+0.025	0.046	-0.015	+0.041	0.076	-0.025
3	6	+0.008		-0.005	+0.013	0.020	-0.008	+0.013	0.023	-0.008	+0.018	0.038	-0.013	+0.030	0.059	-0.018	+0.046	0.091	-0.030
6	10	+0.010		-0.006	+0.015	0.025	-0.010	+0.015	0.030	-0.010	+0.023	0.046	-0.015	+0.036	0.071	-0.023	+0.056	0.112	-0.036
10	18	+0.010		-0.008	+0.018	0.028	-0.010	+0.018	0.036	-0.010	+0.025	0.056	-0.018	+0.041	0.086	-0.025	+0.071	0.142	-0.041
18	30	+0.013		-0.010	+0.020	0.033	-0.013	+0.020	0.044	-0.013	+0.030	0.066	-0.020	+0.051	0.107	-0.030	+0.089	0.178	-0.051
30	50	+0.015		-0.010	+0.025	0.041	-0.015	+0.025	0.051	-0.015	+0.041	0.086	-0.025	+0.064	0.135	-0.041	+0.102	0.216	-0.064
50	80	+0.018		-0.013	+0.030	0.054	-0.018	+0.030	0.059	-0.018	+0.046	0.102	-0.030	+0.076	0.160	-0.046	+0.114	0.254	-0.076
80	120	+0.023		-0.015	+0.036	0.064	-0.023	+0.036	0.074	-0.023	+0.056	0.122	-0.036	+0.102	0.196	-0.056	+0.127	0.292	-0.102
120	180	+0.025		-0.018	+0.041	0.071	-0.025	+0.041	0.089	-0.025	+0.064	0.142	-0.041	+0.114	0.221	-0.064	+0.152	0.343	-0.114
180	250	+0.030		-0.020	+0.046	0.081	-0.030	+0.046	0.107	-0.030	+0.071	0.168	-0.046	+0.127	0.262	-0.071	+0.178	0.419	-0.127
250	315	+0.030		-0.023	+0.051	0.086	-0.030	+0.051	0.119	-0.030	+0.076	0.191	-0.051	+0.152	0.305	-0.076	+0.203	0.483	-0.152
315	400	+0.036		-0.025	+0.056	0.099	-0.036	+0.056	0.150	-0.036	+0.089	0.221	-0.056	+0.152	0.368	-0.089	+0.229	0.584	-0.152

VALUES IN THOUSANDTHS OF AN INCH

Nominal Size Range Inches		Class LN1 Light Press Fit			Class LN2 Medium Press Fit			Class LN3 Heavy Press Fit			Class LN4			Class LN5			Class LN6		
Over	To	Hole Tol. GR6 -0	Maximum Interference	Shaft Tol. GR5 +0	Hole Tol. GR7 -0	Maximum Interference	Shaft Tol. GR6 +0	Hole Tol. GR7 -0	Maximum Interference	Shaft Tol. GR6 +0	Hole Tol. GR8 -0	Maximum Interference	Shaft Tol. GR7 +0	Hole Tol. GR9 -0	Maximum Interference	Shaft Tol. GR8 +0	Hole Tol. GR10 -0	Maximum Interference	Shaft Tol. GR9 +0
0	.12	+0.25	0.40	-0.15	+0.4	0.65	-0.25	+0.4	0.75	-0.25	+0.6	1.2	-0.4	+1.0	1.8	-0.6	+1.6	3.0	-1.0
.12	.24	+0.30	0.50	-0.20	+0.5	0.80	-0.30	+0.5	0.90	-0.30	+0.7	1.5	-0.5	+1.2	2.3	-0.7	+1.8	3.6	-1.2
.24	.40	+0.40	0.65	-0.25	+0.6	1.00	-0.40	+0.6	1.20	-0.40	+0.9	1.8	-0.6	+1.4	2.8	-0.9	+2.2	4.4	-1.4
.40	.71	+0.40	0.70	-0.30	+0.7	1.10	-0.40	+0.7	1.40	-0.40	+1.0	2.2	-0.7	+1.6	3.4	-1.0	+2.8	5.6	-1.6
.71	1.19	+0.50	0.90	-0.40	+0.8	1.30	-0.50	+0.8	1.70	-0.50	+1.2	2.6	-0.8	+2.0	4.2	-1.2	+3.5	7.0	-2.0
1.19	1.97	+0.60	1.00	-0.40	+1.0	1.60	-0.60	+1.0	2.00	-0.60	+1.6	3.4	-1.0	+2.5	5.3	-1.6	+4.0	8.5	-2.5
1.97	3.15	+0.70	1.30	-0.50	+1.2	2.10	-0.70	+1.2	2.30	-0.70	+1.8	4.0	-1.2	+3.0	6.3	-1.8	+4.5	10.0	-3.0
3.15	4.73	+0.90	1.60	-0.60	+1.4	2.50	-0.90	+1.4	2.90	-0.90	+2.2	4.8	-1.4	+4.0	7.7	-2.2	+5.0	11.5	-3.5
4.73	7.09	+1.00	1.90	-0.70	+1.6	2.80	-1.00	+1.6	3.50	-1.00	+2.5	5.6	-1.6	+4.5	8.7	-2.5	+6.0	13.5	-4.0
7.09	9.85	+1.20	2.20	-0.80	+1.8	3.20	-1.20	+1.8	4.20	-1.20	+2.8	6.6	-1.8	+5.0	10.3	-2.8	+7.0	16.5	-4.5
9.85	12.41	+1.20	2.30	-0.90	+2.0	3.40	-1.20	+2.0	4.70	-1.20	+3.0	7.5	-2.0	+6.0	12.0	-3.0	+8.0	19.0	-5.0
12.41	15.75	+1.40	2.60	-1.00	+2.2	3.90	-1.40	+2.2	5.90	-1.40	+3.5	8.7	-2.2	+6.0	14.5	-3.5	+9.0	23.0	-6.0

Table 21 Locational interference fits

Nominal Size Range Millimetres		Class FN1 Light Drive Fit			Class FN2 Medium Drive Fit			Class FN3 Heavy Drive Fit			Class FN4 Shrink Fit			Class FN5 Heavy Shrink Fit		
Over	To	Hole Tol. GR6 −0	Maximum Interference	Shaft Tol. gr5 +0	Hole Tol. H7 −0	Maximum Interference	Shaft Tol. s6 +0	Hole Tol. H7 −0	Maximum Interference	Shaft Tol. t6 +0	Hole Tol. GR8 −0	Maximum Interference	Shaft Tol. gr7 +0	Hole Tol. H8 −0	Maximum Interference	Shaft Tol. t7 +0
0	3	+0.006	0.013	−0.004	+0.010	0.216	−0.006				+0.010	0.024	−0.006	+0.015	0.033	−0.010
3	6	+0.007	0.015	−0.005	+0.013	0.025	−0.007				+0.013	0.030	−0.007	+0.018	0.043	−0.013
6	10	+0.010	0.019	−0.006	+0.015	0.036	−0.010				+0.015	0.041	−0.010	+0.023	0.051	−0.015
10	14	+0.010	0.020	−0.008	+0.018	0.041	−0.010				+0.018	0.046	−0.010	+0.025	0.058	−0.018
14	18	+0.010	0.023	−0.008	+0.018	0.041	−0.010				+0.018	0.046	−0.010	+0.025	0.064	−0.018
18	24	+0.013	0.028	−0.010	+0.020	0.048	−0.013				+0.020	0.053	−0.013	+0.030	0.076	−0.020
24	30	+0.013	0.030	−0.010	+0.020	0.048	−0.013	+0.020	0.053	−0.013	+0.020	0.058	−0.013	+0.030	0.084	−0.020
30	40	+0.015	0.033	−0.010	+0.025	0.061	−0.015	+0.025	0.066	−0.015	+0.025	0.079	−0.015	+0.041	0.102	−0.025
40	50	+0.015	0.036	−0.010	+0.025	0.061	−0.015	+0.025	0.071	−0.015	+0.025	0.086	−0.015	+0.041	0.127	−0.025
50	65	+0.018	0.046	−0.013	+0.030	0.069	−0.018	+0.030	0.082	−0.018	+0.030	0.107	−0.018	+0.046	0.157	−0.030
65	80	+0.018	0.048	−0.013	+0.030	0.074	−0.018	+0.030	0.094	−0.018	+0.030	0.119	−0.018	+0.046	0.183	−0.030
80	100	+0.023	0.061	−0.015	+0.035	0.094	−0.023	+0.035	0.112	−0.023	+0.036	0.150	−0.023	+0.056	0.213	−0.036

Nominal Size Range Inches		Class FN1 Light Drive Fit			Class FN2 Medium Drive Fit			Class FN3 Heavy Drive Fit			Class FN4 Shrink Fit			Class FN5 Heavy Shrink Fit		
Over	To	Hole Tol. GR6 −0	Maximum Interference	Shaft Tol. GR5 +0	Hole Tol. GR7 −0	Maximum Interference	Shaft Tol. GR6 +0	Hole Tol. GR7 −0	Maximum Interference	Shaft Tol. GR6 +0	Hole Tol. GR7 −0	Maximum Interference	Shaft Tol. GR6 +0	Hole Tol. GR8 −0	Maximum Interference	Shaft Tol. GR7 +0
0	.12	+0.25	0.50	−0.15	+0.40	0.85	−0.25				+0.40	0.95	−0.25	+0.60	1.30	−0.40
.12	.24	+0.30	0.60	−0.20	+0.50	1.00	−0.30				+0.50	1.20	−0.30	+0.70	1.70	−0.50
.24	.40	+0.40	0.75	−0.25	+0.60	1.40	−0.40				+0.60	1.60	−0.40	+0.90	2.00	−0.60
.40	.56	+0.40	0.80	−0.30	+0.70	1.60	−0.40				+0.70	1.80	−0.40	+1.00	2.30	−0.70
.56	.71	+0.40	0.90	−0.30	+0.70	1.60	−0.40				+0.70	1.80	−0.40	+1.00	2.50	−0.70
.71	.95	+0.50	1.10	−0.40	+0.80	1.90	−0.50				+0.80	2.10	−0.50	+1.20	3.00	−0.80
.95	1.19	+0.50	1.20	−0.40	+0.80	1.90	−0.50	+0.80	2.10	−0.50	+0.80	2.30	−0.50	+1.20	3.30	−0.80
1.19	1.58	+0.60	1.30	−0.40	+1.00	2.40	−0.60	+1.00	2.60	−0.60	+1.00	3.10	−0.60	+1.60	4.00	−1.00
1.58	1.97	+0.60	1.40	−0.40	+1.00	2.40	−0.60	+1.00	2.80	−0.60	+1.00	3.40	−0.60	+1.60	5.00	−1.00
1.97	2.56	+0.70	1.80	−0.50	+1.20	2.70	−0.70	+1.20	3.20	−0.70	+1.20	4.20	−0.70	+1.80	6.20	−1.20
2.56	3.15	+0.70	1.90	−0.50	+1.20	2.90	−0.70	+1.20	3.70	−0.70	+1.20	4.70	−0.70	+1.80	7.20	−1.20
3.15	3.94	+0.90	2.40	−0.60	+1.40	3.70	−0.90	+1.40	4.40	−0.70	+1.40	5.90	−0.90	+2.20	8.40	−1.40

Table 22 Force and shrink fits

INDEX